Anna Bajek and Bartosz Tylkowski (Eds.)
Medical Physics

Also of interest

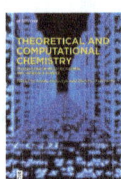
Theoretical and Computational Chemistry.
Applications in Industry, Pharma, and Materials Science
Gulaczyk, Tylkowski (Eds.), 2021
ISBN 978-3-11-067815-4, e-ISBN 978-3-11-067821-5

Chemical Technologies and Processes
Staszak, Wieszczycka, Tylkowski (Eds.), 2020
ISBN 978-3-11-065636-7,
e-ISBN 978-3-11-065627-5

Chemical Synergies.
From the Lab to In Silico Modelling
Bandeira, Tylkowski (Eds.), 2018
ISBN 978-3-11-048135-8, e-ISBN 978-3-11-048206-5

Metals in Wastes
Wieszczycka, Tylkowski, Staszak (Eds.), 2018
ISBN 978-3-11-054628-6, e-ISBN 978-3-11-054706-1

Polymer Engineering
Tylkowski, Wieszczycka, Jastrzab (Eds.), 2017
ISBN 978-3-11-046828-1, e-ISBN 978-3-11-046974-5

Medical Physics

Models and Technologies in Cancer Research

Edited by
Anna Bajek and Bartosz Tylkowski

DE GRUYTER

Editors
Prof. Anna Bajek
The Ludwik Rydygier Collegium Medicum
Nicolaus Copernicus University
Chair of Urology and Andrology
Department of Tissue Engineering
Karlowicza 24
85-092 Bydgoszcz
Poland
a_bajek@wp.pl

Dr. Bartosz Tylkowski
Eurecat, Centre Tecnològic de Catalunya
Marcel·lí Domingo s/n
43007, Tarragona
Spain
bartosz.tylkowski@eurecat.org

ISBN 978-3-11-066229-0
e-ISBN (PDF) 978-3-11-066230-6
e-ISBN (EPUB) 978-3-11-066234-4

Library of Congress Control Number: 2021939065

Bibliographic information published by the Deutsche Nationalbibliothek
The Deutsche Nationalbibliothek lists this publication in the Deutsche Nationalbibliografie; detailed bibliographic data are available on the Internet at http://dnb.dnb.de.

© 2021 Walter de Gruyter GmbH, Berlin/Boston
Cover image: sturti / E+ / gettyimages
Typesetting: Integra Software Services Pvt. Ltd.
Printing and binding: CPI books GmbH, Leck

www.degruyter.com

Preface

Cancer affects everyone regardless of age, gender or social status and represents a tremendous burden for patients, families, and societies at large. It is estimated that the number of people newly diagnosed with cancer every year in Europe could increase from the current 3.5 million to more than 4.3 million by 2035. Modern cancer research is a high-tech undertaking, overlapping with many fields in the physical sciences. This book aims to review the state of the art and to provide the readers with a comprehensive and in-depth understanding of recent developments, technologies and models, which have been investigated and applied in cancer research. By joining efforts of scientists from multi- and-interdisciplinary fields, more people would live without cancer, more cancer patients would be diagnosed earlier, would suffer less, and have a better quality of life after treatment. Chapter 1 describes personalized and targeted therapies based on predictive biomarkers, which are the keywords in the contemporary oncology. Chapters 2 emphasizes recent advancements in cancer chemotherapy while principles and challenges in radiation therapy are detailed in Chapter 3. Chapter 4 provides an overview of the modern cell culture techniques that are currently developing, which allow the introduction of modern models that reflect the organs and physiological system. Currently available cell culture approach is a key aspect of studying these interactions, however, a technology that eliminates the limitations of standard methods is still being sought. Insights into application of natural bioactive substances, natural killer cells, macrophages, and dendritic cells in treatment of various cancer types are sub-topics of Chapter 5 and 6. Chapter 7 deals with non-radioactive imaging strategies for *in vivo* immune cell tracking, while Chapter 8 provides current advancements in design of encapsulation technologies for anticancer drugs delivery. Application of 3D tumor model – a platform for anticancer drug development, which gives an opportunity to supplement and overcome several limitations of currently used 2D cell culture models, is reviewed in Chapter 9.

We believe that the large number of references, to all significant topics mentioned, should make this book useful not only to undergraduates but also to graduate students, medical doctors, patients, and industrial researchers. Having an idea and turning it into a book is as hard as it sounds. The experience is both internally challenging and rewarding. We wish to acknowledge all contributing authors for making this book project a success! Complete thanks to Danka Tylkowska, Jarek Bajek, Krzysztof Roszkowski as well as Bogumiła & Eugeniusz Kaźnica, and Carme Vallverdú Mas who inspired and supported us in the editorial process and helped make this happen.

Ania Bajek and Bartek Tylkowski

Contents

Peface —— V

List of contributing authors —— XIII

Magdalena Wiśniewska, Michał Wiśniewski and Marzena A. Lewandowska
1 Personalized and targeted therapies —— 1
1.1 Introduction —— 1
1.1.1 Predictive biomarkers —— 1
1.1.2 Diagnostic biomarker —— 2
1.1.3 Monitoring biomarker —— 3
1.1.4 Risk (screening) biomarkers —— 3
1.1.5 Pharmacodynamic (response) biomarkers —— 3
1.1.6 Safety biomarker —— 3
1.2 EGFR mutations —— 4
1.3 ALK gene rearrangements —— 5
1.4 ROS1 rearrangements —— 6
1.5 RET rearrangements —— 6
1.6 NTRK gene fusions —— 7
1.7 MET ex14 skipping mutations —— 8
1.8 BRAF-targeted therapies —— 8
1.9 KRAS, NRAS and BRAF wild type in colorectal cancer —— 10
1.10 Overexpression of platelet-derived growth factor receptor β (PDGFRB) in dermatofibrosarcoma protuberans (DFSP) and giant cell fibroblastoma (GCF) —— 11
1.11 KIT and PDGFRA mutations in gastrointestinal stromal tumors —— 12
1.12 Estrogen receptor —— 13
1.13 PIK3CA mutations —— 13
1.14 BRCA mutations —— 14
1.15 HER-2 overexpression —— 14
1.16 PD-L1 (programmed cell death receptor ligand 1) —— 17
1.17 dMMR/MSI status —— 18
References —— 19

Krzysztof Koper, Sławomir Wileński and Agnieszka Koper
2 Advancements in cancer chemotherapy —— 27
2.1 Introduction —— 27
2.2 The role of chemotherapy in cancer treatment —— 28
2.3 The biological basis of chemotherapy —— 28
2.4 Classification of cytostatic drugs —— 28
2.4.1 Alkylating drugs —— 29

2.4.2	Antimethabolites —— 31	
2.4.3	Plant alkaloids and other natural products —— 33	
2.4.4	Platinum derivatives —— 36	
2.5	Cell cycle–nonspecific and cell cycle–specific anticancer drugs —— 36	
2.6	Route of administration of chemotherapy and forms of anticancer drugs —— 37	
2.6.1	Importance of liposomes in clinical practice on the example of doxorubicin —— 41	
2.6.2	Subcutaneous administration —— 42	
2.6.3	Intraperitoneal chemotherapy —— 43	
2.6.4	Intrathecal chemotherapy —— 45	
2.6.5	Transarterial chemoembolisation —— 45	
2.6.6	Electrochemotherapy —— 46	
2.6.7	Implants —— 46	
2.6.8	Gene therapy —— 46	
2.7	Conclusion —— 47	
	References —— 47	

Janusz Winiecki
3   Principles of radiation therapy —— 51
3.7.1   Ionizing radiation interaction with a living matter —— 51
3.7.2   Treatment planning and realization of radiotherapy —— 61
         References —— 79

Karolina Balik, Karolina Matulewicz, Paulina Modrakowska,
Jolanta Kowalska, Xavier Montane, Bartosz Tylkowski and Anna Bajek
4 Advanced cell culture techniques for cancer research —— 81
4.1   Cell-to-cell interactions —— 81
4.1.1   Immunotherapy and cancer research —— 81
4.1.2   Cancer microenvironment —— 82
4.2   Standard methods of cell-to-cell interactions studies —— 83
4.3   Microfluidic devices —— 84
4.3.1   Microfluidic 2D models —— 84
4.3.2   Microfluids 3D models —— 85
4.4   Artificial organs—why do we need them? —— 87
4.4.1   Cell culture analog (CCA) system —— 88
4.4.2   Organoids system —— 90
4.5   "Omic" techniques in biological research —— 92
4.6   Metabolomics —— 93
4.7   Techniques for metabolomics in 2D cell culture —— 95
4.7.1   Experimental design at the level of 2D cell culture —— 95

4.7.2 Experimental design at the level of 3D cell culture — 97
4.8 The direction of *in vitro* cancer research — 98
References — 98

Adrianna Sobolewska, Aleksandra Dunisławska and Katarzyna Stadnicka
5 Natural substances in cancer—do they work? — 103
6.1 Introduction — 103
6.2 Microbiome and cancer — 103
6.2.1 Probiotics and synbiotics in immunotherapy — 105
6.3 Food ingredients and cancer — 116
6.3.1 Resveratrol — 119
6.3.2 Curcumin — 122
6.3.3 Quercetin — 125
6.4 Summary — 127
References — 128

Anna Helmin-Basa, Lidia Gackowska, Sara Balcerowska, Marcelina Ornawka, Natalia Naruszewicz and Małgorzata Wiese-Szadkowska
6 The application of the natural killer cells, macrophages and dendritic cells in treating various types of cancer — 137
6.1 Introduction — 137
6.2 Innate immune cells and the tumor microenvironment — 141
6.3 NK cell–based immunotherapy in treating cancers — 144
6.3.1 Novel strategies for restoring NK cell anti-tumoral activity — 144
6.3.2 Anti-checkpoint receptor monoclonal antibodies — 144
6.3.3 Other antibodies — 148
6.3.4 Agonist antibodies — 149
6.3.5 Modified antibody molecules — 149
6.3.6 Fusion proteins — 150
6.3.7 Immunomodulatory drugs — 150
6.3.8 Cytokines — 151
6.3.9 Adoptive transfer of NK cells — 151
6.3.10 NK cells and combination immunotherapy — 152
6.3.11 Oncolytic virus–based therapy — 153
6.4 Macrophages in cancer treatment — 153
6.4.1 Depleting TAMs — 154
6.4.2 Reprograming TAMs — 155
6.5 DCs in cancer treatment — 156
6.5.1 *Ex vivo* generated DC vaccines — 156
6.5.2 Targeting antigen to DCs *in vivo* — 158
6.6 Conclusions — 159
References — 159

Łukasz Kiraga, Paulina Kucharzewska, Damian Strzemecki, Tomasz P. Rygiel and Magdalena Król
7 Non-radioactive imaging strategies for *in vivo* immune cell tracking — 173
7.1   Introduction — 173
7.2   X-ray computed tomography (CT) — 174
7.2.1   CT contrast agents — 174
7.2.2   CT-based immune cell tracking — 176
7.3   Magnetic resonance imaging (MRI) — 178
7.3.1   Paramagnetic contrast agents (T1 – shortening) — 178
7.3.2   Superparamagnetic contrast agents (T2 – shortening) — 179
7.3.3   "Hot-spot" MRI — 180
7.3.4   Highly-shifted proton (HSP) MRI — 181
7.3.5   CEST MRI — 182
7.3.6   PARACEST MRI — 182
7.4   Optical imaging — 183
7.4.1   Bioluminescence Imaging (BLI) — 183
7.4.2   Fluorescence imaging (FLI) — 184
7.5   Summary — 186
            References — 188

Xavier Montané, Karolina Matulewicz, Karolina Balik, Paulina Modrakowska, Marcin Łuczak, Yaride Pérez Pacheco, Belen Reig Vano, Josep M. Montornés, Anna Bajek and Bartosz Tylkowski
8 Present trends in the encapsulation of anticancer drugs — 193
8.1   Introduction — 193
8.2   Encapsulated anticancer drug delivery systems — 194
8.2.1   Inorganic NPs — 196
8.2.2   Dendrimers — 196
8.2.3   Biopolymeric NPs — 198
8.2.4   Polymeric micelles — 199
8.2.5   Liposomes — 200
8.2.6   Polymersomes — 200
8.2.7   CNTs — 201
8.2.8   QDs — 202
8.3   Hybrid encapsulation systems — 203
8.4   Overview — 204
8.4   Conclusions — 204
            References — 207

Łukasz Kaźmierski and Małgorzata Maj
9 3D tumor model – a platform for anticancer drug development — 213
- 9.1 Need for 3D tumor model (personalized medicine and drug development) — 213
- 9.1.1 3D printed tumor models — 214
- 9.1.2 Need for new *in vitro* models — 214
- 9.2 General understanding of 3D printing technology — 214
- 9.2.1 FDM 3D printing — 215
- 9.2.2 SLA and MSLA 3D printing — 220
- 9.2.3 SLM and DMLS 3D printing — 221
- 9.3 3D *in vitro* model — 221
- 9.3.1 Classical 2D cell culture — 222
- 9.3.2 Spheroid culture — 222
- 9.3.3 Classical 2D versus 3D cell culture approach — 223
- 9.3.4 Spheroid formation methods — 224
- 9.4 Bioprinting — 228
- 9.4.1 Bioprinter types — 228
- 9.4.2 Bioprinter media (bioinks) — 231
- 9.4.3 Workflow and software — 232
- 9.4.4 Typical concerns in bio-printing — 234
- References — 238

Index — 241

List of contributing authors

Magdalena Wiśniewska
Department of Oncology and Brachytherapy,
Collegium Medicum Bydgoszcz,
Nicolaus Copernicus University, Toruń,
Poland; and Department of Clinical Oncology,
Oncology Centre,
Bydgoszcz, Poland

Marzena A. Lewandowska
Department of Thoracic Surgery and Tumors
Collegium Medicum Bydgoszcz
Nicolaus Copernicus University
Toruń, Poland
and Department of Molecular Oncology and
Genetics, Innovative Medical Forum
Oncology Centre
Bydgoszcz, Poland

Michał Wiśniewski
Outpatient Chemotherapy Department
Oncology Centre
Bydgoszcz, Poland

Agnieszka Koper
Department of Oncology
Collegium Medicum Bydgoszcz,
Nicolaus Copernicus University in Toruń,
Toruń, Poland

Krzysztof Koper
Department of Clinical Oncology and
Oncological Nursing
Chair of Oncological Surgery
Collegium Medicum Bydgoszcz,
Nicolaus Copernicus University in Toruń,
Toruń, Poland

Sławomir Wileński
Department of Central Cytostatic
Drug Oncology Centre,
Bydgoszcz, Poland

Janusz Winiecki
Department of Oncology and Brachytherapy,
Collegium Medicum Bydgoszcz,
Nicolaus Copernicus University in Toruń,
Toruń, Poland; and
Department of Medical Physics.
Oncology Centre, Bydgoszcz, Poland

Anna Bajek
Department of Tissue Engineering
Chair of Urology and Andrology
Collegium Medicum Bydgoszcz,
Nicolaus Copernicus University in Toruń,
Toruń, Poland

Karolina Balik
Department of Tissue Engineering
Chair of Urology and Andrology
Collegium Medicum Bydgoszcz,
Nicolaus Copernicus University in Toruń,
Toruń, Poland

Karolina Matulewicz
Department of Tissue Engineering
Chair of Urology and Andrology
Collegium Medicum Bydgoszcz,
Nicolaus Copernicus University in Toruń,
Toruń, Poland

Paulina Modrakowska
Department of Tissue Engineering
Chair of Urology and Andrology
Collegium Medicum Bydgoszcz,
Nicolaus Copernicus University in Toruń,
Toruń, Poland

Jolanta Kowalska
Department of Biological Pest Control and
Organic Agriculture
Institute of Plant
Protection, National Research Institute,
Poznan, Poland

Xavier Montane
Departament de Química Analítica i Química Orgànica
Universitat Rovira i Virgili,
Tarragona, Spain

Bartosz Tylkowski
Chemical Technologies Unit,
Eurecat Centre Tecnològic de Catalunya,
Tarragona, Spain

Katarzyna Stadnicka
Department of Animal Biotechnology and Genetics,
UTP University of Science and Technology
Bydgoszcz, Poland

Adrianna Sobolewska
Department of Anatomy,
Collegium Medicum Bydgoszcz,
Nicolaus Copernicus University in Toruń,
Toruń, Poland

Aleksandra Dunisławska
Department of Animal Biotechnology and Genetics
UTP University of Science and Technology
Bydgoszcz, Poland

Anna Helmin-Basa
Department of Immunology
Collegium Medicum Bydgoszcz,
Nicolaus Copernicus University,
Toruń, Poland

Lidia Gackowska,
Department of Immunology
Collegium Medicum Bydgoszcz,
Nicolaus Copernicus University in Toruń,
Toruń, Poland

Sara Balcerowska
Department of Immunology
Collegium Medicum Bydgoszcz,
Nicolaus Copernicus University in Toruń,
Toruń, Poland

Marcelina Ornawka
Department of Immunology
Collegium Medicum Bydgoszcz,
Nicolaus Copernicus University in Toruń,
Toruń, Poland

Natalia Naruszewicz
Department of Immunology
Collegium Medicum Bydgoszcz,
Nicolaus Copernicus University in Toruń,
Toruń, Poland

Małgorzata Wiese-Szadkowska
Department of Immunology
Collegium Medicum Bydgoszcz,
Nicolaus Copernicus University in Toruń,
Toruń, Poland

Magdalena Król
Department of Cancer Biology
Institute of Biology
Warsaw University of Life Sciences
Warsaw, Poland
and Cellis AG
Zurich, Switzerland

Łukasz Kiraga
Department of Cancer Biology
Institute of Biology
Warsaw University of Life Sciences
Warsaw, Poland
and Cellis AG
Zurich, Switzerland

Paulina Kucharzewska
Department of Cancer Biology
Institute of Biology
Warsaw University of Life Sciences
Warsaw, Poland
and Cellis AG
Zurich, Switzerland

Damian Strzemecki
Cellis AG
Zurich, Switzerland

Tomasz P. Rygiel
Cellis AG
Zurich, Switzerland
and Department of Immunology
Medical University of Warsaw
Warsaw, Poland

Marcin Łuczak
Samorządowa Szkoła Podstawowa nr 1 im. 68
Wrzesińskiego Pułku Piechoty
we Wrześni
Września, Poland

Yaride Pérez Pacheco
Departament d'Enginyeria Química,
Universitat Rovira I Virgili
Tarragona, Spain

Belen Reig-Vano
Department of Chemical Engineering,
Universitat Rovira i Virgili
Tarragona, Spain

Josep M. Montornés
Chemical Technologies Unit,
Eurecat, Centre Tecnològic de Catalunya,
Tarragona, Spain

Łukasz Kaźmierski
Department of Tissue Engineering
Chair of Urology and Andrology
Collegium Medicum Bydgoszcz,
Nicolaus Copernicus University in Toruń,
Toruń, Poland

Małgorzata Maj
Department of Tissue Engineering
Chair of Urology and Andrology
Collegium Medicum Bydgoszcz,
Nicolaus Copernicus University in Toruń,
Toruń, Poland

Magdalena Wiśniewska, Michał Wiśniewski and
Marzena A. Lewandowska

1 Personalized and targeted therapies

Abstract: Biomarker is defined as indicator of normal or pathogenic biological process or response to an intervention or exposure. There are several categories of biomarkers but predictive biomarkers play the most important role in the treatment of neoplasms. In some cancers there may be more than one potential biomarker, and their identification determines the treatment of the patient. Identification of predictive biomarkers allows the development of novel targeted therapies resulting in tailored treatment. In this chapter we discuss most important predictive biomarkers used in contemporary oncology for which there is approved therapies.

Keywords: biomarker, predictive biomarkers, tailored treatment, targeted therapies

1.1 Introduction

Biomarkers are critical to the rational development of medical diagnostics and therapeutics [1]. Biomarker is defined as indicator of normal or pathogenic biological process or response to an intervention or exposure, according to Food and Drug Administration – National Institute of Health (FDA-NIH) Working Group [2].

There are several categories of biomarkers according to FDA-NIH classification. Predictive biomarkers can be used to identify response to exposure to a therapy or an environmental agent. The response could be a relief of symptoms, improvement in survival, or an adverse effect [2]. Predictive biomarkers are essential to modern personalized therapy in contemporary oncology and are the topic of this chapter. Other categories of biomarkers according to FDA-NIH classification include diagnostic, prognostic, monitoring, safety and pharmacodynamic biomarkers.

1.1.1 Predictive biomarkers

Predictive biomarkers are the most important in the treatment of neoplastic diseases. Their presence allows predicting in which group of patients the therapy will be effective. In some cancers there may be more than one potential biomarker, and their identification determines the treatment of the patient. Table 1.1 represents the broad range of biomarkers, which are analyzed in non–small cell lung cancer to

---

This article has previously been published in the journal Physical Sciences Reviews. Please cite as: Wiśniewska, M., Wiśniewski, M., Lewandowska, M. A. Personalized and targeted therapies *Physical Sciences Reviews* [Online] 2021, 6 DOI: 10.1515/psr-2019-0057

qualify patient for appropriate therapy. There are also biomarkers whose presence is associated with the effectiveness of a given therapy regardless of the type of tumor (so-called organ-independent biomarkers), examples of such biomarkers are mutations in the NTRK or high levels of MSI. In this chapter, predictive biomarkers for which there are approved therapies in the treatment of solid tumors will be discussed.

Table 1.1: Predictive biomarkers in non-small cell lung cancer.

Disease	Biomarker	Drug
Non-small cell lung cancer	EGFR	Erlotinib
		Gefitinib
		Osimertinib
		Afatinib
		Dacomotinib
	ALK	Crizotinib
		Ceritinib
		Lorlatinib
		Alectinib
		Brigantinib
	ROS1	Crizotinib
		Ceritinib
		Lorlatinib
		Alectinib
		Brigantinib
	BRAF	Dabrafenib
		Vumurafenib
	MEK	Capmatinib
	RET	Selpercatinib
	PD-L1	Pembrolizumab
	NTRK	Entrectinib
		Larotrektinib
	No biomarker	Chemotherapy

1.1.2 Diagnostic biomarker

Diagnostic biomarker is used as an indicator of a presence of a disease or condition to assess a subtype of the disease, e.g. profiling of gene expression may be used to distinguish subgroups of patients with diffuse large B-cell lymphoma and different gene signatures of malignant cells. Another example of diagnostic biomarker could be measurement of glomerular filtration rate [GFR] in diagnosis of patients with chronic kidney disease [3, 4].

1.1.3 Monitoring biomarker

Monitoring biomarker is an indicator of disease or medical condition status. It may also indicate an effect of a drug of environmental factor. Examples of monitoring biomarkers are prostate-specific antigen (PSA) in patients with prostate cancer or cancer antigen 125 (CA 125) in patients with ovarian cancer. Both biomarkers are used to assess disease status or burden [5, 6].

1.1.4 Risk (screening) biomarkers

Risk (screening) biomarkers indicate the potential for developing a disease or medical condition in an individual who does not currently have clinically apparent disease or the medical condition. Example of risk biomarkers are mutations in genes related to increased risk of developing cancer as BRCA 1/2 (Breast Cancer genes 1 and 2) mutations which are inked with breast cancer [7, 8].

1.1.5 Pharmacodynamic (response) biomarkers

Pharmacodynamic (response) biomarkers indicate a biological response that may be seen in a patient treated with a drug or an environmental factor. Example of a such biomarker is a standardized uptake value (SUV) measured by PET/CT (Positon Emission Tomography/Computed Tomography) with 18-FG-glucose marked contrast, used as a response biomarker when assessing a response in a patients treated with chemotherapy for diffuse large B-cell lymphoma or Hodgkin lymphoma [9, 10].

1.1.6 Safety biomarker

Safety biomarker is measured for determining a risk of toxicity or an adverse effect. It may be assessed after or before an exposure to a drug or an environmental factor. Safety biomarker example cloud be a serum creatinine for monitoring renal toxicity in a patients treated with a potentially nephrotoxic drug [3].

1.2 EGFR mutations

The epidermal growth factor (EGF) receptor (EGFR) is one of transmembrane proteins that take part in the actions of a family of growth factors [11]. EGFR mutations are found in about 50% of Asian NSCLC patients and in about 10% of patients with NSCLC in Western countries [12]. The most common EGFR gene mutations are exon 19 deletions (Figure 1.1) and the point mutation L858R at exon 21 (about 45 and 40% of patients, respectively), activating the tyrosine kinase domain in EGFR and may be detected by qPCR methods [13]. The 'uncommon' EGFR mutations occur in 10–18% EGFR mutations. They usually are 20 insertions, exon 18 point mutations or complex mutations [14, 15].

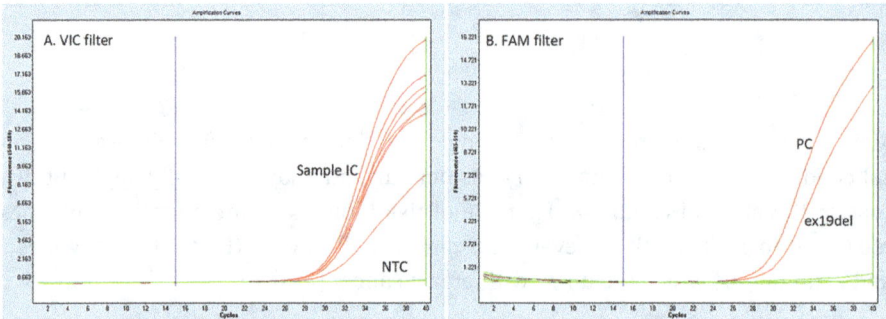

Figure 1.1: Deletion in EGFR exon 19 detected in DNA isolated from an FFPE tissue sample. Somatic mutation analysis performed by qPCR using EGFR mutation analysis kit (entrogen). A – amplification curve of sample internal control (IC); B – amplification curve of deletion in exon 19 and positive control (PC).

EGFR mutations have predictive effect on tyrosine kinase inhibitors (TKIs) treatment. Patients with these mutations better respond to erlotinib, gefitinib, afatinib and osimertinib. The clinical trial assessing efficacy of erlotinib in patients with NSCLC with activating EGFR mutations showed longer median progression-free survival (PFS) in the erlotinib group, as in the standard chemotherapy group (9.7 vs. 5.2 months respectively) [16]. Also another agent, gefitinib showed significantly longer PFS compared with the chemotherapy group (median PFS 9.2 vs. 6.3 months) [17]. In the study assessing the results of chemotherapy compared with afatinib, median PFS was 6.9 and 13.6 months, respectively [18]. In about 60% of patients at the time of progression during the treatment with erlotinib, afatinib or gefitinib may occur the p.Thr790Met point mutation (T790M) in EGFR gene. Osimertinib is active in T790M mutated NSCLC patients with response rate 66% and PFS 11.0 months [19]. Treatment with osimertinib of EGFR mutated NSCLC patients in first line setting results with median PFS 18.9 versus 10.2 months for standard tyrosine kinase inhibitors [20].

1.3 ALK gene rearrangements

ALK (anaplastic lymphoma kinase) is a gene encoding a receptor tyrosine kinase belonging to the insulin receptor. Oncogenic translocations in ALK may occur in different type of neoplasms and lead to a constitutively active tyrosine kinase having an oncogenic potential [21]. The most common translocation is the small inversion in the short arm of chromosome 2 causing a fusion between the 5′ portion of the EML4 gene and the 3′ portion of the ALK gene [22].

Gold standard for ALK rearrangement detection in tumor tissues is molecular cytogenetics analysis using fluorescent *in situ* hybridization (Figure 1.2), however ALK rearrangements may be identified using reverse transcription-polymerase chain reaction as well [22, 23]. Prescreening with immunohistochemistry assay for ALK fusions also can be done and is approved by FDA [24]. NGS is also appropriate to assess for ALK fusions [25].

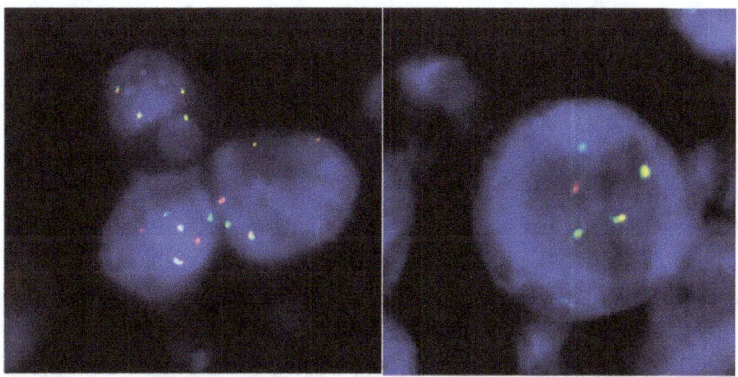

Figure 1.2: Molecular cytogenetic analysis of *ALK* rearangement by fluorescence *in situ* hybridization (Abbott Molecular). Fused signals (green and red or seen as yellow) indicates correct *ALK locus*; splited, separated red and green signals on the upper left site indicates rearrangement *ALK locus*.

Many studies have shown that cancer cells with confirmed *ALK* gene rearrangements are sensitive to ALK inhibition [26]. ALK translocations are reported in NSCLC, anaplastic large cell lymphomas or neuroblastomas [26, 27].

ALK gene rearrangements are present in 2–7% of NSCLC. Patients with these rearrangements are usually younger, light or never smokers and have adenocarcinoma histological type [28]. Co-expression of ALK rearrangements with EGFR or KRAS mutations is hardly ever showed, meaning that ALK is a separate oncogenic driver [22].

The first drug approved by FDA in ALK gene rearrangements positive metastatic NSCLC patients was crizotinib. Crizotinib is a tyrosine kinase inhibitor of ALK, ROS1 and MET. The median PFS in patients treated with at least one chemotherapy regimen before was 7.7 months in the crizotinib group and 3.0 months in the chemotherapy group. The response rates were 65% in the crizotinib patients and 20% in chemotherapy patients. In first line treatment with crizotinib PFS was also significantly longer with crizotinib group than with chemotherapy group (10.9 vs. 7.0 months). Objective response rates were 74 and 45%, respectively. The most common adverse events in patients receiving with crizotinib were visual disorder, gastrointestinal tract side effects, and elevated liver aminotransferase levels [29, 30]. Alectinib is now treatment of choice in ALK rearrangements positive metastatic NSCLC as it showed superior efficacy and lower toxicity than crizotinib [31]. Other recommended drugs are ceritinib and brigatinib (Table 1.1).

1.4 ROS1 rearrangements

ROS proto-oncogene 1 (ROS1) encodes a receptor thyrosine kinase similar to ALK and other members of the insulin-receptor family [32]. ROS1 has been shown to undergo genetic rearrangements in many of human cancers, including NSCLC, ovarian cancer, gastric cancer and colorectal cancer. ROS1 rearrangements lead to the synthesis of fusion proteins in which the kinase domain of ROS1 becomes constitutively active and enhances cellular proliferation. ROS1 rearrangements may be present in about 1–2% of patients with NSCLC, especially in EGFR or KRAS mutations and ALK rearrangements negative and usually in adenocarcinomas [33]. Similar to ALK fusions, ROS1 rearrangements are tested using FISH (Figure 1.3), however NGS and immunohistochemistry is also possible [23].

The recommended treatment options for ROS1 rearrangements – positive metastatic NSCLC patients are crizotinib, ceritinib and entrectinib. Crizotinib showed high efficacy with median PFS 19.2 months and the objective response rate of 72% [34]. Entrectinib is highly active in patients with ROS1 rearrangements-positive NSCLC, including those with CNS disease – overall response rate was 77% and median duration of response lasted for 24.6 months. Most common side effects were weight increase and neutropenia [35]. For ceritinib objective response rate was 62%, duration of response was 21 months and disease control rate was 81% [36]. Lorlatinib is recommended second-line treatment option with overall objective response rates 41% [37].

1.5 RET rearrangements

RET is a thyrosin kinase receptor that enhances cell proliferation and differentiation. RET rearrangements in NSCLC occur between the RET gene and other domains, especially kinesin family 5B and coiled coil domain containing-6 leading to overexpression of the RET protein. RET rearrangements occur in about 0.7–2% NSCLC patients, more often in adenocarcinoma histologic type, never-smokers and younger age [38].

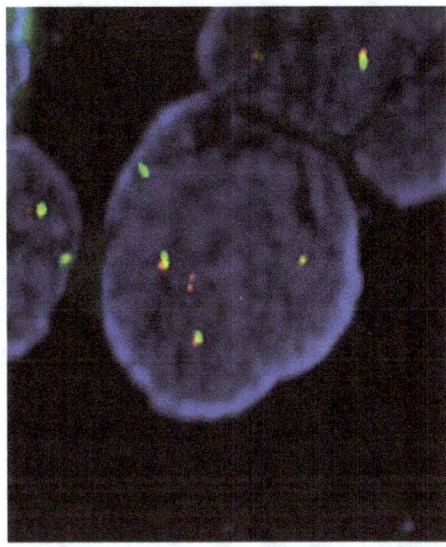

Figure 1.3: Molecular cytogenetic analysis of ROS rearrangement by fluorescence *in situ* hybridization (Abbott Molecular). Fused signals (green and red or seen as yellow) indicates correct *ROS locus*; splited, separated red signal in the middle of nuclei and green signal on the upper left site indicates rearrangement *ROS locus*.

RET rearrangements may be detected using molecular cytogenetic FISH analysis or by RT-PCR or NGS assays.

Selpercatinib is treatment option for patients with RET rearrangements-positive, advanced NSCLC, advanced or metastatic RET-mutant medullary thyroid cancer and RET fusion-positive thyroid cancer. The efficacy of this agent was confirmed in a multicenter, open-label, multi-cohort clinical trial where the overall response rate in previously untreated RET fusion-positive advanced NSCLC patients was 88% [39].

Cabozantinib is a multikinase inhibitor with activity against RET. The studies confirmed efficacy of cabozantinib in patients with RET-rearranged advanced lung cancers with overall response rate of 28% what defines RET rearrangements as actionable drivers in patients with lung cancers [40].

1.6 NTRK gene fusions

NTRK1, NTRK2, and NTRK3 are genes encoding three transmembrane protein receptors: TrkA, TrkB, and TrkC belonging to the tropomyosin receptor kinase (Trk) family. Gene fusions involving NTRK1, 2, and 3 and their partner genes act like oncogenic drivers for solid tumors [41]. NTRK gene fusions occur in about 0.2% of NSCLC patients across genders, ages, smoking histories, and histologies. NTRK gene fusions do not have concurrent alterations in KRAS, EGFR, ALK and ROS1 [42]. NTRK gene fusions occur in 0.2–1% patients with colorectal cancer [41]. NTRTK fusions may be detected using FISH, RT-PCR, NGS and IHC assays. NTRK testing may be considered in patients negative for the main biomarkers.

Available therapeutic options for patients with *NTRK* gene fusions-positive advanced solid tumors are larotrectinib and entrectinib. The trail assessing efficacy of larotrectinib in adults and children who had tumors with these gene fusions showed the overall response rate of 75%. Larotrectinib had marked and durable antitumor activity regardless of the age of the patient or of the tumor type. Observed adverse events were mainly of grade 1 [43]. Entrectinib was well tolerated and induced clinically significant, durable systemic and CNS responses with ORR 57%, median duration of response was 10 months. The most common treatment-related adverse events of grade 3 and 4 were increased weight and anemia [44].

1.7 MET ex14 skipping mutations

c-Met is the tyrosine kinase receptor for hepatocyte growth factor. The known genomic alternations in the MET gene may include exon 14 skipping mutations, gene copy number gain or amplification and MET protein coexpression. Depending on what we are looking for molecular analysis using real-time PCR (exon skipping), MLPA (gain or amplification) or FISH analysis (gain or amplification) can be implemented [13, 23].

The receptor developing as a result of the MET mutations has increased c-Met signaling and oncogenic potential [45, 46]. MET gene alternations may occur in 3–4% of advanced NSCLC, significantly more often in elderly patients in contrast to ALK and ROS1 rearrangements. MET mutations do not usually coexist with EGFR, ROS1, BRAF and ALK genetic variants [45, 46].

Capmatinib is an oral MET phosphorylation inhibitor. Results of the GEOMETRY Mono-1 trial of capmatinib in advanced NSCLC showed responses in 39.1% of previously treated patients and in 71.4% of treatment-naive patients. Median duration of response was 9.72 and 8.41 months, respectively [46]. Patients with METex14 skipping mutations have a modest response for immunotherapy (about 16%) therefore they may be candidates for targeted therapies or chemotherapy before considering immunotherapy [47].

1.8 BRAF-targeted therapies

The RAS–RAF–MEK–ERK–MAP kinase pathway mediates cell response to growth signals [48]. BRAF is an intracellular signaling kinase in the mitogen-activated protein kinase (MAPK) pathway [49].

Activating mutations in the BRAF oncogene occur in 41–55% of advanced melanomas. The most common result of this mutations is substitution of the valine at amino acid position 600 to glutamic acid in the BRAF protein (V600E), changing the kinase into an active conformation. Occasionally patients have V600 K mutations, other genotypes are very rare [50]. BRAF mutations (usually V600E) occur in 3% of patients with lung adenocarcinoma, typically in current or former smokers. The incidence of BRAF mutations other than V600E is higher in lung cancer than in melanoma [51]. BRAF V600 mutations occur also in patients with colorectal cancers (CRC) in 6% of cases [52]. The most commonly used method to assess BRAF mutations is real-time PCR (Figure 1.4).

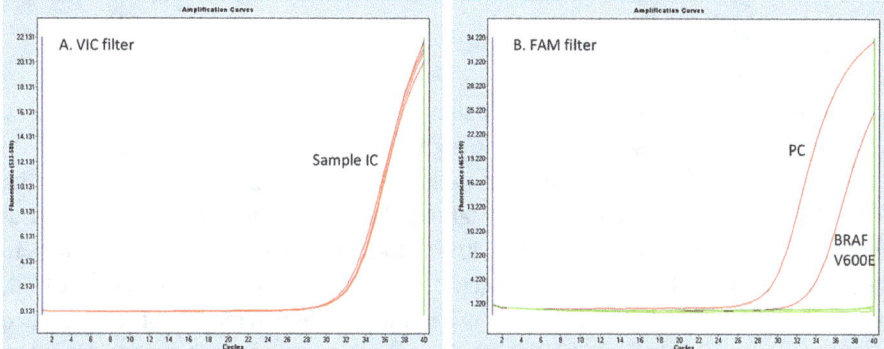

Figure 1.4: BRAF V600E mutation in tissue sample. Somatic mutation analysis performed using qPCR BRAF mutation analysis kit II (Entrogen). A – amplification curves of sample internal control (IC); B – amplification curves of positive control (PC) and sample with BRAF V600E mutation.

BRAF inhibitors, vemurafenib and dabrafenib have been reported to have clinical activity in patients with BRAF V600-mutated, unresectable stage III and stage IV melanoma. In the BRIM-3 trial, median overall survival was significantly longer in the vemurafenib group than in the dacarbazine group (13.6 vs. 9.7 months) [53]. In the BREAK-3 trial assessing efficacy of dabrafenib, median overall survival was 18.2 versus 15.6 months for dabrafenib group and dacarbazine group, respectively [54]. Common toxic events with vemurafenib and dabrafenib are rash, fatigue, and joint pain [55].

Clinical trials showed that in patients with BRAF V600–mutated metastatic melanoma the addition of MEK inhibitor cobimetinib to vemurafenib caused a significant improvement in PFS. The median PFS was 9.9 months in the combination group and 6.2 months in the monotherapy group and the rates of complete or partial response were 68 and 45%, respectively [55]. The addition of trametinib to dabrafenib was also more effective than monotherapy. Median overall survival was 25.1 months in the combination group versus 18.7 months in the dabrafenib only group [56].

In patients with advanced NSCLC harboring BRAF V600E mutations recommended treatment option is dabrafenib and trametinib combination with overall response rate 63% response durations ≥6 months in 64% of responders in previously treated patients.

In treatment naïve patients, ORR was 61% with response durations ≥6 months in 59% of responders.

In previously treated patients with metastatic CRC receiving a combination of BRAF inhibitor encorafenib and cetuximab median overall survival was 8.4 months in encorafenib/cetuximab arm versus 5.4 months in arm treated with chemotherapy alone [57]. The combination therapy was recently approved by FDA.

1.9 KRAS, NRAS and BRAF wild type in colorectal cancer

KRAS is the human homolog of the Kirsten rat sarcoma-2 virus oncogene encoding a GTP-binding protein self-inactivating signal transducer by cycling from GDP- to GTP-bound states in response to stimulation of a cell surface receptor, including EGFR. KRAS can harbor oncogenic mutations that provide to synthesis of constitutively active protein [58]. Mutations within KRAS in codons 12 and 13 resulting in the activation of RAS/RAF signaling occur in 35–42% of colorectal cancer (CRC) [58, 59]. Presence of *KRAS* tumor mutation is a known predictive marker of resistance to EGFR-targeted therapies in CRC [59]. Confirmed mutations in exons 2, 3 and 4 NRAS gene allow to separate an additional group of patients that do not benefit from anti-EGFR therapy [60]. The third gene used in predicting response from anti-EGFR therapy is BRAF. The mutated BRAF protein product is constitutively active and probably allows omitting inhibition of EGFR by anti-EGFR agents [61].

Due to actual knowledge, only patients with KRAS, NRAS and BRAF wild type (not mutated, Figure 1.5) CRC benefit from anti-EGFR therapy [62]. There are two anti-EGFR agents available, panitumumab and cetuximab. In the PRIME trial, assessing the efficacy of panitumumab with FOLFOX4 chemotherapy in first line treatment of patients with RAS wild type, PFS was 10.1 months in panitumumab–FOLFOX4 arm versus 7.9 months in chemotherapy alone arm. Overall survival was 26.0 months in the panitumumab–FOLFOX4 group versus 20.2 months in the FOLFOX4-alone group [63]. The results were similar in the trial with first-line panitumumab with FOLFIRI chemotherapy in RAS-wild type patients [64]. The CRYSTAL study confirmed that patients with KRAS not-mutated advanced CRC benefit from the combination therapy of anti-EGFR agent cetuximab and FOLFIRI in first-line treatment [52]. Cetuximab and panitumumab was also efficient as monotherapy in third-line treatment [65, 66]. The addition of cetuximab or panitumumab in the BRAF-mutated patients did not significantly improve PFS comparing with control treatment regimens [62].

1.10 Overexpression of platelet-derived growth factor receptor β (PDGFRB)

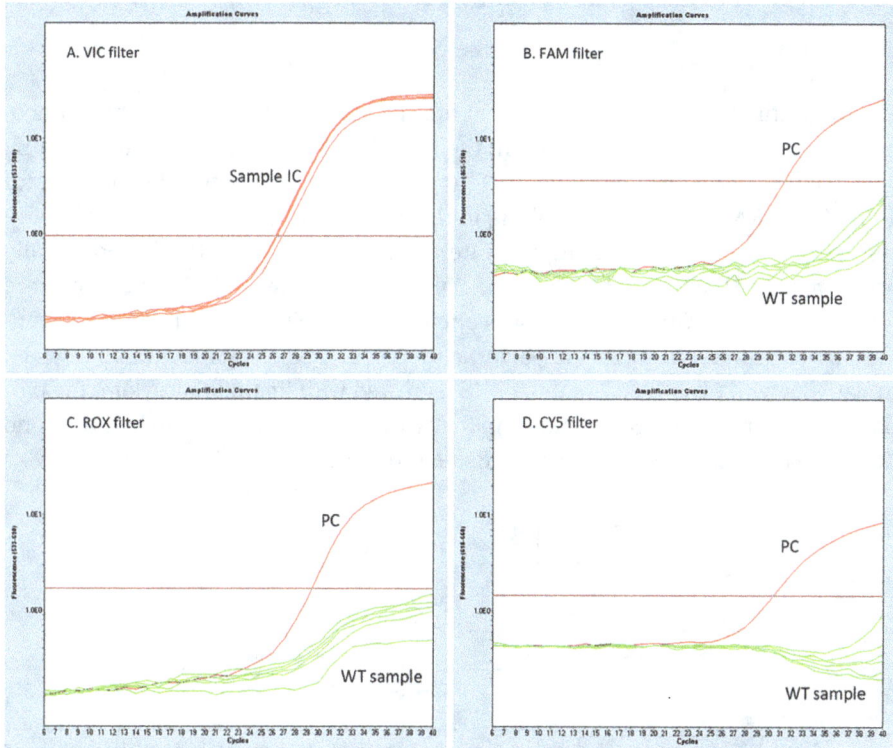

Figure 1.5: Wild type sample in KRAS, NRAS, BRAF, AKT1 and PIK3CA genes. Somatic mutation analysis performed using qPCR colorectal cancer mutation detection panel (entrogen). IC – amplification curves of sample internal control in 6 reactions; PC – amplification curve of positive control.

1.10 Overexpression of platelet-derived growth factor receptor β (PDGFRB) in dermatofibrosarcoma protuberans (DFSP) and giant cell fibroblastoma (GCF)

DFSP and GCF are recurrent, infiltrative skin tumors which are genetically characterized by chromosomal rearrangements involving chromosomes 17 and 22, causing the fusion of the collagen type Iα1 (COLIA1) gene and the platelet-derived growth factor B-chain (PDGFB) gene. The product of this fusion gene is COL1A1/PDGF-B fusion protein that is processed to mature PDGF-BB taking part in DFSP and GCF tumor development and growth. Imatinib mesylate is a specific inhibitor of many protein-tyrosine kinases, including the PDGFRs. Imatinib mesylate has shown efficacy in treatment of unresectable, recurrent and metastatic DFSP in adult patients [67, 68]. Objective response rate was approaching 50% [69].

1.11 KIT and PDGFRA mutations in gastrointestinal stromal tumors

Gastrointestinal stromal tumors (GISTs) are the most common mesenchymal neoplasms of the gastrointestinal tract. In the majority of GISTs are detected activating mutations of the genes encoding KIT or platelet-derived growth factor receptor alpha (PDGFRA) [Figure 1.6], resulting in the constitutive activation of protein tyrosine kinase signaling. KIT mutations are found in 70–85% of GISTs and usually occur in exon 11 and 9, exon 13 and 17 mutations are rare. PDGFRA mutations are found in 5–15% of GISTs and are not concurrent with KIT mutations [70]. The mutant KIT or PDGFRA isoforms are potential therapeutic targets for imatinib mesylate. In patients with exon 11 KIT-mutated GISTs treated with imatinib mesylate, the partial response rate was 83.5% and in patients with exon 9 KIT-mutated GISTs or no KIT/PDGFRA-mutated GISTs it was 47.8 and 0.0% respectively [71].

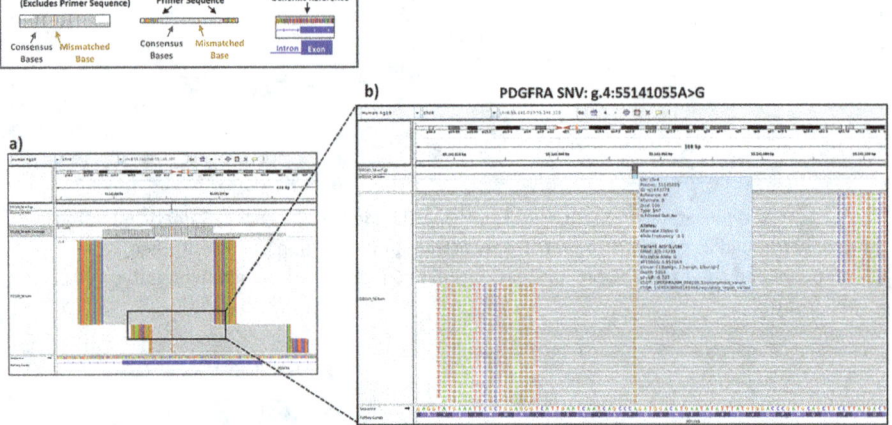

Figure 1.6: Usage of new generation sequencing as a tool for comprehensive 16 gene panel assay to detect and amplify clinically relevant hotspot mutations in solid tumors.(a) Processed sequencing data from Entrogen's THSP NGS assay, as shown in IGV. The aligned region covers all of exon11 (shown in blue) of PDGFRA. (b) Close-up of a PDGFRA SNV called at position g.4:55141055A > G (c.1701A > G; p.Pro592). This is a synonymous variant (silent), and likely, benign (dbSNP ID: rs1873778).

The analysis of Heinrich et al. assessing response of sunitinib in imatinib mesylate-resistant GISTs has shown that sunitinib was more efficient in patients with KIT exon 9 mutations (overall response rates 58%) than in patients with KIT exon 11 mutations (34%). PFS was also longer for patients with KIT exon 9-mutated GISTs than for patients with KIT exon 11-mutated GISTs [72].

KIT mutations may occur in patients with malignant melanoma, usually in mucosal and acral subtypes [73]. In phase II clinical trial imatinib was efficient in patients with KIT-mutated metastatic melanoma with (overall response rate of 23.3%) [74].

1.12 Estrogen receptor

Estrogen receptors alpha (ERα) and beta (ERβ) are nuclear transcription factors that are taking part in many human cell regulation processes. The processes include regulating specific genes transcription in the cell nucleus. Estrogen receptors may be detected in many cells and tissues, for example breast, uterus, ovary, bone, testicles, prostate, liver, or immune system [75].

Estrogen receptor status in breast cancer is a known predictive marker of response to endocrine therapy. Besides, patients ER-negative benefit more from adjuvant therapy of early breast cancer, therefore ER receptor expression may help in qualifying to adjuvant treatment [76].

Tamoxifen is the oral synthetic antiestrogen. Many clinical trials have shown its efficacy in breast cancer patients both advanced and early as an adjuvant treatment [77]. In ATLAS study in women with ER-positive disease, adjuvant tamoxifen administered up to 10 years reduced the risk of breast cancer recurrence, breast cancer mortality and overall mortality [78].

Estrogens in postmenopausal women are synthesized in other tissues than ovary, including connective tissue, skin, liver or muscles. The aromatase is the enzyme converting testosterone into estradiol and androstenedione into estrone. Oral aromatase inhibitors (anastrozole, letrozole, and exemestane) inhibit over 98% of estrogen synthesis and have become the treatment of choice in adjuvant endocrine therapy in postmenopausal ER-positive breast cancer patients. Following their successful implementation for the treatment of metastatic breast cancer, the 'third-generation' have now become standard adjuvant endocrine treatment for postmenopausal estrogen receptor-positive breast cancers [79]. Another agent used in breast cancer treatment is fulvestrant, a selective estrogen receptor down-regulator blocking proliferation of breast cancer cells with confirmed efficacy in advanced ER-positive breast cancer patients [80]. Endocrine therapy has shown good clinical effects in metastatic ER-positive uterine cancer [81].

1.13 PIK3CA mutations

Activating mutations in the gene PIK3CA occur in about 40% of patients with HR-positive, HER2-negative breast cancer have, causing hyperactivation of the alpha isoform (p110α) of phosphatidylinositol 3-kinase (PI3K). Alpelisib is a selective p110α inhibitor. For addition of alpelisib to fulvestrant in previously treated PIK3CA-mutated,

HR-positive, HER2-negative advanced breast cancer has been reported prolonged PFS (11 vs 5.7 months in fulvestrant only group). Most common adverse events were hyperglycemia and rash [82].

1.14 BRCA mutations

BRCA1 and BRCA2 are tumor-suppressor genes encoding proteins taking part in the repair of DNA double-strand breaks. The poly(adenosine diphosphate–ribose) polymerase (PARP) family of enzymes is involved in the repair of DNA single-strand breaks. In BRCA1 and BRCA2, are autosomal dominant breast cancer susceptibility genes. Germ line mutations of BRCA1 and BRCA2 genes occur often in families of high risk of breast and ovarian cancers [83]. BRCA1 is localized on chromosome 17q and BRCA2 on chromosome 13q [84]. BRCA1 mutations are found in about 2.4% and BRCA2 mutations in 2.3% of breast cancer patients [83].

Olaparib is the oral PARP inhibitor with confirmed efficacy in patients with recurrent ovarian cancer harboring BRCA mutation (Figure 1.7) and in patients with metastatic breast cancer with confirmed germline BRCA mutation [120]. In clinical trial, assessing the efficacy of olaparib in metastatic HER2-negative, BRCA-mutated breast cancer who had received no more than two previous chemotherapy regimens median PFS was significantly longer in the olaparib group than in the standard-therapy group (7.0 vs. 4.2 months) [85]. Olaparib is also effective in patients with platinum-sensitive, relapsed ovarian cancer harboring BRCA1/2 mutation [86]. Phase 3 trial has shown efficacy of olaparib in metastatic pancreatic cancer germline BRCA mutation positive patients who had not progressive disease during first-line platinum-based chemotherapy. The median PFS was longer in the olaparib group than in the placebo group (7.4 vs. 3.8 months) [87].

The other PARP inhibitor is talazoparib. The efficacy of this agent was confirmed in advanced breast cancer germline BRCA1/2-mutated patients. Talazoparib compared with standard chemotherapy has shown benefit in PFS (8.6 vs. 5.6 months, respectively) [88].

Another PARP inhibitor, efficient in recurrent ovarian cancer with confirmed BRCA-mutation or high percentage of genome-wide loss of heterozygosity is rucaparib. Median PFS in was 16.6 months in the rucaparib group versus 5.4 months in the placebo group. In patients with homologous recombination deficient cancers it was 13.6 versus 5.4 months [89].

1.15 HER-2 overexpression

HER2 (ErbB2, Neu) is a member of the EGFR family of receptor tyrosine kinases. These receptors are involved in cell proliferation and differentiation and their

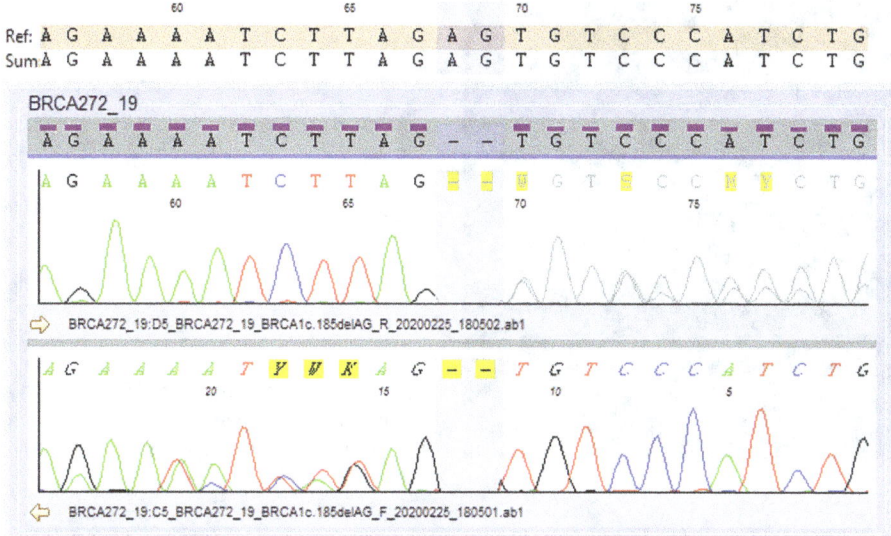

Figure 1.7: Point mutation in BRCA1 gene detected in tissue sample (FFPE) and confirmed in DNA isolated from blood as germline variant. Shaded violet column on chromatogram indicates single nucleotide substitution c.181T > G (p.Cys61Gly). Mutation analysis by sanger sequencing performed using SeqStudio genetic analyzer (BRCA1 NM_007294.3:c.181T > G) (applied Biosystems).

activation alternations take part in development of many neoplasms. About 20–30% of breast cancer patients have confirmed HER2 overexpression what is correlated with poorer prognosis [90]. Amplification of the HER2 oncogene leads to increased level of expression of HER2 on the breast-tumor cells (Figure 1.8) and HER-2 overexpression can accelerate tumorigenesis [91].

Trastuzumab is a recombinant monoclonal antibody against HER2 receptor with confirmed clinical activity in HER-2 overexpressed breast cancer. The clinical trials have shown that trastuzumab provides benefit in overall survival and disease-free survival in women with early breast cancer that overexpresses HER-2. In HER-2 overexpressed breast cancer patients after radical treatment administration of trastuzumab for one year significantly reduced the risk of a disease-free survival event and death compared with observation only [92]. Trastuzumab is effective also in metastatic HER2-positive breast cancer. In the analysis of Slamon et al. trastuzumab added to chemotherapy significantly improved overall survival (25.1 vs. 20.3 months) [93]. Trastuzumab is also effective HER2-positive advanced gastric or gastro-esophageal junction cancer. The study assessing adding trastuzumab to chemotherapy showed median overall survival of 13.8 months in combination group versus 11.1 months in chemotherapy alone group [94].

Another anti-HER antibody is pertuzumab which is used in combination with trastuzumab. This combination is more effective due to more complete signaling blockade (so called dual blockade). Dual blockade of HER-2 receptor combined with

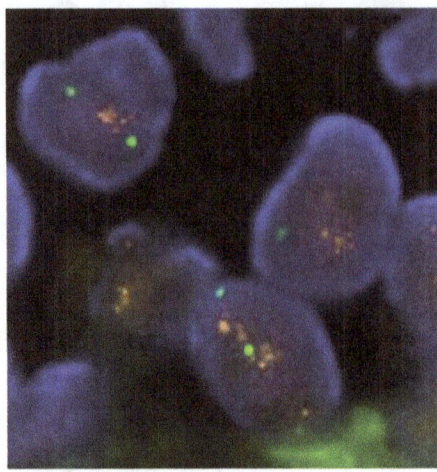

Figure 1.8: Genetic analysis of HER2 gene amplification by fluorescence *in situ* hybridization (Abbott Molecular).

chemotherapy is efficient in neoadjuvant treatment of breast cancer increasing pathological complete response by 15% [95] and also in the adjuvant setting improving rate of invasive disease free survival response (92.0% in the combination group versus 90.2% in trastuzumab only group in 3 years period for patients with regional node involvement) [96]. The efficacy of pertuzumab and trastuzumab combination has been also reported for metastatic HER-2 positive breast cancer (median overall survival was 56.5 months in the combination group, versus 40.8 months trastuzumab only group in patients treated for advanced breast cancer in first line) [97].

Lapatinib is a tyrosine kinases of HER2 and epidermal growth factor receptor type 1 inhibitor. The clinical benefit of lapatinib combined with capecitabine chemotherapy has been reported in advanced HER2-positive breast cancer patients that progressed during trastuzumab therapy. The clinical trial showed median PFS 8.4 months in the combination-therapy group versus 4.4 months in the capecitabine only group [98].

Trastuzumab emtansine (T-DM1) is an antibody–drug combination of trastuzumab and the cytotoxic drug emtansine (DM1), a microtubule inhibitor. T-DM1 binding to HER-2 receptor leads to cellular internalization of the complex and reveals cytotoxic component inhibiting microtubules. T-DM1 has been reported to be more effective than capecitabine plus lapatinib or treatment of the physician's choice in HER2-positive advanced breast cancer patients treated before with trastuzumab and chemotherapy [99]. In the trial of Krop et al. involving pretreated HER2-positive advanced breast cancer patients PFS was significantly improved with T-DM1 comparing with physician's choice treatment (median 6.2 vs. 3.3 months respectively) [100]. The effect of T-DM1 was confirmed also in patients who diagnosed with residual invasive disease in the breast or axilla in pathological report after radical surgery and after neoadjuvant treatment containing trastuzumab [99].

Margetuximab is an anti-HER2 monoclonal antibody binding to the same receptor as trastuzumab and causing comparable antiproliferative effect. Margetuximab potentiates adaptive immunity in treated patients, including enhanced clonality of T-cells and induction of HER2-specific T- and B-cell responses. SOPHIA trial including patients with HER2-positive metastatic breast cancer after prior anti-HER2 therapies has shown that margetuximab with chemotherapy significantly prolonged PFS versus trastuzumab with chemotherapy (median PFS 5.8 vs. 4.9 months, respectively) [101].

1.16 PD-L1 (programmed cell death receptor ligand 1)

Programmed death 1 receptor (PD-1) is a protein that is expressed on T cells, B cells, NK cells, activated monocytes, and dendritic cells. PD-1 role is to prevent autoimmunity by inhibiting activity of T-cells in peripheral tissues. PD-1 binds to its ligands PD-L1 and PD-L2. PD-L1 is surface protein, expressed on B and T lymphocytes, myeloid and dendritic cells, as well as on tumor cells [102, 103].

Within the tumor microenvironment, PD-1 on T cells binds to PD-L1 or PD-L2 on tumor cells and inhibits an antitumor immune response by dampening T-cell receptor signaling [102–104]. Expression of PD-L1 or PD-L2 by tumor cells contributes to tumor evasion or immune destruction by inactivating T cells [105]. Blockade of PD-1 is hypothesized to reactivate an antitumor immune response in the tumou microenvironment [102].

PD-L1 expression varies across tumor types and thresholds and scoring systems to determine PD-L1 positivity can vary between antibody assays [106, 107]. Studies in several tumor types have shown that PD-L1 expression may be linked to the clinical efficacy of immunotherapy [104, 108–111]. Role of PD-L1 as a prognostic and predictive biomarker varies across tumor types and continues to be evaluated [108, 112].

Anticancer drugs that interfere with PD-1 pathway are antibodies which by binding to PD-1 or PD-L1 prevent T-cell inactivation and restore immune response. Studies in several cancer types have shown that PD-L1 expression may be correlated with clinical efficacy of immunotherapy [106, 107, 109]. However, only few antibodies require assessing of PD-L1 status for treatment. One is atezolizumab, human anti-PD-L1 antibody, approved for treatment of metastatic triple negative breast cancer with expression of PD-L1 on tumor-infiltrating immune cells of 1% or more. Atezolizumab in combination with nab-paclitaxel chemotherapy prolonged overall survival for 7 months and progression free survival for 2.5 months comparing to chemotherapy alone [111]. Another drug requiring certain level of expression before treatment initiation is pembrolizumab, an anti-PD-1 human antibody. Pembrolizumab monotherapy in metastatic non-small cell lung cancer patients with PD-L1 tumor proportion score of 50% or greater and no driver mutations resulted with improvement in survival of 11.7 and increased response rate comparing to standard chemotherapy [113].

Efficacy of many other antibodies binding to PD-1 or PD-L1 increases along with higher level of PD-L1 expression but efficacy is still present even with no expression of PD-L1 is detected. Therefore PD-L1 status is not an ideal biomarker for all cancers and drugs targeting PD-1 pathway [110, 114, 115].

1.17 dMMR/MSI status

In the colorectal carcinogenesis are involved three molecular pathways: the chromosomal instability pathway, the mutator pathway (microsatellite instability, MSI) and the epigenetic instability pathway. MSI is the result of DNA mismatch repair (dMMR) system dysfunction [116]. Microsatellites are defined as repetitive units in DNA. Normally, the insertions or deletions in these regions are repaired by the MMR system and deficiency of this system causes dMMR. MMR gene products involve MLH1, MSH2, MSH6 and PMS2. A germ line mutation in one of these genes, usually MLH1 or MSH2, causes dMMR in patients with Lynch syndrome occurring in 0.8–5% of all CRCs [117]. CRC has been classified by the presence of insertion-deletions in microsatellites containing loci into three groups: microsatellite instability high (MSI-H), microsatellite instability low (MSI-L), and microsatellite stable (MSS) [118].

High levels of MSI are observed in about 5% of CRC patients. CRCs with MSI-H have a significant lymphocyte infiltration and high expression of immune checkpoints what makes them sensitive to PD-1 inhibitors. The KEYNOTE-177 randomized trial of first-line pembrolizumab versus physician's chemotherapy for dMMR or MSI-H mCRC showed statistically significant better PFS for pembrolizumab group [119].

Take home message:
1. Biomarker is defined as indicator of normal or pathogenic biological process or response to an intervention or exposure.
2. There are several categories of biomarkers: predictive, diagnostic, prognostic, monitoring, safety and pharmacodynamics. Predictive biomarkers play the most important role in the treatment of neoplastic diseases.
3. Identification of predictive biomarkers allows the development of novel targeted therapies resulting in tailored treatment.

Author contributions: All the authors have accepted responsibility for the entire content of this submitted manuscript and approved submission.
Research funding: This research was supported by funds for statutory research from the Ludwik Rydygier Collegium Medicum Nicolaus Copernicus University (UMK CM 2018 WL 103 and UMK CM 2020 WL 183).
Conflict of interest statement: none.

References

1. Califf RM. Biomarker definitions and their applications. *Exp Biol Med* 2018;243:213–21.
2. FDA-NIH Biomarker Working Group. *BEST (biomarkers, endpointS, and other tools) resource*. MD: Silver Spring; 2016.
3. Wasung ME, Chawla LS, Madero M. Biomarkers of renal function, which and when? *Clin Chim Acta* 2015;438:350–7.
4. Scott DW, Wright GW, Williams PM, Lih C-J, Walsh W, Jaffe ES, et al. Determining cell-of-origin subtypes of diffuse large B-cell lymphoma using gene expression in formalin-fixed paraffin-embedded tissue. *Blood* 2014;123:1214–17.
5. Sandler HM, Eisenberger MA. Assessing and treating patients with increasing prostate specific antigen following radical prostatectomy. *J Urol* 2007;178:S20–24.
6. Gundogdu F, Soylu F, Erkan L, Tatli O, Mavi S, Yavuzcan A. The role of serum CA-125 levels and CA-125 tissue expression positivity in the prediction of the recurrence of stage III and IV epithelial ovarian tumors (CA-125 levels and tissue CA-125 in ovarian tumors). *Arch Gynecol Obstet* 2011;283:1397–402.
7. Thorlacius S, Struewing JP, Hartge P, Olafsdottir GH, Sigvaldason H, Tryggvadottir L, et al. Population-based study of risk of breast cancer in carriers of BRCA2 mutation. *Lancet* 1998;352:1337–9.
8. Struewing JP, Hartge P, Wacholder S, Baker SM, Berlin M, McAdams M, et al. The risk of cancer associated with specific mutations of BRCA1 and BRCA2 among Ashkenazi Jews. *N Engl J Med* 1997;336:1401–8.
9. Wahl RL, Jacene H, Kasamon Y, Lodge MA. From RECIST to PERCIST: evolving Considerations for PET response criteria in solid tumors. *J Nucl Med* 2009;50(1 Suppl):122S–50S.
10. Seshachalam A, Karpurmath SV, Rathnam K, Raman SG, Janarthinakani M, Prasad K, et al. Does interim PET scan after 2 cycles of ABVD predict outcome in Hodgkin lymphoma? Real-world evidence. *J Glob Oncol* 2019;5:1–13.
11. Jorissen RN, Walker F, Pouliot N, Garrett TP, Ward CW. Epidermal growth factor receptor: mechanisms of activation and signalling. *Exp Cell Res* 2003;284:31–53.
12. Hirsch FR, Bunn PAJr. EGFR testing in lung cancer is ready for prime time. *Lancet Oncol* 2009;10:432–3.
13. Lewandowska MA, Czubak K, Klonowska K, Jozwicki W, Kowalewski J, Kozlowski P. The use of a two-tiered testing strategy for the simultaneous detection of small EGFR mutations and EGFR amplification in lung cancer. *PloS One* 2015;10:e0117983.
14. O'Kane GM, Bradbury PA, Feld R, Leighl NB, Liu G, Pisters KM, et al. Uncommon EGFR mutations in advanced non-small cell lung cancer. *Lung Canc* 2017;109:137–44.
15. Zhang YL, Yuan JQ, Wang KF, Fu X-H, Han X-R, Threapleton D, et al. The prevalence of EGFR mutation in patients with non-small cell lung cancer: a systematic review and meta-analysis. *Oncotarget* 2016;7:78985–93.
16. Rosell R, Carcereny E, Gervais R, Vergnenegre A, Massuti B, Felip E, et al. Erlotinib versus standard chemotherapy as first-line treatment for European patients with advanced EGFR mutation-positive non-small-cell lung cancer (EURTAC): a multicentre, open-label, randomised phase 3 trial. *Lancet Oncol* 2012;13:239–46.
17. Mitsudomi T, Morita S, Yatabe Y, Negoro S, Okamoto I, Tsurutani J, et al. Gefitinib versus cisplatin plus docetaxel in patients with non-small-cell lung cancer harbouring mutations of the epidermal growth factor receptor (WJTOG3405): an open label, randomised phase 3 trial. *Lancet Oncol* 2010;11:121–8.

18. Sequist LV, Yang JC, Yamamoto N, O'Byrne K, Hirsh V, Mok T, et al. Phase III study of afatinib or cisplatin plus pemetrexed in patients with metastatic lung adenocarcinoma with EGFR mutations. *J Clin Oncol* 2013;31:3327–34.
19. Mok TS, Wu YL, Ahn MJ, Garassino MC, Kim HR, Ramalingam SS, et al. Osimertinib or platinum-pemetrexed in EGFR T790M-positive lung cancer. *N Engl J Med* 2017;376:629–40.
20. Soria JC, Ohe Y, Vansteenkiste J, Reungwetwattana T, Chewaskulyong B, Lee KH, et al. Osimertinib in untreated EGFR-mutated advanced non-small-cell lung cancer. *N Engl J Med* 2018;378:113–25.
21. Ducray SP, Natarajan K, Garland GD, Turner SD, Egger G. The transcriptional roles of ALK fusion proteins in tumorigenesis. *Cancers* 2019;11.
22. Solomon B, Varella-Garcia M, Camidge DR. ALK gene rearrangements: a new therapeutic target in a molecularly defined subset of non-small cell lung cancer. *J Thorac Oncol* 2009;4:1450–4.
23. Chrzanowska NM, Kowalewski J, Lewandowska MA. Use of fluorescence in situ hybridization (FISH) in diagnosis and tailored therapies in solid tumors. *Molecules* 2020;25.
24. Lindeman NI, Cagle PT, Beasley MB, Chitale DA, Dacic S, Giaccone G, et al. Molecular testing guideline for selection of lung cancer patients for EGFR and ALK tyrosine kinase inhibitors: guideline from the College of American Pathologists, International Association for the Study of Lung Cancer, and Association for Molecular Pathology. *J Thorac Oncol* 2013;8:823–59.
25. Weickhardt AJ, Aisner DL, Franklin WA, Varella-Garcia M, Doebele RC, Camidge DR. Diagnostic assays for identification of anaplastic lymphoma kinase-positive non-small cell lung cancer. *Cancer* 2013;119:1467–77.
26. Morris SW, Kirstein MN, Valentine MB, Dittmer K, Shapiro D, Saltman D, et al. Fusion of a kinase gene, ALK, to a nucleolar protein gene, NPM, in non-Hodgkin's lymphoma. *Science* 1994;263:1281–4.
27. Osajima-Hakomori Y, Miyake I, Ohira M, Nakagawara A, Nakagawa A, Sakai R. Biological role of anaplastic lymphoma kinase in neuroblastoma. *Am J Pathol* 2005;167:213–22.
28. Kwak EL, Bang YJ, Camidge DR, Shaw AT, Solomon B, Maki RG, et al. Anaplastic lymphoma kinase inhibition in non-small-cell lung cancer. *N Engl J Med* 2010;363:1693–703.
29. Shaw AT, Kim DW, Nakagawa K, Seto T, Crinó L, Ahn MJ, et al. Crizotinib versus chemotherapy in advanced ALK-positive lung cancer. *N Engl J Med* 2013;368:2385–94.
30. Solomon BJ, Mok T, Kim DW, Wu Y-L, Nakagawa K, Mekhail T, et al. First-line crizotinib versus chemotherapy in ALK-positive lung cancer. *N Engl J Med* 2014;371:2167–77.
31. Peters S, Camidge DR, Shaw AT, Gadgeel S, Ahn JS, Kim D-W, et al. Alectinib versus crizotinib in untreated ALK-positive non-small-cell lung cancer. *N Engl J Med* 2017;377:829–38.
32. Acquaviva J, Wong R, Charest A. The multifaceted roles of the receptor tyrosine kinase ROS in development and cancer. *Biochim Biophys Acta* 2009;1795:37–52.
33. Davies KD, Doebele RC. Molecular pathways: ROS1 fusion proteins in cancer. *Clin Canc Res* 2013;19:4040–5.
34. Shaw AT, Ou SH, Bang YJ, Camidge DR, Solomon BJ, Salgia R, et al. Crizotinib in ROS1-rearranged non-small-cell lung cancer. *N Engl J Med* 2014;371:1963–71.
35. Drilon A, Siena S, Dziadziuszko R, Barlesi F, Krebs MG, Shaw AT, et al. Entrectinib in ROS1 fusion-positive non-small-cell lung cancer: integrated analysis of three phase 1-2 trials. *Lancet Oncol* 2020;21:261–70.
36. Lim SM, Kim HR, Lee JS, Lee KH, Lee Y-G, Min YJ, et al. Open-label, multicenter, phase II study of ceritinib in patients with non-small-cell lung cancer harboring ROS1 rearrangement. *J Clin Oncol* 2017;35:2613–18.
37. Killock D. Lorlatinib in ROS1-positive NSCLC. *Nat Rev Clin Oncol* 2020;17:7.

38. Ferrara R, Auger N, Auclin E, Besse B. Clinical and translational implications of RET rearrangements in non-small cell lung cancer. *J Thorac Oncol* 2018;13:27–45.
39. Drilon A, Oxnard G, Wirth L, Solomon B. PL02.08 registrational results of LIBRETTO-001: a phase 1/2 trial of LOXO-292 in patients with RET fusion-positive lung cancers. *J Thorac Oncol* 2019;14:S6–7.
40. Drilon A, Rekhtman N, Arcila M, Wang L, Ni A, Albano M, et al. Cabozantinib in patients with advanced RET-rearranged non-small-cell lung cancer: an open-label, single-centre, phase 2, single-arm trial. *Lancet Oncol* 2016;17:1653–60.
41. Gatalica Z, Xiu J, Swensen J, Vranic S. Molecular characterization of cancers with NTRK gene fusions. *Mod Pathol* 2019;32:147–53.
42. Farago AF, Taylor MS, Doebele RC, Zhu VW, Kummar S, Spira AI, et al. Clinicopathologic features of non-small-cell lung cancer harboring an NTRK gene fusion. *JCO Precis Oncol* 2018;2018.
43. Drilon A, Laetsch TW, Kummar S, DuBois SG, Lassen UN, Demetri GD, et al. Efficacy of larotrectinib in TRK fusion-positive cancers in adults and children. *N Engl J Med* 2018;378:731–9.
44. Doebele RC, Drilon A, Paz-Ares L, Siena S, Shaw AT, Farago AF, et al. Entrectinib in patients with advanced or metastatic NTRK fusion-positive solid tumours: integrated analysis of three phase 1-2 trials. *Lancet Oncol* 2020;21:271–82.
45. Awad MM, Oxnard GR, Jackman DM, Savukoski DO, Hall D, Shivdasani P, et al. MET exon 14 mutations in non-small-cell lung cancer are associated with advanced age and stage-dependent MET genomic amplification and c-Met overexpression. *J Clin Oncol* 2016;34:721–30.
46. Vansteenkiste JF, Van De Kerkhove C, Wauters E, Van Mol P. Capmatinib for the treatment of non-small cell lung cancer. *Expert Rev Anticancer Ther* 2019;19:659–71.
47. Mazieres J, Drilon A, Lusque A, Mhanna L, Cortot AB, Mezquita L, et al. Immune checkpoint inhibitors for patients with advanced lung cancer and oncogenic driver alterations: results from the IMMUNOTARGET registry. *Ann Oncol* 2019;30:1321–8.
48. Davies H, Bignell GR, Cox C, Stephens P, Edkins S, Clegg S, et al. Mutations of the BRAF gene in human cancer. *Nature* 2002;417:949–54.
49. Dhillon AS, Hagan S, Rath O, Kolch W. MAP kinase signalling pathways in cancer. *Oncogene* 2007;26:3279–90.
50. Long GV, Menzies AM, Nagrial AM, Haydu LE, Hamilton AL, Mann GJ, et al. Prognostic and clinicopathologic associations of oncogenic BRAF in metastatic melanoma. *J Clin Oncol* 2011;29:1239–46.
51. Paik PK, Arcila ME, Fara M, Sima CS, Miller VA, Kris MG, et al. Clinical characteristics of patients with lung adenocarcinomas harboring BRAF mutations. *J Clin Oncol* 2011;29:2046–51.
52. Van Cutsem E, Kohne CH, Lang I, Folprecht G, Nowacki MP, Cascinu S, et al. Cetuximab plus irinotecan, fluorouracil, and leucovorin as first-line treatment for metastatic colorectal cancer: updated analysis of overall survival according to tumor KRAS and BRAF mutation status. *J Clin Oncol* 2011;29:2011–9.
53. McArthur GA, Chapman PB, Robert C, Larkin J, Haanen JB, Dummer R, et al. Safety and efficacy of vemurafenib in BRAF(V600E) and BRAF(V600K) mutation-positive melanoma (BRIM-3): extended follow-up of a phase 3, randomised, open-label study. *Lancet Oncol* 2014;15:323–32.
54. Hauschild A, Grob J, Demidov L, Jouary T. An update on BREAK-3, a phase III, randomized trial: dabrafenib (DAB) versus dacarbazine (DTIC) in patients with BRAF V600E-positive mutation metastatic melanoma (MM). *J Clin Oncol* 2013;31:9013.
55. Larkin J, Ascierto PA, Dreno B, Atkinson V, Liszkay G, Maio M, et al. Combined vemurafenib and cobimetinib in BRAF-mutated melanoma. *N Engl J Med* 2014;371:1867–76.

56. Long GV, Stroyakovskiy D, Gogas H, Levchenko E, de Braud F, Larkin J, et al. Dabrafenib and trametinib versus dabrafenib and placebo for Val600 BRAF-mutant melanoma: a multicentre, double-blind, phase 3 randomised controlled trial. *Lancet* 2015;386:444–51.
57. Kopetz S, Grothey A, Yaeger R, Van Cutsem E, Desai J, Yoshino T, et al. Encorafenib, binimetinib, and cetuximab in BRAF V600e-mutated colorectal cancer. *N Engl J Med* 2019;381:1632–43.
58. Amado RG, Wolf M, Peeters M, Van Cutsem E, Siena S, Freeman DJ, et al. Wild-type KRAS is required for panitumumab efficacy in patients with metastatic colorectal cancer. *J Clin Oncol* 2008;26:1626–34.
59. Roth AD, Tejpar S, Delorenzi M, Yan P, Fiocca R, Klingbiel D, et al. Prognostic role of KRAS and BRAF in stage II and III resected colon cancer: results of the translational study on the PETACC-3, EORTC 40993, SAKK 60-00 trial. *J Clin Oncol* 2010;28:466–74.
60. Sorich MJ, Wiese MD, Rowland A, Kichenadasse G, McKinnon RA, Karapetis CS. Extended RAS mutations and anti-EGFR monoclonal antibody survival benefit in metastatic colorectal cancer: a meta-analysis of randomized, controlled trials. *Ann Oncol* 2015;26:13–21.
61. Wan PT, Garnett MJ, Roe SM, Lee S, Niculescu-Duvaz D, Good VM, et al. Mechanism of activation of the RAF-ERK signaling pathway by oncogenic mutations of B-RAF. *Cell* 2004;116:855–67.
62. Pietrantonio F, Petrelli F, Coinu A, Di Bartolomeo M, Borgonovo K, Maggi C, et al. Predictive role of BRAF mutations in patients with advanced colorectal cancer receiving cetuximab and panitumumab: a meta-analysis. *Eur J Canc* 2015;51:587–94.
63. Douillard JY, Oliner KS, Siena S, Tabernero J, Burkes R, Barugel M, et al. Panitumumab-FOLFOX4 treatment and RAS mutations in colorectal cancer. *N Engl J Med* 2013;369:1023–34.
64. Geredeli C, Yasar N. FOLFIRI plus panitumumab in the treatment of wild-type KRAS and wild-type NRAS metastatic colorectal cancer. *World J Surg Oncol* 2018;16:67.
65. Kim TW, Elme A, Kusic Z, Park JO, Udrea AA, Kim SY, et al. A phase 3 trial evaluating panitumumab plus best supportive care vs best supportive care in chemorefractory wild-type KRAS or RAS metastatic colorectal cancer. *Br J Canc* 2016;115:1206–14.
66. Karapetis CS, Khambata-Ford S, Jonker DJ, O'Callaghan CJ, Tu D, Tebbutt NC, et al. K-ras mutations and benefit from cetuximab in advanced colorectal cancer. *N Engl J Med* 2008;359:1757–65.
67. Sjoblom T, Shimizu A, O'Brien KP, Pietras K, Dal Cin P, Buchdunger E, et al. Growth inhibition of dermatofibrosarcoma protuberans tumors by the platelet-derived growth factor receptor antagonist STI571 through induction of apoptosis. *Canc Res* 2001;61:5778–83.
68. McArthur GA. Molecularly targeted treatment for dermatofibrosarcoma protuberans. *Semin Oncol* 2004;31:31–6.
69. Rutkowski P, Van Glabbeke M, Rankin CJ, Ruka W, Rubin BP, Debiec-Rychter M, et al. Imatinib mesylate in advanced dermatofibrosarcoma protuberans: pooled analysis of two phase II clinical trials. *J Clin Oncol* 2010;28:1772–9.
70. Yoo C, Ryu MH, Jo J, Park I, Ryoo B-Y, Kang Y-K. Efficacy of imatinib in patients with platelet-derived growth factor receptor alpha-mutated gastrointestinal stromal tumors. *Cancer Res Treat* 2016;48:546–52.
71. Heinrich MC, Corless CL, Demetri GD, Blanke CD, von Mehren M, Joensuu H, et al. Kinase mutations and imatinib response in patients with metastatic gastrointestinal stromal tumor. *J Clin Oncol* 2003;21:4342–9.
72. Heinrich MC, Maki RG, Corless CL, Antonescu CR, Harlow A, Griffith D, et al. Primary and secondary kinase genotypes correlate with the biological and clinical activity of sunitinib in imatinib-resistant gastrointestinal stromal tumor. *J Clin Oncol* 2008;26:5352–9.

73. Curtin JA, Busam K, Pinkel D, Bastian BC. Somatic activation of KIT in distinct subtypes of melanoma. *J Clin Oncol* 2006;24:4340–6.
74. Guo J, Si L, Kong Y, Flaherty KT, Xu X, Zhu Y, et al. Phase II, open-label, single-arm trial of imatinib mesylate in patients with metastatic melanoma harboring c-Kit mutation or amplification. *J Clin Oncol* 2011;29:2904–9.
75. Paterni I, Granchi C, Katzenellenbogen JA, Minutolo F. Estrogen receptors alpha (ERalpha) and beta (ERbeta): subtype-selective ligands and clinical potential. *Steroids* 2014;90:13–29.
76. Allred DC, Carlson RW, Berry DA, Burstein HJ, Edge SB, Goldstein LJ, et al. NCCN Task Force Report: estrogen receptor and progesterone receptor testing in breast cancer by immunohistochemistry. *J Natl Compr Canc Netw* 2009;7(6 Suppl):S1–21; quiz S22–23.
77. Powles TJ. Efficacy of tamoxifen as treatment of breast cancer. *Semin Oncol* 1997;24:S1–48; S41–54.
78. Davies C, Pan H, Godwin J, Gray R, Arriagada R, Raina V, et al. Long-term effects of continuing adjuvant tamoxifen to 10 years versus stopping at 5 years after diagnosis of oestrogen receptor-positive breast cancer: ATLAS, a randomised trial. *Lancet* 2013;381:805–16.
79. Lonning PE, Eikesdal HP. Aromatase inhibition 2013: clinical state of the art and questions that remain to be solved. *Endocr Relat Canc* 2013;20:R183–201.
80. Lee CI, Goodwin A, Wilcken N. Fulvestrant for hormone-sensitive metastatic breast cancer. *Cochrane Database Syst Rev* 2017;1:CD011093.
81. Decruze SB, Green JA. Hormone therapy in advanced and recurrent endometrial cancer: a systematic review. *Int J Gynecol Canc* 2007;17:964–78.
82. Andre F, Ciruelos E, Rubovszky G, Campone M, Loibl S, RugoHS, et al. Alpelisib for PIK3CA-mutated, hormone receptor-positive advanced breast cancer. *N Engl J Med* 2019;380:1929–40.
83. Malone KE, Daling JR, Doody DR, Hsu L, Bernstein L, Coates RJ, et al. Prevalence and predictors of BRCA1 and BRCA2 mutations in a population-based study of breast cancer in white and black American women ages 35 to 64 years. *Canc Res* 2006;66:8297–308.
84. Ford D, Easton DF, Stratton M, Narod S, Goldgar D, Devilee P, et al. Genetic heterogeneity and penetrance analysis of the BRCA1 and BRCA2 genes in breast cancer families. The Breast Cancer Linkage Consortium. *Am J Hum Genet* 1998;62:676–89.
85. Robson M, Im SA, Senkus E, Xu B, Domchek SM, Masuda N, et al. Olaparib for metastatic breast cancer in patients with a germline BRCA mutation. *N Engl J Med* 2017;377:523–33.
86. Pujade-Lauraine E, Ledermann JA, Selle F, Gebski V, Penson RT, Oza AM, et al. Olaparib tablets as maintenance therapy in patients with platinum-sensitive, relapsed ovarian cancer and a BRCA1/2 mutation (SOLO2/ENGOT-Ov21): a double-blind, randomised, placebo-controlled, phase 3 trial. *Lancet Oncol* 2017;18:1274–84.
87. Golan T, Hammel P, Reni M, Van Cutsem E, Macarulla T, Hall MJ, et al. Maintenance olaparib for germline BRCA-mutated metastatic pancreatic cancer. *N Engl J Med* 2019;381:317–27.
88. Litton JK, Rugo HS, Ettl J, Hurvitz SA, Gonçalves A, Lee K-H, et al. Talazoparib in patients with advanced breast cancer and a germline BRCA mutation. *N Engl J Med* 2018;379:753–63.
89. Coleman RL, Oza AM, Lorusso D, Aghajanian C, Oaknin A, Dean A, et al. Rucaparib maintenance treatment for recurrent ovarian carcinoma after response to platinum therapy (ARIEL3): a randomised, double-blind, placebo-controlled, phase 3 trial. *Lancet* 2017;390:1949–61.
90. Cho HS, Mason K, Ramyar KX, Stanley AM, Gabelli SB, Denney DW, et al. Structure of the extracellular region of HER2 alone and in complex with the Herceptin Fab. *Nature* 2003;421:756–60.
91. Burstein HJ. The distinctive nature of HER2-positive breast cancers. *N Engl J Med* 2005;353:1652–4.

92. Cameron D, Piccart-Gebhart MJ, Gelber RD, Procter M, Goldhirsch A, de Azambuja E, et al. 11 years' follow-up of trastuzumab after adjuvant chemotherapy in HER2-positive early breast cancer: final analysis of the HERceptin Adjuvant (HERA) trial. *Lancet* 2017;389:1195–205.
93. Slamon DJ, Leyland-Jones B, Shak S, Fuchs H, Paton V, Bajamonde A, et al. Use of chemotherapy plus a monoclonal antibody against HER2 for metastatic breast cancer that overexpresses HER2. *N Engl J Med* 2001;344:783–92.
94. Bang YJ, Van Cutsem E, Feyereislova A, Chung HC, Shen L, Sawaki A, et al. Trastuzumab in combination with chemotherapy versus chemotherapy alone for treatment of HER2-positive advanced gastric or gastro-oesophageal junction cancer (ToGA): a phase 3, open-label, randomised controlled trial. *Lancet* 2010;376:687–97.
95. Gianni L, Pienkowski T, Im YH, Roman L, Tseng L-M, Liu M-C, et al. Efficacy and safety of neoadjuvant pertuzumab and trastuzumab in women with locally advanced, inflammatory, or early HER2-positive breast cancer (NeoSphere): a randomised multicentre, open-label, phase 2 trial. *Lancet Oncol* 2012;13:25–32.
96. von Minckwitz G, Procter M, de Azambuja E, Zardavas D, Benyunes M, Viale G, et al. Adjuvant pertuzumab and trastuzumab in early HER2-positive breast cancer. *N Engl J Med* 2017;377:122–31.
97. Swain SM, Baselga J, Kim SB, Ro J, Semiglazov V, Campone M, et al. Pertuzumab, trastuzumab, and docetaxel in HER2-positive metastatic breast cancer. *N Engl J Med* 2015;372:724–34.
98. Geyer CE, Forster J, Lindquist D, Chan S, Romieu CG, Pienkowski T, et al. Lapatinib plus capecitabine for HER2-positive advanced breast cancer. *N Engl J Med* 2006;355:2733–43.
99. von Minckwitz G, Huang CS, Mano MS, Loibl S, Mamounas EP, Untch M, et al. Trastuzumab emtansine for residual invasive HER2-positive breast cancer. *N Engl J Med* 2019;380:617–28.
100. Krop IE, Kim SB, Gonzalez-Martin A, LoRusso PM, Ferrero J-M, Smitt M, et al. Trastuzumab emtansine versus treatment of physician's choice for pretreated HER2-positive advanced breast cancer (TH3RESA): a randomised, open-label, phase 3 trial. *Lancet Oncol* 2014;15:689–99.
101. Rugo HS, Im SA, Cardoso F, Cortes J. Abstract nr GS1-02: phase 3 SOPHIA study of margetuximab + chemotherapy vs trastuzumab + chemotherapy in patients with HER2+ metastatic breast cancer after prior anti-HER2 therapies: second interim overall survival analysis. In: *Proceedings of the 2019 San Antonio breast cancer symposium; 2019 Dec 10–14.* AACR, San Antonio, TX; 2019.
102. Pardoll DM. The blockade of immune checkpoints in cancer immunotherapy. *Nat Rev Canc* 2012;12:252–64.
103. Ceeraz S, Nowak EC, Noelle RJ. B7 family checkpoint regulators in immune regulation and disease. *Trends Immunol* 2013;34:556–63.
104. Keir ME, Butte MJ, Freeman GJ, Sharpe AH. PD-1 and its ligands in tolerance and immunity. *Annu Rev Immunol* 2008;26:677–704.
105. McDermott DF, Atkins MB. PD-1 as a potential target in cancer therapy. *Cancer Med* 2013;2:662–73.
106. Udall M, Rizzo M, Kenny J, Doherty J, Dahm S, Robbins P, et al. PD-L1 diagnostic tests: a systematic literature review of scoring algorithms and test-validation metrics. *Diagn Pathol* 2018;13:12.
107. Kluger HM, Zito CR, Turcu G, Baine MK, Zhang H, Adeniran A, et al. PD-L1 studies across tumor types, its differential expression and predictive value in patients treated with immune checkpoint inhibitors. *Clin Canc Res* 2017;23:4270–9.

108. Brown JA, Dorfman DM, Ma FR, Sullivan EL, Munoz O, Wood CR, et al. Blockade of programmed death-1 ligands on dendritic cells enhances T cell activation and cytokine production. *J Immunol* 2003;170:1257–66.
109. Xu Y, Wan B, Chen X, Zhan P, Zhao Y, Zhang T, et al. The association of PD-L1 expression with the efficacy of anti-PD-1/PD-L1 immunotherapy and survival of non-small cell lung cancer patients: a meta-analysis of randomized controlled trials. *Transl Lung Cancer Res* 2019;8:413–28.
110. Robert C, Ribas A, Schachter J, Arance A, GrobJ-J, Mortier L, et al. Pembrolizumab versus ipilimumab in advanced melanoma (KEYNOTE-006): post-hoc 5-year results from an open-label, multicentre, randomised, controlled, phase 3 study. *Lancet Oncol* 2019;20:1239–51.
111. Schmid P, Rugo HS, Adams S, Schneeweiss A, Barrios CH, Iwata H, et al. Atezolizumab plus nab-paclitaxel as first-line treatment for unresectable, locally advanced or metastatic triple-negative breast cancer (IMpassion130): updated efficacy results from a randomised, double-blind, placebo-controlled, phase 3 trial. *Lancet Oncol* 2020;21:44–59.
112. Freeman GJ, Long AJ, Iwai Y, Bourque K, Chernova T, Nishimura H, et al. Engagement of the PD-1 immunoinhibitory receptor by a novel B7 family member leads to negative regulation of lymphocyte activation. *J Exp Med* 2000;192:1027–34.
113. Reck M, Rodriguez-Abreu D, Robinson AG, Hui R, Csőszi T, Fülöp A, et al. Updated analysis of KEYNOTE-024: pembrolizumab versus platinum-based chemotherapy for advanced non-small-cell lung cancer with PD-L1 tumor proportion score of 50% or greater. *J Clin Oncol* 2019;37:537–46.
114. Gandhi L, Rodriguez-Abreu D, Gadgeel S, Esteban E, Felip E, De Angelis F, et al. Pembrolizumab plus chemotherapy in metastatic non-small-cell lung cancer. *N Engl J Med* 2018;378:2078–92.
115. Motzer RJ, Tannir NM, McDermott DF, Arén Frontera O, Melichar B, Choueiri TK, et al. Nivolumab plus ipilimumab versus sunitinib in advanced renal-cell carcinoma. *N Engl J Med* 2018;378:1277–90.
116. Venderbosch S, Nagtegaal ID, Maughan TS, Smith CG, Cheadle JP, Fisher D, et al. Mismatch repair status and BRAF mutation status in metastatic colorectal cancer patients: a pooled analysis of the CAIRO, CAIRO2, COIN, and FOCUS studies. *Clin Canc Res* 2014;20:5322–30.
117. Koopman M, Kortman GA, Mekenkamp L, Ligtenberg MJL, Hoogerbrugge N, Antonini NF, et al. Deficient mismatch repair system in patients with sporadic advanced colorectal cancer. *Br J Canc* 2009;100:266–73.
118. Vilar E, Tabernero J. Molecular dissection of microsatellite instable colorectal cancer. *Canc Discov* 2013;3:502–11.
119. Andre T, Shiu K, Kim TW, Jensen BV, Jensen LH, Punt CJA, et al. Pembrolizumab versus chemotherapy for microsatellite instability-high/mismatch repair deficient metastatic colorectal cancer: the phase 3 KEYNOTE-177 study. *J Clin Oncol* 2020;38:LBA4.
120. Szczerba E, Kamińska K, Mierzwa T, Misiek M, Kowalewski J, Lewandowska MA. BRCA1/2 mutation detection in the tumor tissue from selected Polish patients with breast cancer using next generation sequencing. *Genes* 2021;12:519. https://doi.org/10.3390/genes12040519.

Krzysztof Koper, Sławomir Wileński and Agnieszka Koper
2 Advancements in cancer chemotherapy

Abstract: Chemotherapy is in most cases a method of systemic treatment of malignant tumors with cytostatic drugs. Although modern methods such as immunotherapy or targeted therapy are used more and more often nowadays, the role of chemotherapy in oncology is still significant. It can be used as an independent treatment method or in combination with other oncological therapies. The action of chemotherapy is closely linked to the cell cycle of the tumor. Advances in technology allow the introduction of different pharmaceutical forms of the same drug. Worse prognosis of metastatic tumors justifies the need to search for new, more effective treatment methods. The main problem of chemotherapy is the occurrence of adverse events. Reducing the frequency and severity of side effects is possible primarily by changing the technique of implementation of chemotherapy administration. These principles are fulfilled by new, increasingly popular therapeutic methods, such as: Perioperative Hyperthermic Intraperitoneal Chemotherapy (HIPEC), Pressurized Intraperitoneal Aerosol Chemotherapy (PIPAC) or transarterial chemoembolization (TACE). The dynamic development of knowledge concerning cytostatic drugs, including targeting the tumor cell with the form of the drug, allows us to assume that in the future this direction will increase the effectiveness and safety of anticancer therapy.

Keywords: anticancer drug forms, chemotherapy, drug administration routes, liposome, micellar solubilization.

2.1 Introduction

Chemotherapy is a method of systemic treatment of malignant tumors with cytostatic drugs. However, some cytostatic drugs can be applied topically, e.g., they are introduced into the body cavities. An example is peritoneal chemotherapy. Cytostatic drugs, especially classic ones, are usually drugs with a narrow therapeutic index and do not show selective anticancer effects, i.e., they damage not only cancer cells but also healthy cells. Hence, cancer chemotherapy is a difficult treatment that requires consideration of pharmacokinetics, mechanism of action, dosage and frequency of use. Usually multi-drug schemes are applied at intervals of 21–28 days. The usage of 2–3 drug schemes reduces the risk of cancer resistance to the therapy. Due to the lack of selectivity of cytostatic drugs, normal cells such as bone marrow cells, gastrointestinal epithelium or reproductive cells are also destroyed.

This article has previously been published in the journal Physical Sciences Reviews. Please cite as: Koper, K., Wileński S., Koper, A. Advancements in cancer chemotherapy *Physical Sciences Reviews* [Online] 2021, 6 DOI: 10.1515/psr-2020-0206

https://doi.org/10.1515/9783110662306-002

2.2 The role of chemotherapy in cancer treatment

Chemotherapy can be used as a stand-alone treatment method or in combination with other oncological therapies. Chemotherapy is currently used in several situations:
1. Inductive treatment—the aim is to reduce tumor mass. Such an action is intended to enable local treatment to be undertaken or significantly reduce its range, e.g., used in breast cancer or head and neck tumors.
2. Neoadjuvant treatment—used entirely as a treatment improving the effects of therapy before the planned radical treatment, e.g., in breast or bladder cancer.
3. Supplementary treatment (adjuvant treatment), applied after local methods (surgery or radiotherapy). It is aimed at reducing the risk of recurrence or distant metastases [1].

2.3 The biological basis of chemotherapy

The effect of chemotherapy is closely related to the cell cycle of the cancer. Cytostatic drugs damage mainly the cells in the cell cycle. They can also affect the processes of transition of cells from one phase to another. Cytostatics in the schemes should be selected so as to affect different phases of the cell cycle.
  Cell cycle phases:
- **M** phase—the phase of mitosis, in which cell division occurs;
- **G1** phase—the stage of cell growth up to the size of the stem cell; in this phase, the differentiation of cytoplasmic structures occurs, the synthesis of RNA, the synthesis of some proteins;
- **S** phase—phase of replication, i.e., DNA synthesis;
- **G2** phase—synthesis of proteins involved in the cell division, which for example are part of the karyokinetic spindle;
- **G0** phase—resting phase.

2.4 Classification of cytostatic drugs

Cytostats are classified according to the mechanism of action and chemical structure of the drug and the phase of the cell cycle in which they operate. Some drugs in different classifications are classified into different groups.
  The classification of cytostatic drugs in terms of mechanism of action and chemical structure is based on the Anatomical Therapeutic Chemical Classification System (ATC) according to WHO Collaborating Centre for Drug Statistics Methodology [2, 3]:

2.4.1 Alkylating drugs

The mechanism of action is based on connecting highly reactive alkyl radicals of the drug to numerous functional groups with negative charge in the acid molecules of DNA and RNA and proteins (Figures 2.1, 2.2). They cause disorders of the structure of deoxyribonucleic acid and disturb the life processes of the cell. They act on all phases of the cell cycle. This group of drugs includes nitrogen iperite derivatives, nitrosourea derivatives, sulfonic acid esters and triazenes.

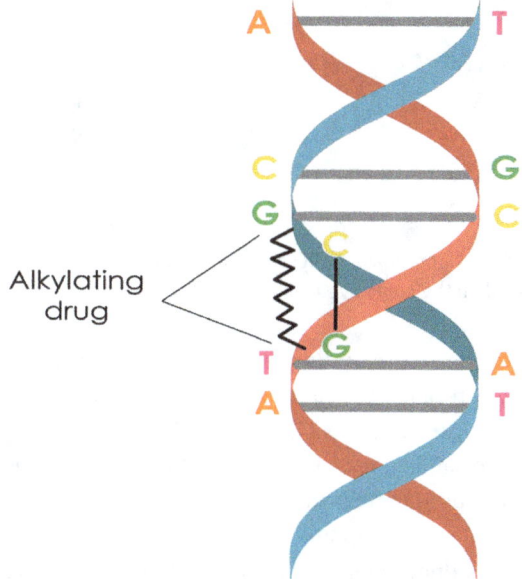

Figure 2.1: Mechanism of action of alkylating drugs; based on the study by Krzakowski and Wyrwicz [1].

2.4.1.1 Derivatives of nitrogen iperite
Chlormethine, cyclophosphamide, ifosfamide, chlorambucil, melphalan, trophosphamide, prednimustine, bendamustine.

Cyclophosphamide and ifosfamide are still commonly used in cancer therapy to the present day. Cyclophosphamide is administered not only to oncological patients, but also in case of organ transplants. In oncology, cyclophosphamide has a wide range of anticancer effects. It is used in monotherapy or in combined treatment in case of: lymphoma, leukemia, multiple myeloma, breast cancer, ovarian cancer, sarcoma and small cell lung cancer. These medicines have significant local and general toxicity. They damage the bone marrow, cause nausea and vomiting, hair loss and pulmonary fibrosis. After application of ifosfamide and cyclophosphamide, hemorrhagic cystitis is

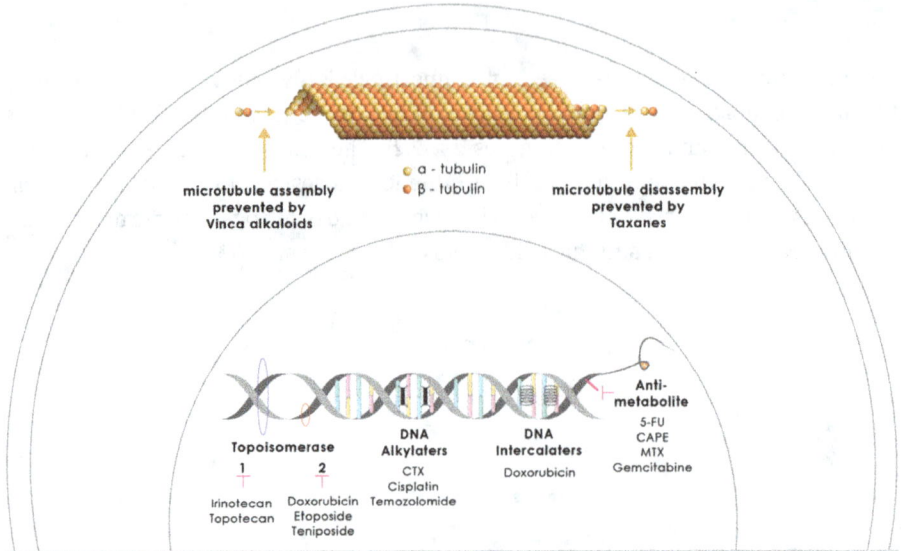

Figure 2.2: Mechanism of chemotherapy. 5-FU: 5-fluorouracil; CAPE: capecitabine, CTX: cyclophosphamide, MTX: methotrexate; based on the study by Krzakowski and Wyrwicz [1].

a characteristic complication. This complication may be prevented by administering mesna (uromitexane). Uromitexane, which binds the metabolites of both drugs, is a protective agent for anticancer treatment against the toxicity of cyclophosphamide and ifosfamide. When taking high doses of ifosfamide, neurological symptoms such as polyneuropathy may occur. Bendamustine is used to treat subsequent lines of lymphoma. Medicines of this group cause deep and late myelosuppression.

In addition, they can cause liver and kidney damage and pulmonary fibrosis. Toxic symptoms may occur with a long delay even after a few weeks after chemotherapy.

2.4.1.2 Derivatives of nitrosourea
Carmustine, lomustine, streptozotocin.

Medicines from this group are used to treat brain tumors (medulloblastoma, astrocytoma, ependymoma, glioblastoma) and metastases to the brain.

2.4.1.3 Sulfonic acid esters
Busulfan, treosulfan.

Busulfan is an organic chemical compound from the sulfonate group. This cytostatic agent is mainly used in oncology to treat chronic myelogenous leukemia. Busulfan may cause: pulmonary fibrosis, bone marrow damage, skin discoloration, gynaecomastia, damage to gastrointestinal mucosa.

2.4.1.4 Triazenes

Dacarbazin, temozolomide.

Dacarbazin is a cytostatic agent with alkylating properties of the triazene group. It is used to treat lymphomas, melanoma, sarcomas and solid tumors in children. Temozolomide, which is a derivative of dacarbazin, was introduced to treatment in 1999. It is used to treat malignant gliomas in monotherapy and in combination with radiotherapy. Dacarbazin can cause myelosuppression, nausea and vomiting, face redness, hair loss. Temozolomide causes: myelosuppression, nausea and vomiting.

2.4.2 Antimethabolites

Antimethabolites inhibit cell division by building themselves into the nuclear material as structural analogues of natural metabolites in cellular biological systems (Figures 2.2,3), act thanks to their chemical structure similar to the metabolites, killing cells in phase S. These drugs are most effective in the treatment of tumors with a high growth fraction. They include folic acid analogues, purine analogues, pyrimidine analogues.

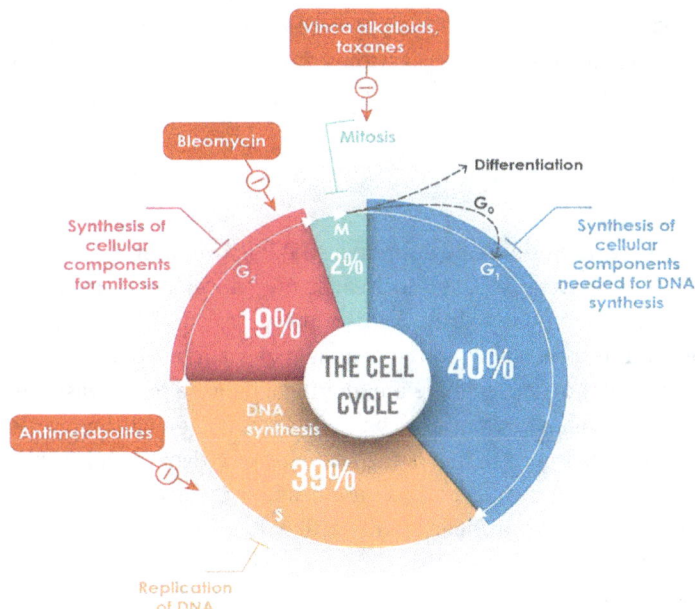

Figure 2.3: Classification of cytostatic agents by cell cycle phase in which they act; based on the studies by Krzakowski and Wyrwicz and Orzechowska-Juzwenko [1], [3].

2.4.2.1 Folic acid analogues

Methotrexate, pemetrexed, pralatrexate.

Methotrexate belongs to the compounds that are folic acid antagonists. It is not only an anticancer drug, but also a disease modulating and immunosuppressive drug. In oncology, it has been used in the treatment of solid head and neck cancers, chorionic epithelioma, acute lymphatic leukemia and acute myeloid leukemia. Pemetrexed belongs to a newer generation of folic acid analogs. This drug is recommended in malignant pleural mesothelioma and in combination with cisplatin for the treatment of nonsmall cell lung cancer as a first-line treatment in the locally advanced stage or with metastases. Methotrexate damages: bone marrow, gastrointestinal mucosa, nausea, vomiting, diarrhea, kidney damage and hair loss. Side effects from the central nervous system most often occur after administration of the drug intrathecally.

2.4.2.2 Purine analogues

Mercaptopurine, thioguanine, fludarabine, cladribine, clofarabine, nelarabine.

Currently, drugs from this group are less commonly used to treat cancer. Fludarabine is recommended in the treatment of chronic B-cell lymphocytic leukemia (CLL). Cladribine is used in the treatment of chronic lymphatic leukemia, hairy cell leukemia, acute myeloid leukemia and non-Hodgkin's lymphoma. Drugs in this group damage the bone marrow, kidneys, liver and cause gastrointestinal disorders. Respiratory toxicity in the form of interstitial pneumonia occurs. Administered in high doses are neurotoxic.

2.4.2.3 Pyrimidine analogues

5-fluorouracil, capecitabine, floxuridine, tegafur, gemcitabine, cytarabine.

5-fluorouracil is a chemotherapist used to treat many cancers, mainly of the gastrointestinal tract. Capecitabine is a drug that is a precursor of 5-fluorouracil, a drug administered orally. It is used mainly in the treatment of gastrointestinal cancers and breast cancer. Side effects of capecitabine are: diarrhea, damage to the gastrointestinal mucous membranes, nausea, vomiting, hand-foot syndrome and damage to the bone marrow. Gemcitabine is a drug that is the analogue of 2′-deoxycytidine. It is used to treat bladder, pancreas, lungs, ovary and breast cancer. Cytosine Arabinoside is a modified nucleoside. It is an anticancer drug that has been used to treat Hodgkin's lymphoma, leukemia and non-Hodgkin's lymphoma. 5-fluorouracil damages the bone marrow and gastrointestinal mucosa, and when administered in high doses can damage the nervous system. Gemcitabine causes: myelosuppression of the bone marrow, nausea and vomiting and flu-like symptoms. Cytosine Arabinoside causes: damage to the bone marrow, nausea, vomiting, diarrhea, damage to the gastrointestinal mucosa, conjunctivitis.

2.4.3 Plant alkaloids and other natural products

Anticancer drugs may be of natural origin—the common feature of this group of drugs is the fact that their active precursors are isolated from natural substances: plants, fungi and bacteria. They include:
- cytotoxic antibiotics, i.e., anthracyclines and other anticancer antibiotics;
- podophyllotoxin derivatives;
- drugs affecting the mitotic spindle;
- camptothecin derivatives.

2.4.3.1 Alkaloids of Catharanthus Roseus and their analogues
Vinblastine, vincristine, vindesine, vinorelbine, vinflunine.

Plant alkaloids (obtained from Catharanthus Roseus) show antimitotic properties. The chemical structure is similar, but the range of anticancer action varies. They belong to the group of phase-specific drugs acting on the M phase of the cell cycle. The mechanism of action is to inhibit cell division in the metaphase stage (Figure 2.2).

Vincristine is used in the treatment of lymphomas, sarcomas, breast cancer and small cell lung cancer. Vinblastine is a cytostatic agent that can be recommended in the treatment of such diseases as: Hodgkin's disease, non-Hodgkin's lymphoma, chronic lymphatic leukemia, testicular carcinoma, epithelioma, Kaposi's sarcoma, breast cancer, bladder cancer, kidney cancer and nonsmall cell lung cancer. Whereas, vinorelbine can be used to treat nonsmall cell lung cancer and advanced breast cancer. Vincristine is neurotoxic, i.e., it causes: peripheral neuropathy, sensory disorders, especially of the tips of the toes of the upper limbs, weakening of muscle strength, atrophy of reflexes. In addition, it causes adverse effects on the autonomic nervous system: constipation, intestinal obstruction, bladder atony with urination disorders. Vinblastine causes: bone marrow damage, nausea, vomiting, diarrhea, hair loss, phlebitis. Neurological symptoms appear less frequently than after vincristine. Vinorelbine and vindesine can damage veins, cause peripheral neuropathy, myelosuppression, nausea and vomiting, liver damage.

2.4.3.2 Podophilotoxin derivatives
Etoposide, teniposide.

The mechanism of action is related to the influence on enzymes involved in DNA biosynthesis. Etoposide has indications for use in small-cell lung cancer, testicular cancer, Hodgkin's lymphoma and non-Hodgkin's lymphomas, ovarian cancer, gestational trophoblastic disease and acute myeloid leukemia. Teniposide inhibiting topoisomerase II has indications for the treatment of acute lymphoblastic leukemia, non-Hodgkin's lymphomas and brain tumors. The etoposide damages the bone marrow, causes: nausea, vomiting, diarrhea, oral mucositis, hair loss. Teniposide—the most common adverse symptoms are also: damage to the bone marrow, nausea, vomiting, diarrhea and hair loss.

2.4.3.3 Taxoids

Paclitaxel, docetaxel, cabazitaxel.

They prevent cell division by dysfunction of the mitotic spindle (Figures 2.2,3). Their antimitotic effects are different from those of the periwinkle alkaloids. They inhibit the depolymerization of microtubules and stabilize them.

Taxoids are the leading new generation of drugs used in chemotherapy. Paclitaxel is administered in many cancers such as breast cancer, ovarian cancer and lung cancer. Docetaxel has been used to treat non-small cell lung cancer, prostate cancer, stomach cancer, head and neck cancer and breast cancer. Paclitaxel causes: neutropenia, peripheral sensory and motor neuropathies, heart rhythm disorders, hair loss, nausea, vomiting, anaphylactic reactions. Docetaxel causes side effects like paclitaxel. During treatment, peripheral oedemas may additionally occur.

2.4.3.4 Cytotoxic antibiotics and related substances

Anthracycline antibiotics and related substances: doxorubicin, daunorubicin, epirubicin, idarubicin, valrubicin, amrubicin, mitoxantrone, pixantrone.

Their anticancer action consists of binding to DNA and creating free radicals, interrupting and stabilizing the double helix of DNA, preventing the transcription of DNA code into RNA and inhibiting RNA synthesis (Figure 2.2). The compounds are isolated from the Streptomyces fungi. This group of drugs includes: anthracyclines I (doxorubicin and daunorubicin) and II generation (epirubicin, idarubicin) and other antibiotics.

The most commonly used cytotoxic antibiotic is doxorubicin. It is mainly recommended for the treatment of breast cancer, Hodgkin's lymphoma and non-Hodgkin's lymphomas. Furthermore, doxorubicin has a recognized use in the treatment of such diseases as: ovarian cancer, endometrial cancer, small cell lung cancer, bladder cancer, thyroid cancer, stomach cancer, prostate cancer, liver cancer as well as acute lymphoblastic and myeloblastic leukemia, multiple myeloma, Wilms' tumor, soft tissue sarcoma, neuroblastoma and squamous cell carcinoma of the head and neck. The range of action of epirubicin is similar to that of doxorubicin described earlier. It is worth mentioning the fact of lower toxicity in the case of epirubicin, especially in terms of cardiotoxicity. Idarubicin is a synthetic antibiotic with antimitotic and cytostatic effects for the treatment of acute nonlymphoblastic and lymphoblastic leukemia, multiple myeloma and advanced breast cancer. Due to its significant cardiotoxic effect, daunorubicin, in turn, is currently used only in the induction treatment of acute leukemia. Doxorubicin is cardiotoxic and causes myelosuppression. It damages the gastrointestinal mucosa and causes: nausea, vomiting, hair loss. Epirubicin—the side effects are the same as during treatment with doxorubicin, but it is less cardiotoxic. Daunorubicin—the most common toxic complications are: damage to the bone marrow, myocardium, liver, gastrointestinal tract, hair loss, phlebitis, stomatitis and fever.

Currently, mitoxantron is a drug rarely used in oncology. If no other treatment is possible, this drug can be administered in patients with non-Hodgkin's lymphoma, acute myelogenous leukemia, advanced breast cancer and castration-resistant prostate cancer. Mitoxantron causes: myelosuppression of the bone marrow, nausea, vomiting, myocardial damage, neuropathy.

Actinomycin: dactinomycin.

Dactinomycin is a cytostatic antibiotic used in the treatment of childhood cancers (Wilms tumor, Ewing sarcoma), soft tissue sarcomas, germ cell tumors of the testis and ovary, and gestational trophoblastic disease (GTD). Dactinomycin is considered to be one of the strongest anticancer drugs. Dactinomycin damages the bone marrow, which symptoms include thrombocytopenia or pancytopenia. It also causes nausea, vomiting, hair loss and dermatitis.

Other cytotoxic antibiotics: bleomycin, mitomycin, plicamycin, ixabepilone.

The most commonly used drug in this group is bleomycin. This drug is used to treat various cancers, such as Hodgkin's lymphoma, non-Hodgkin's lymphoma, testicular cancers or head and neck cancer. Mitomycin is recommended in such diagnoses as: stomach cancer, breast cancer, lung cancer, cervical cancer, head and neck cancer and bladder cancer. Mitomycin is also used in the treatment of peritoneal carcinoma in the course of gastric and colorectal cancer. With the use of this drug the Hyperthermic Intraperitoneal Chemotherapy (HIPEC) is then performed. Bleomycin can cause: toxic pneumonia and fibrosis, skin discoloration, hair loss, stomatitis, nausea and vomiting. When administered intramuscularly, it often causes an increase in body temperature. Mitomycin causes: damage to the bone marrow, nausea and vomiting, damage to the gastrointestinal mucosa, hair loss, skin lesions, lung fibrosis.

2.4.3.5 Camptothecin derivatives
Irinotecan, topotecan.

Drugs from this group act as topoisomerase I inhibitors. Irinotecan is a semisynthetic analogue of the natural camptothecin alkaloid obtained from the *Camptotheca acuminata* tree. This drug, in combination with 5-fluorouracil, is used to treat gastrointestinal cancers. Topotecan is the first registered topoisomerase I inhibitor that can be administered orally. The recommendation for its administration is the diagnosis of advanced ovarian, cervical or small cell lung cancer (SCLC). Irinotecan causes: bone marrow damage, nausea, vomiting, appetite loss, diarrhea, hair loss. Topotecan damages the bone marrow, causes nausea and vomiting.

2.4.3.6 Asparaginase

Asparaginase is an enzyme that breaks down asparagine, obtained from strains of *Escherichia coli* or *Ervinia carotovora*. It is used to treat acute lymphoblastic leukemia.

The drug causes: anaphylactic reactions, chills, fever, nausea and vomiting, hyperglycemia, liver damage.

2.4.4 Platinum derivatives

Cisplatin, carboplatin, oxaliplatin, satraplatin, polyplatillen.

Platinum derivatives are the basic group of drugs used in oncology. Cisplatin limits the growth of cancer (Figure 2.2), enhances cancer immunogenicity and increases sensitivity to radiation therapy. Cisplatin is used in monotherapy or combined therapy with other cytostatic agents to treat: ovarian cancer, cervical cancer, lung cancer, head and neck cancer and testicular cancer. Cisplatin is characterized by high efficacy and high number of adverse effects. Carboplatin was introduced after cisplatin as another drug from this group. It was used in chemotherapy for ovarian cancer and small cell lung cancer. The main recommendation for use of oxaliplatin is gastrointestinal cancers. It has been proven that oxaliplatin combined with 5-fluorouracil in adjuvant treatment of stage III colorectal cancer improves survival compared to 5-fluorouracil monotherapy. Cisplatin is highly nephrotoxic (hydration is necessary to prevent nephrotoxicity). It causes peripheral neuropathy, severe nausea and vomiting, bone marrow damage, and hearing damage. Carboplatin is better tolerated by patients and has a higher therapeutic index compared to cisplatin. It is less nephrotoxic and neurotoxic, but it causes more hematopoietic system suppression than cisplatin.

2.5 Cell cycle–nonspecific and cell cycle–specific anticancer drugs

Due to the phase of the cycle in which the drugs act, we categorize them into (Table 2.1):
- Specific to the cycle phase (active in a certain phase);
- Nonspecific to the cycle phase (active in different phases).

Drugs specific to the cycle phase work in a certain phase. Drugs act up to a certain dose limit, the exceeding of which does not increase the effectiveness of the drug. Drugs that are nonspecific to the cycle phase are characterized by a linear relationship between dose and effect. This results in the destruction of more cancer cells with an increase in the drug dose [4].

Table 2.1: Classification of cytostatic drugs according to their effect on the cell cycle [1, 3, 4].

Phase-specific drugs	Phase nonspecific drugs
Antimethabolites (S phase)	Alkylating drugs
Anticancer antibiotics (S/G2/M phases)	
Drugs from the group of podophyllotoxin derivatives (G2 phase)	
Plant alkaloids (M phase)	
Taxoids (M phase)	
Drugs from the group of camptotecin derivatives (G2 phase)	

2.6 Route of administration of chemotherapy and forms of anticancer drugs

The classification of anticancer drugs according to their pharmaceutical form is not only intended to provide a technological approach but also to demonstrate certain clinical implications in the use of these drugs. Currently, clinical oncology has a wide variety of chemical substances that are used in cancer therapy at different stages of treatment. The classic cytostatic drugs, despite the increasing presence of targeted drugs, are still an important therapeutic option for oncological patients. The effectiveness of these drugs is limited by a narrow therapeutic index, which directly affects the safety of the therapy. Anticancer chemotherapeutics use cell structures that are characteristic of cells that divide rapidly as a molecular target and therefore the toxicity profile of this group of drugs is mainly focused on tissues whose functions result from the possibility of rapid proliferation. These include, among others, bone marrow, mucous membranes, hair follicle bulbs. However, this mechanism does not close the possibility of other effects of anticancer drugs on the patient's body, moreover, the side effects of this group of drugs may affect almost all organs and systems.

The form in which the drug is to be administered to the patient and the route of administration of this drug are the resultant of the physicochemical and pharmacological properties of the drug. There are situations in the clinic where the drug form directly influences the toxicity profile. The classification of the most commonly used form of anticancer drug in clinical practice is presented in Table 2.2.

According to the general trend, the oral route of administration is the preferred one. However, not all medicinal products can be administered by this route. Limitations result from the chemical properties of the active substance (e.g., protein structure, which decomposes in the environment of the gastrointestinal tract), irritating properties, as well as unsatisfactory bioavailability of the drug (percentage of the dose that is absorbed from the gastrointestinal tract into the blood and the time

Table 2.2: The classification of the most commonly used form of anticancer drug in clinical practice.

Anticancer drug forms		Examples
Internal-usage drugs for oral use	Modified and unmodified release rate tablets, including coated tablets	Cyclophosphamide, capecitabine
	Capsules	Vinorelbine
Parenteral drugs	Solution for injections and infusions	5-fluorouracil
	Powders for solution preparation for injection and infusion	Cyclophosphamide, dacarbazine, bleomycin
	Concentrates for injection and infusion (aqueous and anhydrous)	Cisplatin, paclitaxel, etoposide
	Suspensions	Cytosine arabinoside for intrathecal administration
	Liposome suspensions	Liposomal doxorubicin, liposomal irinotecan
	Microspheres	Various cytostatic drugs incorporated into the surface of lactic acid copolymers for use in chemoembolization
	Implants	Carmustine

during which this process occurs). A separate aspect is the patient's willingness to cooperate with the doctor, the ability to follow the recommendations and the patient's persistence in taking medicines. These problems are avoided in case of parenteral administration.

The most commonly used route of administration of anticancer drugs is the intravenous route, which guarantees 100% bioavailability and eliminates the risk of drug mistakes related to the extent of cooperation between the patient and the doctor. However, the intravenous route of administration carries a risk associated with the invasiveness of the administration, e.g., the risk of septic complications, complications associated with the action of the drug at the place of administration (e.g., extravasation), allergic complications.

Due to the narrow therapeutic index of anticancer drugs, which forces the individualization of applied doses of drugs for a specific patient, single doses are prepared in hospital conditions immediately before administration. Medicines produced by the industry in multidose packs in various forms (Table 2.2) must undergo stages leading to the preparation of a product, which is usually a mixture of the active substance and excipients that can be safely administered to the vascular bed. These drugs are usually administered as infusions, i.e., sterile aqueous solutions or emulsions of one or more medicinal substances intended for parenteral use in the form of a drip infusion. In the context of anticancer drugs, this is an important form of drug, since most cytostatic agents and monoclonal antibodies are only administered as infusion fluid.

Preparations administered in the form of infusions require previous preparation from drugs that appear in different forms:
- Solution for infusion;
- Powder (lyophilisate) for solution for infusion;
- Concentrate for solution for infusion;
- Powder (lyophilisate) to prepare the concentrate;
- Liposome concentrates for the preparation of the intravenous infusion.

Factors directly affecting the safety of the therapy with regard to the drug itself are: the type of solvent (infusion fluid) used, its volume, durability of the prepared product, storage conditions, method of administration, possibility of interaction, type of immediate packaging used.

In clinical practice, two types of infusion solutions are used for dilution: 0.9% sodium chloride solution for the intravenous infusion and 5% glucose solution for the intravenous infusion. In addition, it is possible to dilute drugs in other liquids, such as Ringer's solution, a mixture of 0.9% sodium chloride solution with 5% glucose solution, 0.45% sodium chloride solution, 0.05 mol/L sodium bicarbonate solution.

The choice of a suitable solution depends on the durability of the final product resulting in particular from the stability of the active substance in the presence of electrolytes (Cl^-) and the effect of the acidic environment (glucose).

Special sterile and apyrogenic disposable sets are used to administer drugs in the infusion fluid. The set includes an infusion fluid reservoir and transfusion apparatus. The choice of appropriate equipment depends on the type of drug being prepared also in the context of the material of which the equipment is made. Interactions between, e.g., paclitaxel (in a dissolved form in polyoxyethylated castor oil) and packaging containing poly (vinyl chloride) are described, which consists of flushing out hepatotoxic phthalates from the material under the influence of the solvent— Cremophor (polyoxyethylated castor oil) present in the drug. In this case, it is recommended to use packaging that is at least PVC-free in direct contact with the drug [5].

The physicochemical stability of drug solutions is a fundamental aspect with regard to both the efficacy and safety of pharmacotherapy, and the following factors influence stability:
- Product (Producer);
- Solvent;
- Diluter;

- Concentration of the solution;
- Storage conditions (temperature, presence of light);
- Storage time

The development of technology allows the introduction of various pharmaceutical forms of the same drug. Usually the factor determining the search for new technological options for the same molecule is the unsatisfactory toxicity profile or effectiveness of the product. An example is paclitaxel, a drug introduced in the early 1990s. The molecule of this drug is virtually insoluble in water (<0.1 g/L), so solubilizers have been used to make this drug suitable for use in clinical practice. Polyoxyethylated castor oil, also known as Cremophor, was used for this purpose. The presence of Cremophor in the formulation significantly affects the properties of the product. It is a nonionic amphiphilic compound and one of the few surfactants approved for use in parenteral drugs. Compared to other compounds of this type, it is the strongest solubilizer, which is highly toxic and causes numerous side effects (hypersensitivity reactions from urticaria to full-blown anaphylactic shock) [6, 7].

The drug concentrate applied to the solvent, which can be either a 0.9% sodium chloride solution or a 5% glucose solution, forms a characteristic micellar system (Figure 2.4). Micellar solubilization significantly affects the pharmacokinetics of the drug, as paclitaxel placed in micelles is less available for tissue distribution and bile excretion. To achieve a physicochemical stable micellar system, the final concentration of paclitaxel must be between 0.3 and 1.2 mg/mL. Paclitaxel tends to precipitate over time in an aqueous environment, so the solution is administered through a filter with a pore diameter of no more than 0.22 μm.

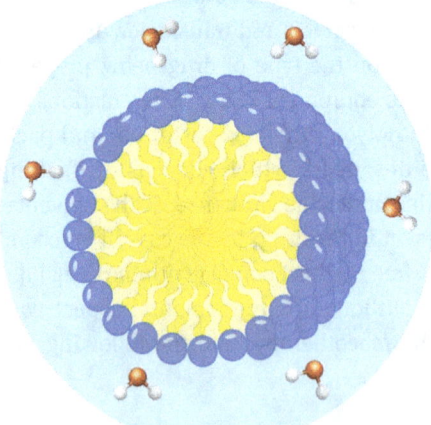

Figure 2.4: Micellar system: Inside, there is a paclitaxel covered with a hydrophobic part of Cremophor. The top layer of micelles is hydrophilic, thus physically compatible with the solvent; based on the study by Krämer and Heuser [5].

Attempts to obtain a permanent preparation of paclitaxel in the form of a conjugate with proteins resulted in the introduction of paclitaxel in the form of nanosuppression. The product does not contain Cremophor, and is stabilized by the presence of albumins, which gives a colloidal nanosuppression. The absence of Cremophor significantly reduces side effects of the drug, increases the biological half-life and improves the distribution of the drug. Extensive research is currently being conducted into further options for the administration of paclitaxel, including intravenous administration using polymeric micelles, administration as a prodrug or oral administration using other excipients [8–11].

2.6.1 Importance of liposomes in clinical practice on the example of doxorubicin

Liposomes are spherical, submicron bubbles with a diameter of 0.02–5 μm, whose membranes are made up of a double layer of amphiphilic molecules or several double layers to form a single or multichamber bubble, respectively. The hydrophilic lipid components of the double layer are directed toward the aqueous phase, while the lipid components of both layers are directed toward each other, forming the inner hydrophobic layer of this membrane [12, 13].

One of the uses of liposomes is that they can be a carrier of drugs used in targeted therapy. The diameter of liposomes for parenteral purposes must be less than 0.2 μm. During storage, liposomes cannot increase their diameter.

After intravenous administration, the liposome suspension is destabilized by the fatty exchange of the membrane-forming phospholipids under the influence of plasma lipoproteins, mainly HDL. The addition of cholesterol prevents lipid exchange and has a beneficial stabilizing effect.

Liposomes are characterized by a strong tropism. They accumulate mainly in the reticular-endothelial system, in the liver, spleen, lymphatic nodes and bone marrow. This is due to the adsorption of various plasma proteins on the liposome surface, the so-called opsonins, and then the liposome uptake by tissue cells, primarily by tissue macrophages. These cells recognize liposomes by their opsonin specificity and eliminate them as foreign bodies. In order to increase the selectivity of liposomes and direct them to diseased tissues and organs, the liposome surface is covered with hydrophilic polymers. Polyethylene glycols (PEG) are most commonly used, i.e., about 2000. Covering the liposome with PEG coating prolongs the time of their circulation in the bloodstream, extends the biological half-life, which can be 20 h.

Most solid tumors develop in the vascular system, which, like the reticular-endothelial system, shows porosity resulting from intensive, chaotic angiogenesis. This allows the penetration of liposomes with cytostatic agents into tumor cells. Anticancer drugs in liposomal form include, among others, doxorubicin and daunorubicin. Liposomal products containing doxorubicin occur in a nonpegylated and pegylated form. They differ in pharmacokinetics and are not bioequivalent because in the nonpegylated preparation doxorubicin is closed in liposomes "ex tempore" as a citrate. The size of the nonpegylated liposomes is about 180 nm. These are oligolamellar vesicles with many double lipid layers, superimposed one on top of the other (OLV). Gradient pH: inside the liposome pH about 4 (citrate buffer), outside the liposome pH about 7 (carbonate buffer) [14, 15].

Pegylated liposomal doxorubicin is stabilized by the gradient of ammonium ions and "precipitation" in the sulfate liposome. It contains doxorubicin hydrochloride encapsulated in liposomes whose surface is covered with methoxypolyethylene glycol (MPEG), which protects the liposomes from being detected by the mononuclear phagocytes system, which causes the drug to stay in the bloodstream longer. The size of the liposomes is about 100 nm. These are large, unilamellar vesicles (LUV). Stabilizing factor: Ammonium ion gradient. There is glucose outside the liposome, ammonium sulfate inside (Figure 2.5) [16, 17].

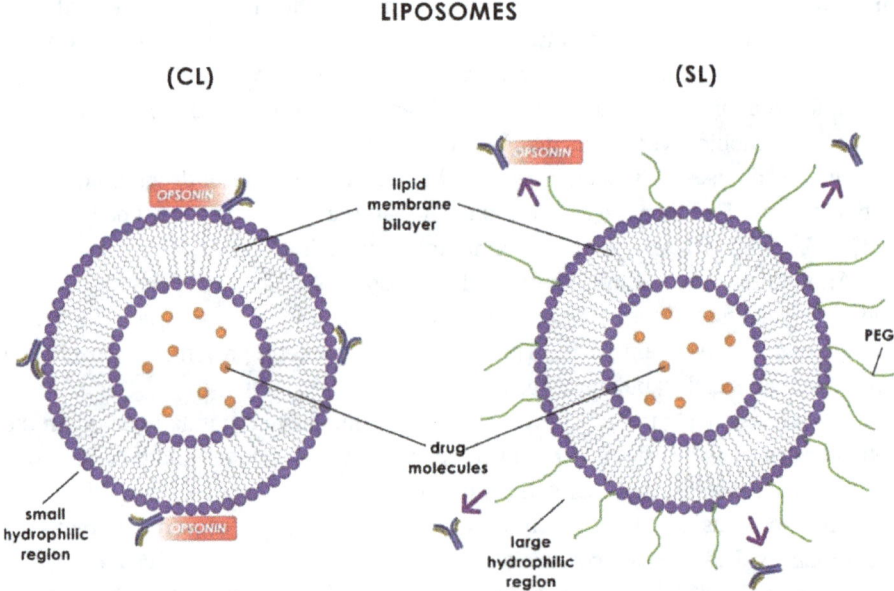

Figure 2.5: Comparison of doxorubicin formulations in the form of nonpegylated (CL) and pegylated (SL) liposomes; based on the studies by Abraham et al. and Niu et al. [13], [16].

2.6.2 Subcutaneous administration

In order to reduce local complications of intravenous administration as well as to increase patient comfort, a method of subcutaneous administration was developed for some drugs. This route of administration is an alternative to intravenous administration and is possible when the drug can be closed in a small volume. A key factor in enabling this route of administration is to ensure that the pharmacokinetic parameters (including, in particular, bioavailability) of subcutaneous administration guarantee the efficacy and safety of the administration at the appropriate level. Currently, the subcutaneous route of administration of anticancer drugs is used for such drugs as: methotrexate, cytarabine. In the preparation of these drugs hyaluronidase is sometimes used. This enzyme increases the dispersion and absorption of simultaneously administered drugs [18–20].

2.6.3 Intraperitoneal chemotherapy

The bad prognosis in the case of metastatic tumors justifies the need to look for new, more effective treatments. This issue has contributed to the development of intraperitoneal chemotherapy (IPC) techniques. Randomized studies have confirmed a significant increase in disease-free survival time and total survival after this method compared to standard intravenous chemotherapy. However, due to unacceptable complications, the effect of this method of chemotherapy on the improvement of treatment results is limited. Only half of the patients undergoing therapy usually fail to complete the entire planned treatment. Reducing the frequency of complications is possible by changing the technique of chemotherapy delivery. Perioperative hyperthermic intraperitoneal chemotherapy (HIPEC) fulfills these requirements. In its assumptions, this procedure combines two fundamental methods of treatment used in oncology—surgery and chemotherapy (Photos 2.1,2). The fact that both procedures (surgery and chemotherapy) are performed almost simultaneously may have a positive effect on the final result of the treatment.

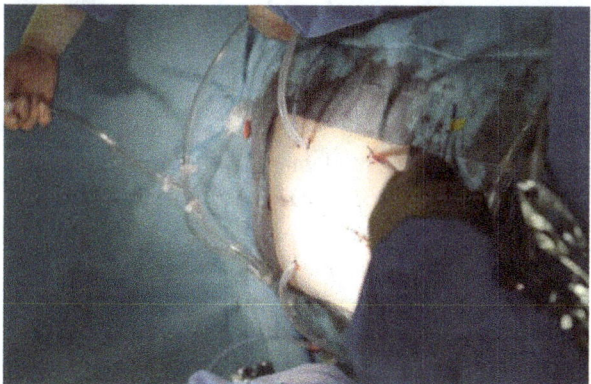

Photo 2.1: HIPEC treatment.

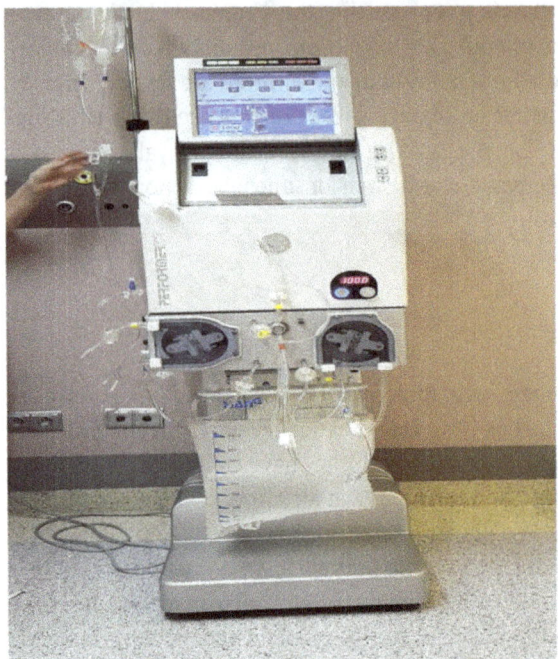

Photo 2.2: HIPEC treatment device.

This method is used primarily in patients who have been found to have metastases to the peritoneum in the course of: colorectal cancer, stomach cancer, pseudomyxoma peritonei, mesothelioma and ovarian cancer. HIPEC treatment may also be effective in treating metastases of colorectal cancer or stomach cancer to the ovary. The rationale for the use of cytoreductive procedures and perfusion intraperitoneal chemotherapy in hyperthermia in the treatment of various cancers, including ovarian cancer, is the fact that the peritoneum is the most common site of recurrence (15–40%). It is worth noting that on the basis of currently available randomized and observational multicenter studies, HIPEC cannot be treated as a method replacing systemic chemotherapy. It is only a way to supplement the golden standard of combined treatment in ovarian cancer, where a combination of surgery and intravenous chemotherapy (carboplatin + paclitaxel) still remains.

In ovarian cancer, when total cytoreduction is achieved in the pelvic area, retroperitoneal space, and in the area of the wall or mesentery of the small intestine there are foci of cancer no more than 2–3 mm, the only way to improve the outcome of surgical treatment seems to be to use perioperative Hyperthermic intraperitoneal chemotherapy. The chemotherapist circulating in the peritoneal cavity is previously heated to 41–42 degrees Celsius. The main advantage of peritoneal chemotherapy is its ability to achieve significantly higher concentrations of the chosen agent in the locoregional region, which results in increased efficiency. The key factor determining the effectiveness of HIPEC procedure is hyperthermia. It has been shown that cancer tissue is more sensitive to elevated temperature than healthy tissue. This condition makes the heated fluid circulating in the peritoneal cavity more toxic to the tumor mass. In assessing the effectiveness of peritoneal chemotherapy in hyperthermia we have both non-randomized and randomized research projects. It is believed that the total survival rate was significantly longer in the group with HIPEC treatment applied.

PIPAC ("Pressurized Intraperitoneal Aerosol Chemotherapy") is an innovative combination of laparoscopy techniques with a modern way of delivering the drug mainly in patients with cancer spread to the peritoneum, most often such cancers as gastric, ovarian and colorectal cancer. It consists in spraying chemotherapy in the form of an aerosol with a laparoscopic technique, straight into the abdominal cavity. In this way, the chemotherapist reaches the cancer cells directly. The PIPAC method is used as a palliative treatment when cytoreductive treatments in combination with HIPEC cannot be performed. However, more and more often PIPAC is treated as part of the treatment sequence, where after remission of metastases to the peritoneum it is possible to perform cytoreductive treatments and HIPEC.

2.6.4 Intrathecal chemotherapy

Intrathecal chemotherapy is the administration of chemotherapy directly into the spinal canal to allow the cytostatic agents to reach the central nervous system. It is used for tumors located in the brain, spinal cord or medulla oblongata. The existing blood–brain barrier means that most drugs administered intravenously do not pass into the central nervous system and are ineffective.

2.6.5 Transarterial chemoembolisation

Transarterial chemoembolisation (TACE) involves the selective insertion of a catheter into the arteries supplying the cancer and administration of embolization material through its lumen in the form of a cytostatic-soaked lactic acid copolymer (e.g., irinotecan). This allows to obtain a locally high concentration of cytostatic agents within the tumor and cut off the arterial blood flow to the neoplastic lesion. This method is used in patients with nonresective primary liver tumors, biliary duct cancer, as well as metastases, especially from the large intestine. TACE owes its effectiveness to two mechanisms. Most liver tumors are supplied with blood from the arterial side. Embolization of the liver artery branch blocks the blood supply to the tumor and limits its growth until more vessels are formed to supply the tumor. Secondly, the local supply of the cystostatic agent can reach a very high concentration only within the tumor. In this way the cytostatic drugs have a local and not systemic effect. The simultaneous closure of the arteries by the embolization material prevents the cytostatic agent from leaching from the vascular bed of the tumor.

2.6.6 Electrochemotherapy

Another option for the administration of anticancer drugs which is an attempt to increase the strength of the drug without significantly increasing the risk of systemic toxicity is electrochemical therapy. This method is used in nonoperative cases of advanced cancers located in the body shells (skin and subcutaneous tissue). Electrochemotherapy is a combination of chemotherapy and cell membrane electroporation. The phenomenon of electroporation is based on a temporary increase in the permeability of cell membranes under the influence of an appropriately selected electrical impulse. Thanks to the thinning of cell membrane structures of cancer cells, free diffusion of intravenous cytostatic agents (e.g., bleomycin) into the cells is significantly facilitated. In this way, high local concentrations of anticancer drugs are obtained, which increases their cytotoxicity up to several hundred times [21, 22].

2.6.7 Implants

Techniques associated with chemotherapy are also used to treat brain tumors. In the case of the treatment of this tumor, there is an option of inserting an implant into a bed after the resection of the tumor. This implant with a diameter of about 1.5 cm and a thickness of about 1 mm is made of a biodegradable matrix based on polifeprosan, in which the active substance (carmustine) is distributed. Carmustine is released over time in a controlled way as a result of the gradual degradation of the implant, acting cytotoxically. This solution results in a locally high concentration of the drug, which persists for a long time [23].

2.6.8 Gene therapy

Reduction of systemic toxicity of cytostatic drugs with simultaneous escalation of their anticancer effects is reflected in the concept of using the prodrug in combination with the agent releasing the drug from inactive form only in the tumor environment. In terms of this assumption, currently the options using macromolecules as components of the drug delivery system seem to be most interesting. Examples include gene directed enzyme prodrug therapy (GDEPT) as well as Antibody directed enzyme prodrug therapy (ADEPT). The idea is to systemically administer a drug in the form of a prodrug, the transformation of which into an active form depends on the presence of an enzyme that is not naturally present in the patient's body, but is delivered locally within the cancer tissue, e.g., by means of genes encoding this enzyme. This experimental gene therapy is considered in relation to known anti-cancer drugs, including for example alkylating drugs [24].

2.7 Conclusion

Since the introduction of chemotherapy to cancer treatment, the prognosis of patients has improved significantly. Unfortunately, the toxicity of this group of drugs resulting mainly from their nonspecific effects on target structures disqualifies the possibility of full use of their potential. Increasing the dose increases the effectiveness, but significantly reduces the safety of the therapy. Therefore, there are still two main ways to develop classical chemotherapy. The first concerns the search for new molecules using both known and new mechanisms of action. The second way of development is related to research on the form of the drug and ways of its administration. The dynamic development of knowledge on cytostatic drugs, including the targeting the drug form to the target of a cancer cell, allows us to assume that this direction will increase the effectiveness and safety of anticancer therapy in the future.

Author contribution: All the authors have accepted responsibility for the entire content of this submitted manuscript and approved submission.
Research funding: None declared.
Conflict of interest statement: The authors declare no conflicts of interest regarding this article.

References

1. Krzakowski M, Wyrwicz L. Leczenie systemowe. In: Krzakowski, M, Potemski, P, Warzocha, K, Wysocki, P, editors. *Onkologia kliniczna*, 3rd ed. Warsaw, Poland: Via Medica; 2014:107–34 pp.
2. The classification of cytostatic drugs based on the anatomical therapeutic chemical classification system; 2020. Available from: https://www.whocc.no/atc_ddd_index/?code=L&showdescription=no.
3. Orzechowska-Juzwenko K. Leki przeciwnowotworowe. In: Kostowski W, Herman Z, editors. *Podstawy Farmakologii*, 2nd ed. Warsaw, Poland: PZWL; 2001:400–44 pp.
4. DeVita VT, Chu E. Medical oncology. In: DeVita VT, Lewrance TS, Rosenber SA, editors. *Cancer-principales and practice oncology*, 9th ed. Philadelphia, USA: Lippincott Wiliams & Wilkins; 2011:312–21 pp.
5. Krämer I, Heuser A. Paclitaxel – pharmaceutical and pharmacological issues. *Environ Health Perspect USA* 1995;1:37–41.
6. Müller RH, Mäder K, Gohla S. Solid lipid nanoparticles (SLN) for controlled drug delivery – a review of the state of the art. *Eur J Pharm Biopharm* 2000;50:161–77. https://doi.org/10.1016/S0939-6411(00)00087-4.

7. Meng Z, Lv Q, Lu J, Yao H, Lv X, Jiang F, et al.. Prodrug strategies for paclitaxel. *Int J Mol Sci* 2016;17:796. https://doi.org/10.3390/ijms17050796.
8. Miele E, Spinelli GP, Miele E, Tomao F, Tomao S. Albumin-bound formulation of paclitaxel (Abraxane ABI-007) in the treatment of breast cancer. *Int J Nanomed* 2009;4:99–105. https://doi.org/10.2147/ijn.s3061.
9. He Z, Wan X, Schulz A, Bludau H, Dobrovolskaia MA, Stern ST, et al. A high capacity polymeric micelle of paclitaxel: implication of high dose drug therapy to safety and in vivo anti-cancer activity. *Biomaterials* 2016;101:296–309. https://doi.org/10.1016/j.biomaterials.2016.06.002.
10. Rowinsky EK, Donehower RC. Paclitaxel (taxol). *N Engl J Med* 1995;332:1004–14. https://doi.org/10.1056/nejm199504133321507.
11. Chu Z, Chen JS, Liau CT, Wang HM, Lin YC, Yang MH, et al. Oral bioavailability of a novel paclitaxel formulation (Genetaxyl) administered with cyclosporin A in cancer patients. *Anti Canc Drugs* 2008;19:275–81. https://doi.org/10.1097/cad.0b013e3282f3fd2e.
12. Huang JR, Lee MH, Li WS, Wu HC. Liposomal irinotecan for treatment of colorectal cancer in a preclinical model. *Cancers* 2019;11:281. https://doi.org/10.3390/cancers11030281.
13. Abraham SA, Waterhouse DN, Mayer LD, Cullis PR, Madden TD, Bally MB. The liposomal formulation of doxorubicin. *Methods Enzymol* 2005;391:71–97. https://doi.org/10.1016/s0076-6879(05)91004-5.
14. Batist G, Barton J, Chaikin P, Swenson C, Welles L. Myocet (liposome-encapsulated doxorubicin citrate): a new approach in breast cancer therapy. *Expet Opin Pharmacother* 2002;3:1739–51. https://doi.org/10.1517/14656566.3.12.1739.
15. Swenson CE, Perkins WR, Roberts P, Janoff AS. Liposome technology and the development of Myocet (liposomal doxorubicin citrate). *Breast* 2001;10:1–7. https://doi.org/10.1016/s0960-9776(01)80001-1.
16. Niu G, Cogburn B, Hughes J. Preparation and characterization of doxorubicin liposomes. *Methods Mol Biol* 2010;624:211–9. https://doi.org/10.1007/978-1-60761-609-2_14.
17. Chang HI, Yeh MK. Clinical development of liposome-based drugs: formulation, characterization, and therapeutic efficacy. *Int J Nanomed* 2012;7:49–60. https://doi.org/10.2147/IJN.S26766.
18. Balis FM, Mirro JJr, Reaman GH, Evans WE, McCully C, Doherty KM, et al. Pharmacokinetics of subcutaneous methotrexate. *J Clin Oncol* 1998;6:1882–6. https://doi.org/10.1200/JCO.1988.6.12.1882.
19. Anderson KC, Landgren O, Arend RC, Chou J, Jacobs IA. Humanistic and economic impact of subcutaneous versus intravenous administration of oncology biologics. *Future Oncol* 2019;15:3267–81. https://doi.org/10.2217/fon-2019-0368.
20. Bittner B, Richter W, Schmidt J. Subcutaneous administration of biotherapeutics: an overview of current challenges and opportunities. *BioDrugs* 2018;32:425–40. https://doi.org/10.1007/s40259-018-0295-0.
21. Probst U, Fuhrmann I, Beyer L, Wiggermann P. Electrochemotherapy as a new modality in interventional oncology: a review. *Technol Canc Res Treat* 2018;17:1–12. https://doi.org/10.1177/1533033818785329.
22. Campana LG, Marconato R, Valpione S, Galuppo S, Alaibac M, Rossi CR, et al. Basal cell carcinoma: 10-year experience with electrochemotherapy. *J Transl Med* 2017;15:122. https://doi.org/10.1186/s12967-017-1225-5.

23. Kleinberg L. Polifeprosan 20, 3.85% carmustine slow release wafer in malignant glioma: patient selection and perspectives on a low-burden therapy. *Patient Prefer Adherence* 2016;10:2397–406. https://doi.org/10.2147/PPA.S93020.
24. Zhang J, Kale V, Chen M. Gene-directed enzyme prodrug therapy. *AAPS J* 2015;17:102–10. https://doi.org/10.1208/s12248-014-9675-7.

Janusz Winiecki
3 Principles of radiation therapy

Abstract

Introduction: Radiotherapy is one of the basic methods of cancer treatment. Tens of millions of people around the world are exposed to ionizing radiation each year in the hope that it will help fight the disease or slow down its progress. Radiotherapy owes its success mainly to important discoveries in the field of physics, which allowed to understand the essence of the interaction of ionizing radiation with matter, in particular living matter.
Materials: The following study explains which types of radiation have the ability to ionize matter. The difference between the interaction of electrically charged particles and neutral particles was explained. The author briefly described methods of delivering radiation to diseased tissues and how adjacent tissues are protected. The most important physical quantities describing the quality and dose of the delivered radiation were introduced.
Conclusions: Safe use of radiotherapy as one of the methods of oncological treatment requires proficient knowledge of the basics of radiobiology and the physics of nuclear interactions. The study describes the most important steps in the preparation and implementation of radiotherapy, but it is not sufficient to fully understand this method. However, it provides an opportunity to be familiar with the issue in general.

Keywords: dose delivery, ionizing radiation, physical principles of radiotherapy, radiation interaction with matter, techniques in radiotherapy, treatment planning

3.7.1 Ionizing radiation interaction with a living matter

Radiation is a physical phenomenon whose main feature is transport of energy. There are many different types of radiation, and each of these types usually has different physical properties. In practice, only radiation that can penetrate the human tissue and transfer energy to the medium can be used for radiotherapy, but not everyone. Sometimes the absorbed power is simply converted into a heat, which does not cause significant biological or chemical effects. However, the radiation power may also release electrons from the electrically neutral atoms or molecules within the absorbent – if it happens, the radiation is called **ionizing radiation**.

This article has previously been published in the journal Physical Sciences Reviews. Please cite as: Winiecki, J. Principles of radiation therapy *Physical Sciences Reviews* [Online] 2020, 5. DOI: 10.1515/psr-2019-0063

https://doi.org/10.1515/ 9783110662306-003

The ionization destroys molecules by breaking down chemical bounds between atoms. It must lead to changes in chemical and physical properties of the molecules (e.g., DNA, functional proteins). Healthy cells have the ability to repair some of the damages that occur as a result of exposure to ionizing radiation. Irreparable damage is one of the causes of **apoptosis** or permanent changes in the functioning of cells and tissues, including mutations and secondary malignant tumors. Treatment with ionizing radiation should be then proceed in such a way that healthy cells have a chance and time to **repair** the damages.

There is a substantial difference between of how does the ionizing **corpuscular radiation** (a stream of alfa particles, electrons, positons, protons, etc.) and the electromagnetic waves (e.g., gamma and X radiation) interact with the matter. The way in which charged particles interact with a matter is called **direct ionization** [4]. The principle of direct ionization is the electric forces (Coulomb forces) appearing between the approaching particle and the another one i.e., electron or atomic nucleus.

3.7.1.1 Light charged particles interaction with a matter

The electromagnetic field associated with light incident particle (e.g., electron, positon) collides with the electric field of an orbital electron. As a result of collision, the orbital electron leaves the parent atom and begins its own story as a **scattered (secondary) radiation**. Effects of the collision are also visible in the incident particle's behavior. According to the laws of momentum and energy conservation, the particle rapidly changes the direction of movement, which makes its path extremely complicated (Figure 3.7.1.1 b).

Primary radiation gradually loses its kinetic energy by consequently ionizing or excitation of subsequent atoms. The excitation of atoms is observed when the amount of energy delivered to the selected electron is not sufficient to eject it from the shell. On the other hand, high-speed electrons have also a possibility to overcome the barrier created by orbital electrons. The original trajectory of the particle is in this case deflected by electrical forces, which appear between the nucleus and the approaching particle. According to the general theory of electromagnetic radiation, a part of electron's kinetic energy is in this moment converted to the X-radiation. The phenomenon is known (from German language) as **bremsstrahlung**.

It should be noted that one penetrating particle can participate in several consecutive activities. Each collision leads to the loss of a specific portion of energy that is transferred to the secondary radiation. Direct ionization in the implementation of charged particles is a very efficient process. Firstly, because the primary particle can damage many molecules during its lifetime, secondly, damage is also caused by secondary radiation (especially delta electrons).

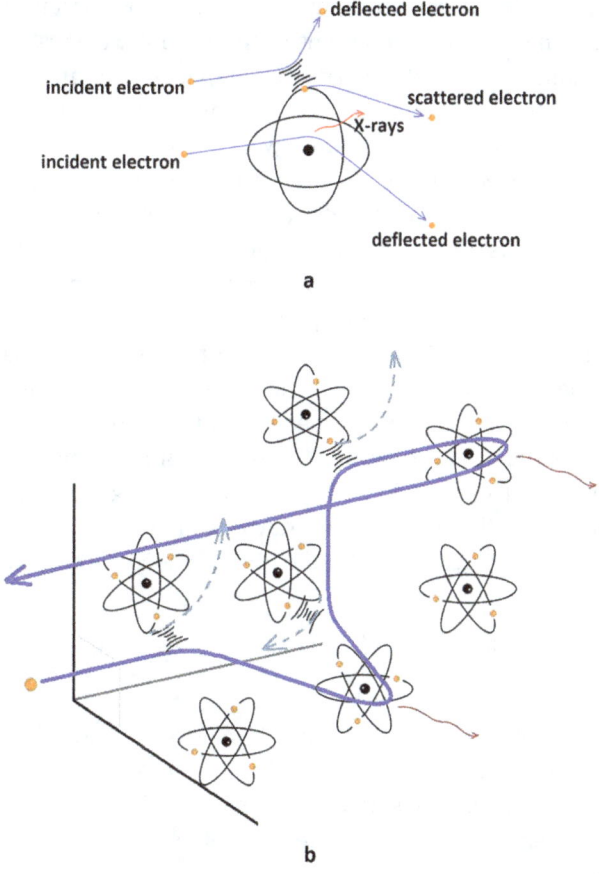

Figure 3.7.1.1: Interaction of electrons with matter: a – single interactions, b – electron path.

3.7.1.2 Photons interaction with a matter

The interaction of high-energy photons (X-rays, gamma) with matter is generally known as **indirect ionization.** This is a much less efficient process than ionization made by electrons: photons do not have an electric charge, and therefore the energy can be transferred only in the case of a direct hit. This, of course, happens relatively rarely, considering how small the target is (orbital electrons, nucleus) in comparison to the whole volume of atom.

The three most important phenomena in radiotherapy during which photons transfer their energy to the medium are: photoelectric effect, Compton scattering (Compton effect), and pair production. The first two effects rely on the interaction with an orbital electron, which absorbs the energy carried out by the electromagnetic wave. If, as a result of collision, the photon disappears and all its energy is delivered to the electron, we talk about **photoelectric effect**. Part of the photon energy is consumed to break the attraction between the electron and nucleus (work function), the rest of energy is converted into initial kinetic energy of recoiled electron (photoelectron).

The **Compton Effect** is a type of scattering when a quantum of light collides with a free or valence electron (an outer shell electron, weakly bound with a nucleus). As the binding energy of the electron is much lower than the energy of X-rays and gamma radiation, the incident photon does not disappear completely when the electron is released. It is scattered, its energy decreases, and wavelength is getting longer. The momentum of colliding photon is divided into two secondary particles: a scattered photon and a rejected electron as shown in Figure 3.7.1.2.

If photons have energy higher than 1.022 MeV, the effect of **pair production** is available. It is the third of the most important interactions between high-energy photons and the matter. The term "pair production" means that the product of this phenomenon is two particles: electron and positron. The pair is formed if photon passes very close to the atomic nucleus (interaction with the electromagnetic field of nucleus). The threshold energy for this effect is 1.022 MeV due to the fact that the generation of one electron (or positron) requires just 0.511 MeV of energy, according to Einstein's theory:

$$0.511 \text{MeV} = m_e \cdot c^2,$$

where, m_e – mass of the electron, c – speed of the light.

Electron and positron have the same mass but opposite electrical charge. The total charge of the particles is zero, because the photon that disappeared also had no electrical charge. The energy that remains is transferred equally to both particles and converted into their kinetic energy. The speed obtained in this way allows the particles to move away from each other.

The probability of occurrence of described phenomena depends strongly on the energy of beam and absorbent density (atomic number). Low energy radiation (0.010 eV–0.100 MeV) interacts almost only in a way known as photoelectric effect – the Compton Effect is possible, but it is less common (less than 10% of cases) as shown in Figure 3.7.1.3 [1]. However, with increasing photon energy, the chance of a Compton Effect also increases. X-rays and gamma with energy close to 1 MeV interact almost exclusively by the Compton Effect, and if the value 1.022 MeV is exceeded, an additional option appears – the pair production effect.

When considering the issue of the interaction of high-energy electromagnetic radiation with matter, it is impossible not to try to explain the difference between X-rays and gamma rays. The simplest criterion is the source and method of generation, although it does not completely explain the difference. It can be said that gamma rays always have an isotope source: they are created as a result of reactions taking place in the atomic nucleus (radioactive decays). There is also one specific type of gamma radiation: annihilation radiation. Its energy is exactly 0.511 MeV,

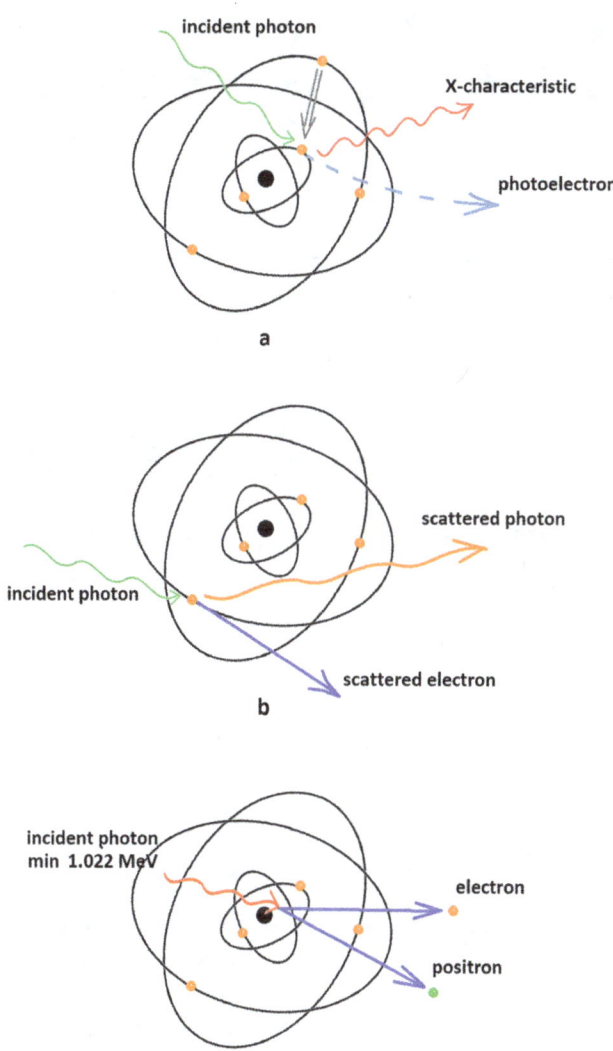

Figure 3.7.1.2: Interaction of photons with matter: a – photoelectric effect, b – Compton effect, c – pair production effect.

and it is created as a result of the collision of a negative electron with a positive positron. Both particles disappear and at the point where it happened, two gamma photons are formed (remember the pair production effect). The phenomenon of annihilation was used in positron emission tomography (PET): positrons are brought by glucose combined with fluorine-18 isotope (FDG – fludeoxyglucose). It is known that glucose is extensively absorbed by cancer cells. In this way, the tumor is recognized by a positron-emitting isotope.

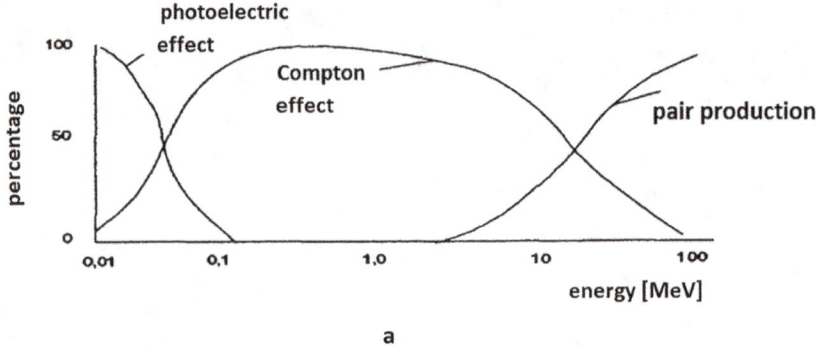

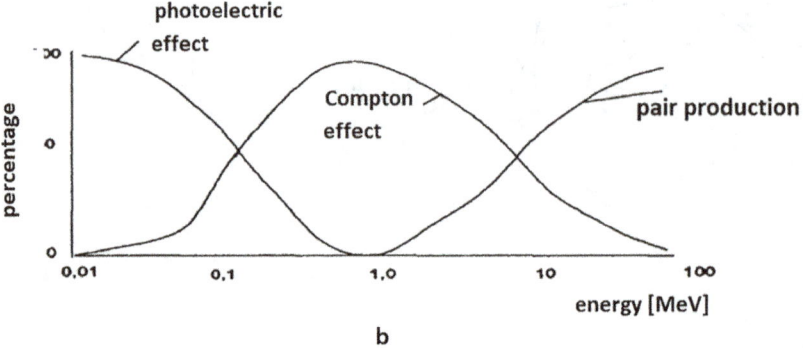

Figure 3.7.1.3: Percentage share of individual phenomena of photon interaction depending on photon energy: a – water, b – copper [1].

X-rays should always be associated with a rapid loss of energy by electrons. Usually, the X-ray tubes and accelerators are mentioned on this occasion (bremsstrahlung effect). X-rays can be produced also as a result of electrons falling from highest shells to the lowest one – this way the so-called characteristic radiation is generated. Therefore, it cannot be simplified by saying that gamma radiation is always natural and X-rays are artificial.

Except of the method of beam generation, there is also a certain quality difference between gamma and X radiation. First of all, it should be made clear that each gamma-emitting isotope generates its own unique set of beams. For example, the well-known and popular in medicine and industry Co-60 isotope is a source of three types of radiation: gamma quanta with energy 1.17 and 1.33 MeV, and beta particles (electrons) with energy 1.48 MeV. Wherever a cobalt unit would be installed (at the bottom of the ocean, Mount Everest peak, or on the surface of the Moon), it will always produce radiation of the same spectrum: several known components regardless of time.

Accelerators and X-ray tubes generate a beam of a broad and continuous spectrum – it consists of a huge amount of energy-different particles (e.g., photons). The energy spread in a case of X-ray tube is a result of the construction of device: X-ray beam appears as a result of electron energy conversion, which is carried out on a special disk (anticathode made from tungsten and molybdenum). The problem is that only some of the electrons stop on the surface of the disk. The others penetrate the target and lose some energy, which is converted into heat – consider Figure 3.7.1.4.

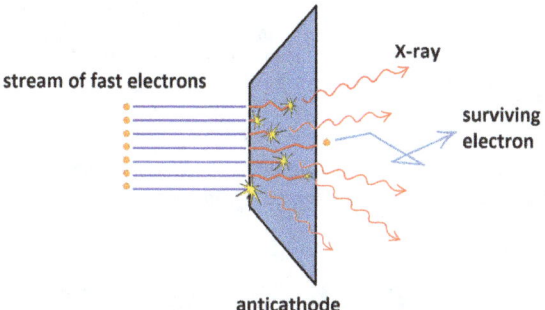

Figure 3.7.1.4: Bremsstrahlung process taking place on the X-ray tube anti-cathode.

Working with X-rays is, from physicist's point of view, more difficult than with gamma rays. As it was said, the probability of occurrence of certain phenomena of interaction with matter depends on the energy of photons. For X-ray generators, we have a lot of photons with different energies in the beam, which interact with matter at the same time. In addition, the spectrum characteristics and intensity of the beam may fluctuate due to the technical condition of the device. For gamma, the spectrum of the beam is constant in time, and the time-relative changes in intensity are well known (law of radioactive decay). Photon beams produced by X-ray tubes or accelerators are determined not by the energy (because it is a set of many different photons), but by the value of the electron-accelerating voltage (e.g., X-100kV, X-6MV).

Each photon usually can be subject to only one activity (no more than a few) in its path, which makes photons inefficient in ionization. As it was said before, charged particles are much more productive. However, their productivity and ease of interaction with tissue means that compared to photons, they lose their energy very quickly. The stream of such particles has no chance to reach a deeper tumor if the source of radiation is outside the patient's body. Typical range of light charged particles in the tissue usually does not exceed 3 cm. The use of **external photon beams** has the sense that intensely ionizing particles are produced inside the patient, in the tumor, and its surroundings (consider main three effects discussed earlier).

Another solution is the temporary implantation of a radioactive source (isotopes or small electronic devices) inside the body, close to the tumor. Treatment with this method is known as **brachytherapy** – Figures 3.7.1.5 and 3.7.1.6.

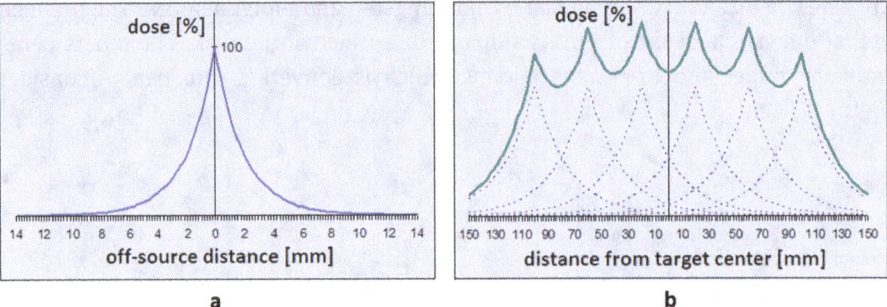

Figure 3.7.1.5: Dose distribution in brachytherapy: a – dose distribution around the point source of radiation, b – dose distribution along the line, along which the source was moving.

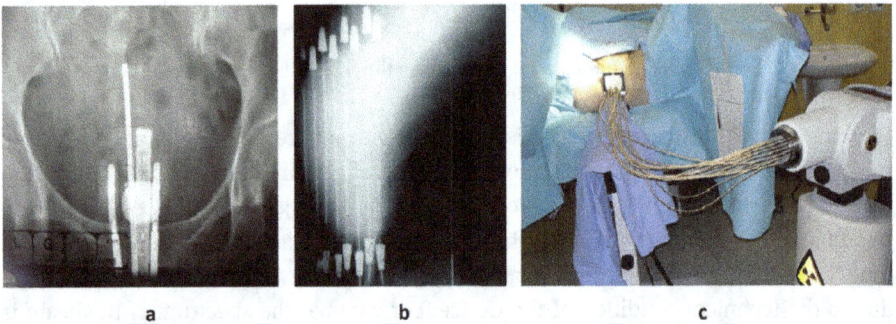

Figure 3.7.1.6: Brachytherapy: a – applicators used for uterine brachytherapy, b – breast treatment, c – prostate treatment.

Chronologically, the oldest method of treatment with radiation is contact brachytherapy. This method began to be used already at the beginning of the 20th century. For example, small radium sources (Ra-228) were used, which for practical reasons were enclosed in casings called radium corks. The corks met the hygiene aspect as they were replaceable. They also protected against burns, because radium sources in addition to radiation emission were also very hot. The source of radiation (the cork with isotope) should be set as close as possible to the tumor: in the mouth, nasal cavity, or the woman's genital tract. For skin diseases, the source was applied directly to its surface. Because the work was done manually, it also caused a direct threat to the operator performing the procedure. Today, brachytherapy uses special applicators (guides, catheters, needles), which are first placed in the target place. The position of the applicators is verified before a treatment using magnetic resonance imaging or computed tomography. Insertion of applicators close to the tumor allows selecting the optimal stopping points for the source and determining irradiation time. After these activities, the source is inserted into the applicators. This approach is known as **afterloading**.

Some clinics around the world specialize in Boron-Neutron Capture Therapy for brain gliomas. The idea of BNCT is to saturate first a glial tissue with boron compounds. This is realized by injecting into the patient's body biological active compounds containing boron ligands. Boron accumulates in the glioma because the tumor absorbs selected chemical compounds with which boron has been combined.

Boron nuclei are not radioactive, but they start to produce alpha particles if they are bombarded with neutrons. The therapist must therefore direct the neutron stream to the target volume where accumulation of boron compounds has been observed.

Alpha particles have enormous ionization capacity because their electrical charge is twice that of electrons. The average range of alpha particles does not exceed 2 mm in the tissue, hence it can be said that therapy is approximately limited only to the volume of tissues impregnated with boron compounds. "Approximately" means that we do not take into account the radiation dose generated by individual interactions of neutrons with absorbent atoms.

Neutrons ionize matter indirectly because they do not have their own electrical charge like photons do not. However, neutrons have a huge mass (about 2000 times greater than the mass of an electron), so they can easily knock out protons or even larger elements from atomic nuclei that they meet on their way (consider nuclear fission or disintegration). When discussing BNCT, the negative effects associated with the dose generated by the neutron beam are often overlooked, although the amount of energy transferred to the tissues is not small at all. The key is that the extra dose from neutrons is dispersed in a large volume due to the incomparably large range of secondary protons, and the therapeutic dose from alphas is exactly delivered to the tumor.

Radiotherapy with protons is also available (proton therapy). Therapeutic protons, as well as electrons, have an electric charge, but they are much heavier (comparable to neutrons). They ionize matter also directly by electrical interactions with a cloud of orbital electrons and with an atomic nucleus, but it is very difficult to change their destination. Collimated proton beams are therefore used for very precise radiotherapy procedures, which will be discussed later. Protons activated during the BNCT procedure are not collimated, and therefore, the dose is dispersed in all directions.

3.7.1.3 The dose

A quantitative explanation of the irradiation should be started by introducing the concept of dose. The absorbed dose – **D** is the amount of average radiation energy absorbed by the matter (mass unit of the matter):

$$D = \Delta E_{abs}/\Delta m$$

Absorption means that energy has been utilized, for example, it has been used to cause local changes in the structure of matter (ionization). It is easy to understand that such a definition is simple to apply to directly ionizing particles. The photons, because they are indirect ionizing particles, rather transfer the energy to the matter than the energy is absorbed by matter. The total amount of energy transferred by photons to the medium and converted to initial kinetic energy of secondary radiation is known as KERMA (Kinetic Energy Released in Matter) Figure 3.7.1.7. In fact, the absorbed dose is deposited later by secondary particles during their collisions with orbital electrons and atomic nuclei, as it has been discussed before. Both, the KERMA and Dose, have the same SI unit, which is 1 gray (1Gy):

$$1Gy = 1J/1kg$$

Note that the amount of dose absorbed by the tissue must be less than KERMA, because secondary electrons lose some of their kinetic energy in bremsstrahlung.

The absorbed dose purely has only the physical sense. It does not reflect the importance of radiobiological effects of exposure. Depending on various types and quality of radiation, the same value of absorbed dose can cause different chemical and biological effects. Therefore, the radiation quality factor – **Q** and **equivalent dose** – **H** have been introduced. Equivalent dose is calculated using the absorbed dose D multiplied by quality factor Q of the radiation "r" as follows:

$$H_r = Q_r \cdot D.$$

The SI unit for equivalent dose is 1 sievert (1Sv). For X-rays, gamma, and electron beams, 1Gy is exactly 1Sv, because the value of Q-factor for these types of radiation is just 1Sv/Gy. For heavy particles (protons, neutrons, alpha particles, and heavy ions), the value of Q reaches 20 Sv/Gy, which can lead to drastic consequences if mistaken.

Furthermore, for radiation protection purposes, the **effective dose** – H_{eff} has been introduced. As the body tissues and organs have different radiosensitivities, the special **weighting factors** – ω_T have been assigned to them. For each organ, ω_T-factor describes the overall risk associated with exposure to ionizing radiation and repair ability. The factors do not have their own SI unit. The values of ω_T recommended by International Commission of Radiation Protection (ICRP) are given in Table 3.7.1.

The value of H_{eff} for a selected tissue "T" is calculated as follows:

$$H_{eff} = \sum \omega_T \cdot D,$$

and for the whole body:

$$H_{eff} = \sum \omega_T \cdot D,$$

Table 3.7.1: Weighting factors for major organs and tissues.

Organ/tissue	ω_T
Gonads	0.25
Breast	0.15
Marrow	0.12
Lung	0.12
Thyroid	0.03
Bone (surface)	0.03
Skin (total)	0.01

Which means that the risks associated with all exposed tissues and organs should be considered? The effective dose is applicable to incidental and accidental exposures rather than to radiation therapy. In diagnostic activities with an ionizing radiation, the effective dose is monitored and included as one of the criteria for quality assurance. Doses of radiation absorbed by patients undergoing radiation therapy significantly exceed those usually registered by radiation protection officers. For example, the annual effective dose limit for a typical public exposure is 6 mSv (0.006 Gy for X-rays or gamma), and the average therapeutic dose in radiotherapy is 40–70 Gy. Usually, for accidental exposures, the risk exists for the whole body, hence the need to take into account the dose absorbed by many organs. However, radiation treatment is limited (excluding several cases such as Total Body Irradiation (TBI) or Total Marrow Irradiation (TMI)) to the selected part of the body where the tumor is located (e.g., head, head–neck, chest, pelvis). Steep dose gradients are also used to reduce the dose in the surrounding tissues, which means that the dose for the surrounding tissues tends to zero.

3.7.2 Treatment planning and realization of radiotherapy

"Radiotherapy planning" is a process that requires the involvement and cooperation of many people representing different medical professions: medical physicists, physicians, technicians, etc. All of them have their own task to perform, on which the success of treatment depends. Before this happens, however, the decision to start ionizing radiation treatment must be made by a Multidisciplinary Tumor Board of physicians.

As mentioned in Section 3.7.1, dose deposition must be performed carefully. Usually the total therapeutic dose is divided into several dozen identical fractions delivered one per day. All therapeutic sessions are identical, their duration is usually a few to several minutes. The patient lies on the therapeutic table in the so-called therapeutic position. As the patient's physical condition may deteriorate during the entire treatment and it will be difficult for him to maintain the therapeutic position, special accessories are used to guarantee repetitive body positioning. A set of immobilizing accessories is selected by the doctor in cooperation with a medical physicist and radiotherapy technician. Sometimes the guarantee of correct positioning during the whole treatment can be obtained only by using an individual thermoplastic mask Figure 3.7.2.1. The accessories should not contain metal parts, which could attenuate the beam and reduce the therapeutic effect.

Although the patient underwent many diagnostic tests, computed tomography (CT) is again performed before radiotherapy begins. This time CT is performed in a therapeutic (not diagnostic) position using immobilizing accessories. Three-dimensional body reconstruction is on the one hand an accurate anatomical map of the patient and on the other hand brings information about the density and chemical composition of tissues. Since the probability of collisions of incident penetrating depends strongly on the concentration of electrons and density of the matter, different amounts of secondary radiation will be generated in different tissues (compare lung density and bone density). Physicists need data on tissue density to calculate the amount of scattered radiation and to determine the value of the dose D absorbed in individual tissues.

The first obligatory step in the treatment planning process is delineation of target (tumor) volumes and volumes of normal tissues that might be irradiated and affect the treatment. The physician is obligated to specify gross tumor volume (GTV), clinical target volume (CTV), planning target volume (PTV), and crucial organs at risk (OARs) following the ICRU (The International Commission on Radiation Units and Measurements) recommendations [1], [2], [3].

Target volumes (GTV, CTV) correspond to the tissues of known or suspected cancer infiltration. The specified diagnostic methods (e.g., MRI, PET) help to recognize morphological and physiological abnormalities that are evidence of tumor location. It is important not only where the tumor is located, but also what its shape and dimensions are.

The CTV is usually larger than GTV because it contains evidenced GTV with the surrounding tissues suspected of hiding a few cancer cells, although this did not have to be proven in diagnostic test. Even if they are not visible, single cancer stem cells (CSCs) can cause disease progression. Sometimes, e.g., in the case of prostate cancer, by definition it is assumed that the whole gland should be irradiated, regardless of whether the cancer cells were found: in its entire volume or only in part of the prostate. So in the case of radiation therapy for prostate cancer CTV = GTV.

The physiological tumor movement (periodic shifts of organs) and the therapeutic beam positioning cause a risk of inaccurate CTV irradiation. During breathing, for example, not only movements of lungs but also shifts of such organs as liver or pancreas are observed. In order to take into account all possible spatial uncertainties of the therapy, the PTV concept has been introduced. The defining safety margins and delineation of the PTV volume require extensive medical knowledge in the field of physiology, because the oscillation amplitudes are usually different in each of the three axes of the patient: vertical transverse and longitudinal.

The relationship between GTV, CTV, and PTV is shown in Figures 3.7.1.7–3.7.2.2a.

As it has been mentioned, delineation of OARs is recommended. Which organ to consider as an OAR depends on the target's location? For brain cancer, these can be eyeballs (or only lenses), optic nerve, brainstem, etc. In turn, for chest (lung) cancer, these are the heart, aorta, and spinal cord. In any case, the skin is protected because of its high sensitivity to radiation. Today, tolerance doses for such organs are already well known. Therefore, every effort should be made during dose distribution optimization to ensure that the acceptable dose limits are not exceeded during the treatment. For sure, the delineating helps to control the dose absorbed by OARs. The dose absorbed by OARs can be minimized by using techniques known as "target tracking" and "respiratory gating" – more on optimizing dose distribution later in the text.

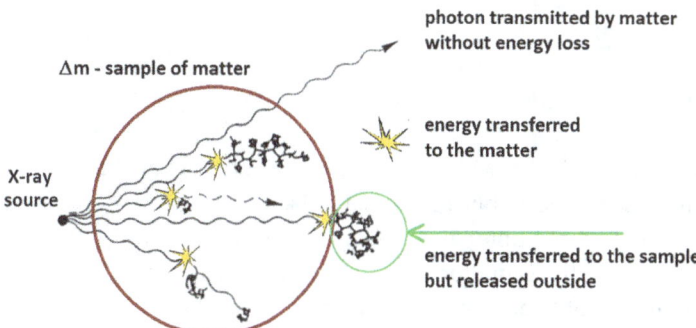

Figure 3.7.1.7: Explanation of KERMA.

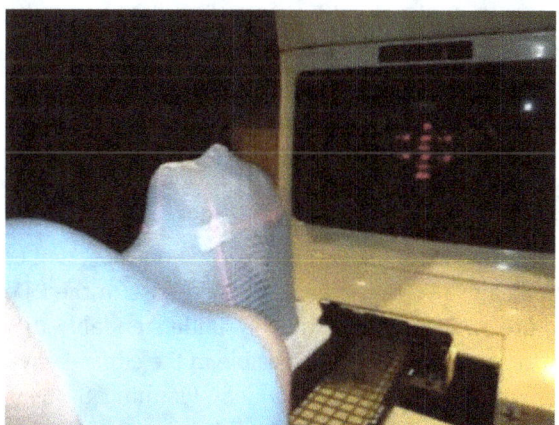

Figure 3.7.2.1: Immobilized Head-and-Neck patient during radiotherapy session.

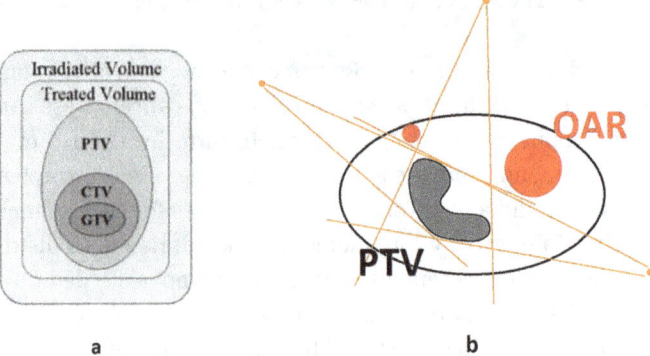

Figure 3.7.2.2: Areas of interest in radiotherapy: a – relations between GTV, CTV and PTV structures ICRU Report 50. Treated Volume is a structure receiving a dose similar to the prescribed for PTV, Irradiated Volume is a structure that has received a measurable dose, b – therapeutic beam settings [2], [3].

3.7.2.1 Dose distribution calculation and optimization

The interaction of ionizing radiation with matter has been thoroughly studied and discussed in numerous scientific publications (some basic information has been provided in Section 3.7.1). The amount of dose absorbed at each point in space (also inside the body) depends on the radiation energy that reaches that point and the quality of a matter at that point (e.g., density and chemical composition).

One must know that during the therapeutic session, the patient's body is every second penetrated by billions of particles. Each of them is independent of the others and reacts independently with the substance it meets. The therapeutic beam is consistently modified and attenuated and as it passes through subsequent layers of tissues. Not only intensity decreases but also modification of radiation quality happens. In the case of X-ray beams, so-called "beam hardening" effect is, for example, observed. It means that low-energy photons disappear quickly. If they occur at greater depths, they are a result of Compton Effect or pair production (discussed in Section 3.7.1). This is not the case with gamma beams, which are usually almost monoenergetic. It is much easier with ionizing particles, because if they go deeper into the tissues, they always have less energy.

Optimization of the dose distribution is, at the first approximation, selection of optimal directions for therapeutic beams to deliver the prescribed dose to the PTV and protect OARs – consider Figure 3.7.2.2b. It is intuitively understandable that calculation of the path of each particle separately is not feasible in the clinical environment. Special computer workstations known as Treatment Planning Systems (TPSs) are developed to enable medical physicists to simulate the dose distribution within a patient's body. In the past, the calculations of treatment time were carried out manually. There was no question of OAR dose reduction – dose delivery to PTV was important of all, so complications in this field were very common.

The TPSs have built-in calculation algorithms to simulate dose distribution generated by selected types of radiation. If nonisotope radiation sources are used in therapy (medical accelerators), the user must enter multiple input data into the system. These parameters are measured by physicists immediately after installing the treatment unit, before working with patients. The basic required parameters are the Bragg curves or percentage depth doses (PDDs), transversal profiles at different depths (TPs), and output factors (OFs). The parameters are usually measured in an automated water phantom (3D beam analyzer), in which the detector can move in all directions due to some technical solutions used.

The PDD curves represent dose distribution along the central axis of the beam (CAX). The curves take into account two components: the impact of dose from primary radiation generated by treatment unit, and the dose from secondary radiation. Since the dimensions of PTV structures may be random in clinical practice, the algorithms usually require many different depth-dose curves corresponding with irradiation fields of different dimensions. Figure 3.7.2.3 shows some typical Bragg curves for the most used radiation beams.

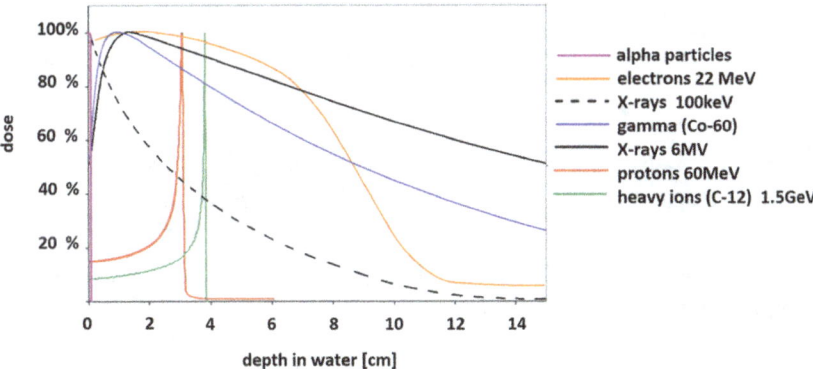

Figure 3.7.2.3: Bragg curves in water for the most commonly used types of radiation.

The Bragg curve for alpha particles is most difficult to see because it coincides with the vertical axis of the diagram. As already discussed, alpha particles have a range of just a millimeter and their curve is most on the left (purple line). The next curve (orange line) is the curve for an electron beam generated by a medical accelerator (9 MeV). Subsequently, photon curves are presented (blue line for gamma from Co-60 isotope, black line for X-6MV, and dashed for X-100kV). The red line is for protons and the green one for heavy ions. The drawings give some additional idea about radiotherapy planning. Electron radiation should be used to treat the surface cancers. By using it, we also have a fantastic opportunity to protect deeper tissues that only receive a low dose from so-called X-ray contamination (last, almost flat section of the Bragg curve). To protect the skin, however, high-energy photon beams should be used, since the dose on the body surface is low and gradually increases (so-called **build-up** effect). The build-up effect is imperceptible in the case of X-rays, whose source is a tube (10–300 kV). Such a type of radiation reaches the highest dose just on the skin and it is forbidden for radiotherapy applications. Note that exactly such radiation is used in CT, X-ray diagnostics, and also for patient positioning verification, which will be discussed later.

Due to the complicated trajectory caused by many events, electrons reach usually the tissues located no deeper than a few centimeters under the skin – consider Figure 3.7.2.4. The most important parameters describing dose distribution (and/or PDD) for electron beams are R_{80}, R_{50}, and R_P (practical range). They indicate at what depth the dose falls down to 80 and 50% of and its maximum value. It should be noted that as the beam energy increases, the electron penetration also improves.

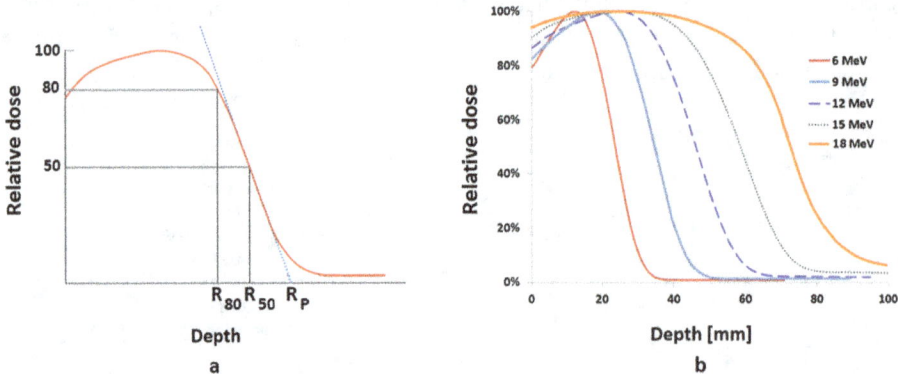

Figure 3.7.2.4: Bragg curves for electron beams: a – illustration of R_{50} and R_P, b – relation between different energy of electrons.

Heavy particles (protons, heavy ions) behave differently than electrons. As it was mentioned, they propagate in matter in straight lines. At the beginning of their paths in the tissue, they are still too fast to be able to ionize atoms that they pass by – the dose absorbed by the matter is really low. However, step by step, they slow down and at some point, when they are slow enough, they suddenly give all their energy to the environment. Because the dose is deposited in a very small area, it was called the Bragg peak. The position of the Bragg peak (depth) depends strongly on the initial energy of protons. In clinical practice, therapeutic beams contain protons with several different energies to obtain the so-called spread Bragg peak. It is a solution enabling delivery of a homogeneous dose to the entire volume of PTV.

Beam profiles contain information about off-axis dose fluctuations (in the plane that is perpendicular to the CAX). The profiles are measured in two perpendicular directions: "in plane" (y) and "cross plane" (x) as shown in Figure 3.7.2.5b. Measurements should be performed at several depths: dmax, 5, 10, 20, and 30 cm. Dmax is a depth, at which the build-up terminates, and the dose reaches the highest value. For the most common radiation beams, they are 15 mm (6 MV) and 29 mm (15 MV). Note that as the depth increases, the shape of the profile changes: it becomes wider (we call it beam divergence) and more and more smooth, devoid of details.

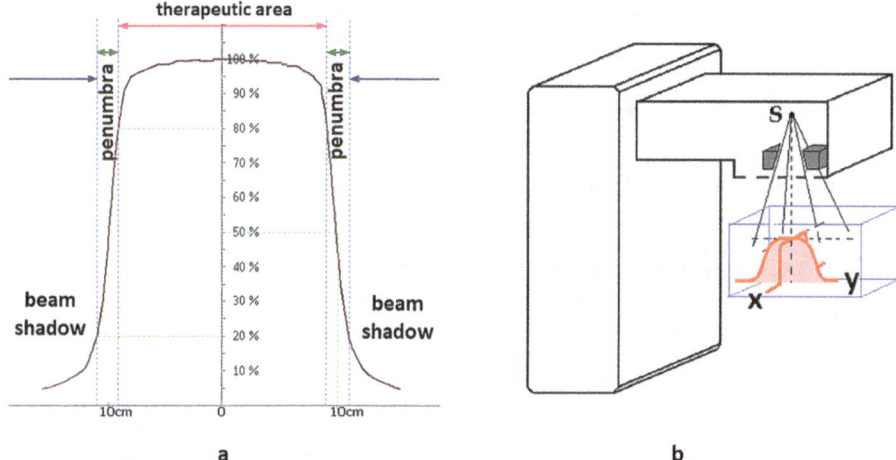

Figure 3.7.2.5: Beam profile: a – sections of the profile, b – scheme of water phantom measurements.

Treatment planning system usually requires profiles for the same fields, for which PDDs were measured. The beam profile contains the following areas: beam shadow, penumbra, and therapeutic area. The shadow of the beam is furthest from CAX and, like the penumbra, occurs on the left and right side of the profile. The penumbra and shadow are caused by the scattering effects, they cause an unwanted dose to the organs adjacent to PTV. The therapeutic area is the central part of the profile and, depending on the design of the treatment unit, has a specific outline. For cobalt units, penumbras and shadows are very unfavorable: in addition to PTV, a large amount of the adjacent tissue is also irradiated. Figures 3.7.2.6 and 3.7.2.7 presents typical profiles for several therapeutic beams. Accelerator beams allow protecting healthy tissues because their penumbra is very narrow. The most common are accelerators with so-called flattening filters (FF). By installing an appropriate filter (metal block in the shape of a cone) in the beam path, the profile of such a beam is almost flat. At a time when computer calculations were rare, the use of flat beams significantly reduced the amount of calculations needed (throughout the therapeutic area, the dose was approximately the same, so it was enough to make calculations only for CAX). Nowadays, flattening filter-free (FFF) accelerators are most often installed.

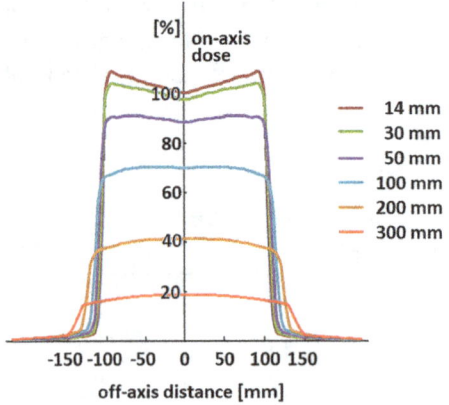

Figure 3.7.2.6: Beam profiles measured for 6 MV photon beam (20 x 20 cm).

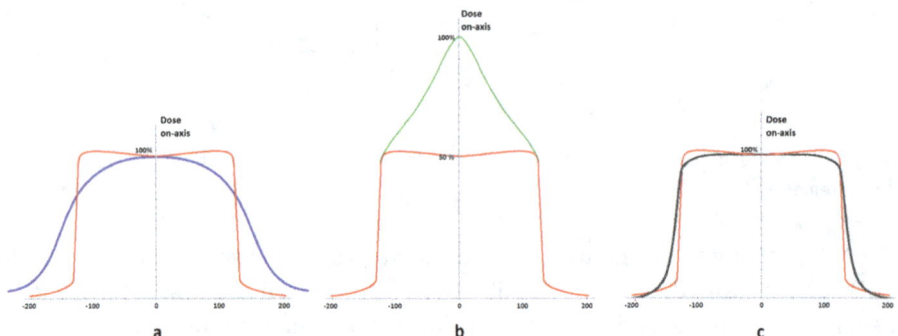

Figure 3.7.2.7: Examples of beam profiles used in radiotherapy: red line – 6 MV photon beam (linear accelerator with a flattening filter), blue line – gamma beam (Co-60), green line – 6 MV FFF photon beam (flattening filter free accelerator), gray line – 6 MeV electron beam (linear accelerator).

As nearly 90% of radiotherapy treatments are currently performed using photons generated by medical accelerators (MV energy range), further considerations regarding treatment planning and dose calculation will be limited to therapy with high-energy X-rays.

Prediction of dose distribution in the medium requires a combination of information contained in the PDD curve and beam profiles. The concept is presented in a simplified way in Figure 3.7.2.8a. Each of the points taken from beam profile should be multiplied by the corresponding value read from the PDD curve and then by OF value (normalization factor depending on the field area). Since the measurements have been performed in a water phantom, the result of calculation is only true for water. The algorithms implemented in TPSs can read the density and chemical composition data of a tissue from patient's CT and take it into account when predicting dose distribution.

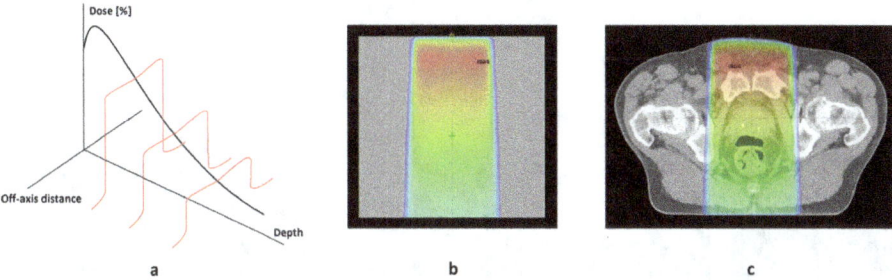

Figure 3.7.2.8: Prediction of dose distribution: a – combination of PDD and beam profile, b – resultant dose distribution in water, c – dose distribution in pelvis.

Modern algorithms use the acquired dosimetry data for detailed spectral analysis of the radiation beam. The accurate knowledge of particles energy together with the knowledge of tissue density and chemical composition enables literal implementation of the laws of physics. As all physical phenomena, in particular the interaction of radiation with matter is described by the theory of probability. The latest algorithms use Monte Carlo methods to predict the paths of individual particles and the probability of energy transfer to the matter.

PTV shapes and sizes can be different for each patient. Even if we look at the same target from different directions (BEV – Beams Eye View), we can see a different projection. It is therefore necessary to adapt the beam shape to the current PTV envelope. In the case of a classic medical accelerator, this is done using a collimator. The device is an integral part of the accelerator located at the point where radiation exits the treatment unit – Figures 3.7.2.9 and 3.7.2.10.

Beam collimation takes place on three sections of the collimator: primary collimator, collimator jaws, and multileaf collimator (MLC). The goal set for the primary collimator is to eliminate the scattered radiation arising in the target, i.e., the place where the kinetic energy of electrons is converted to X-rays. Only those photons remain in the beam that moves in the CAX-like direction. In the next step, the radiation beam passing through the jaw system receives a rectangular shape.

The discovery of an MLC has revolutionized external beam radiotherapy. The MLC consists of a set of metal bars that are driven and controlled independently of

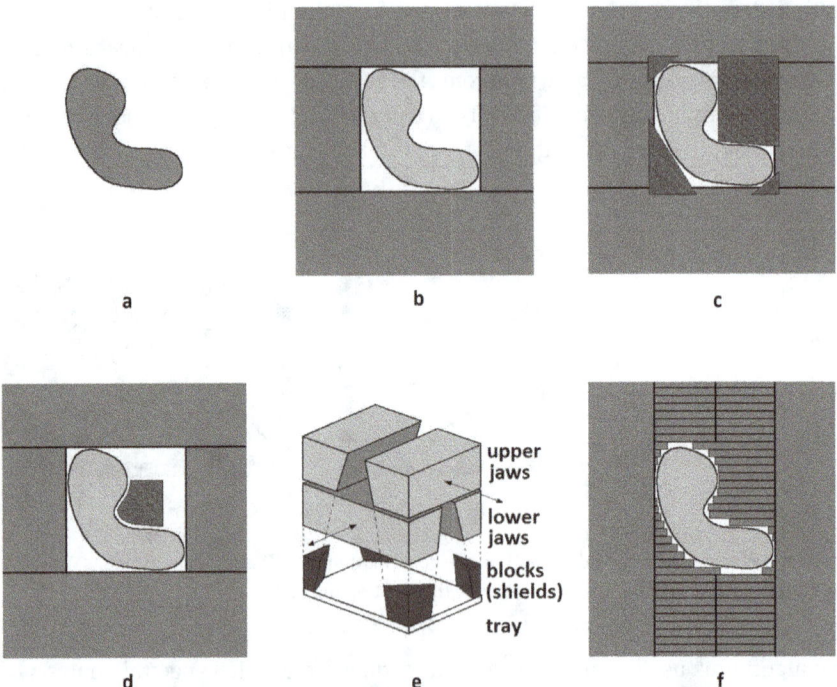

Figure 3.7.2.9: Adjusting the beam shape to the PTV envelope: a – PTV seen from BEV, b – adjustment of jaws positions to the tumor size, c – individual blocks (top view), d – central shield, e – collimation system without MLC (general view), f – concept of using a multi-leaf collimator.

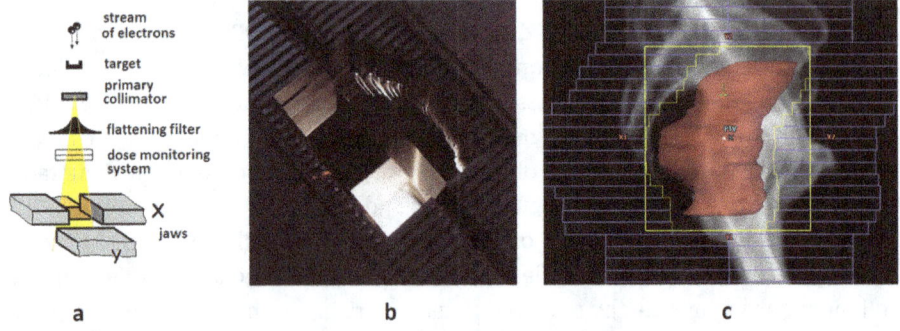

Figure 3.7.2.10: The collimation system of a linear (classic) accelerator: a – general scheme, b – the MLC from the patient's point of view, c – adjusting the MCL aperture to PTV shape.

each other. Tungsten bars are made of one piece of metal so that they fit tightly together. In the so-called static techniques, the leaves reach the expected position (they determine the aperture similar to the shape of a tumor) and remain stationary during the exposure.

The arrangement of the leaves must be selected individually for each therapeutic beam to match the current BEV. Before MLC got into routine practice, healthy tissues were covered with heavy metal blocks (Wood's alloy) mounted on a special tray under the collimator jaws. First of all, using the blocks was hard work for radiotherapy technicians (the shields were heavy and it was necessary to make a separate set of blocks for each beam). Secondly, replacing the shields before each therapeutic beam made the duration of the therapeutic session longer. On the other hand, the shields made of solid metal did not have gaps, so there was no radiation leakage as in the case of MLC. Due to radiation leakage between the adjacent leaves, MLC should be used wisely (e.g., when the spinal cord or optic nerve is protected). In 2001, Oelfke and Bortfeldt described their concept of the so-called

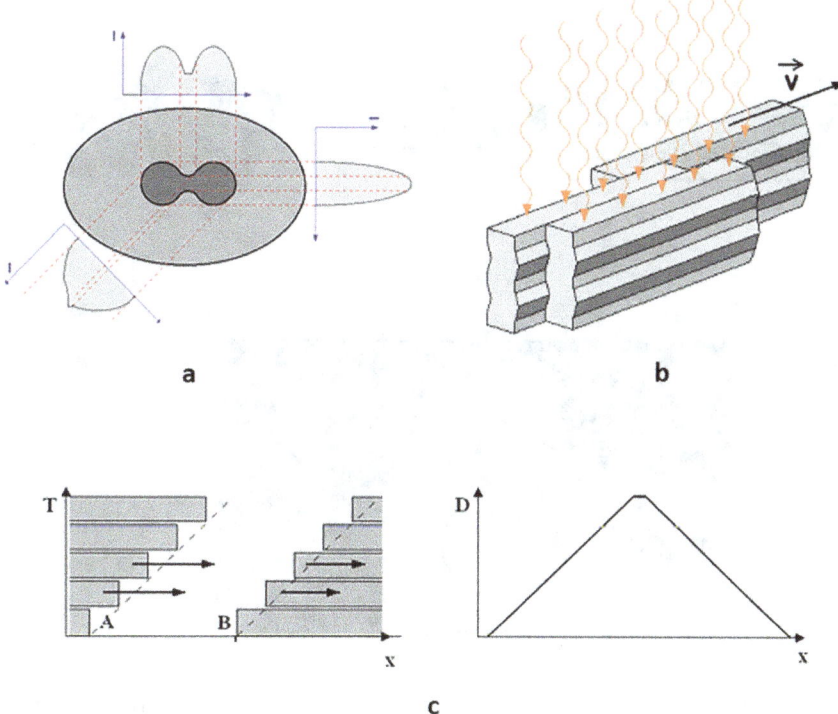

Figure 3.7.2.11: Conception of Intensity Modulated Radiation Therapy: a – intended intensity distribution for selected beams, b – MLC leaves during operation, c – realization of smooth gradients of dose along leaf axis (x).

inverse planning and a new era in radiation therapy has begun: the era of Intensity Modulated Radiation Therapy (IMRT) Figure 3.7.2.1.

To deliver the same dose at different depths, the losses caused by attenuation of the beam by shallower tissues should be compensated. In the IMRT technique, selected fragments (segments) of the therapeutic field are exposed longer than the others. This can be achieved by changing the position of the MLC leaves during irradiation: the tissue is exposed for as long as necessary and then it is shielded by a leaf. Conversion of the expected dose distribution into the desired movement of individual collimator leaves is part of inverse planning and requires the use of algorithms known as MLC optimizers.

In the IMRT technique, the directions of entry of the beams into the patient's body are selected by a medical physicist (treatment planning from physicist perspective), which is analogous to static techniques. The difference is that the leaves do not open the space above PTV, but move according to the designated plan. The use of inverse planning and intentional modification of beam intensity within the therapeutic field gives IMRT a high ability to focus the dose in the target and to

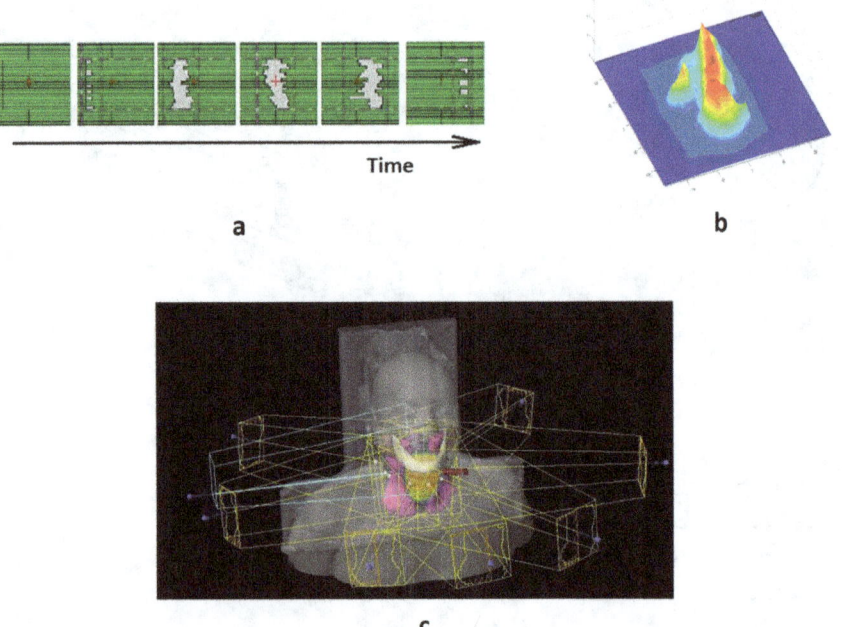

Figure 3.7.2.12: The IMRT technique: a – positions of selected leaves of MLC during IMRT procedure vs. time, b – acquired dose distribution, c – therapeutic beam arrangement.

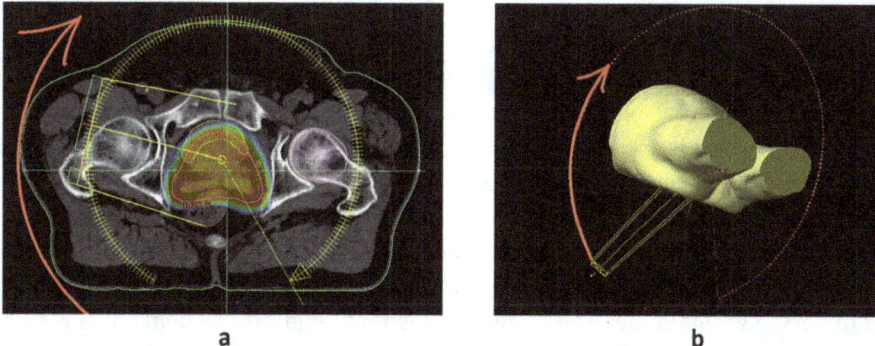

Figure 3.7.2.13: The concept of VMAT technique: a – beam rotation (arc) and dose acquired dose distribution, b – beam rotation overview.

protect the tissue adjacent to it. The concept has been expanded in a VMAT (Volumetric Modulated Arc Therapy) - Figures 3.7.2.12 and 3.7.2.13.

In radiotherapy with external beams, the protection of PTV surrounding tissues is achieved by using of several or more beams entering from different directions. The effect of protection increases as the number of therapeutic beams increases. The trend is noticeable in both static and dynamic techniques such as IMRT. As a result of this observation, rotational techniques appeared in radiotherapy. Already in the days of cobalt units, rotational therapies were carried out with the arm rotating around the patient. From a theoretical point of view, rotational therapy is a treatment involving the use of a huge number of beams. The beams are arranged radially and differ slightly from each other by the angles of entry. VMAT is the most advanced variation of rotational technique because intensity modulation and rotation of the unit's arm around the patient occur simultaneously.

VMAT can be performed by specially designed devices, such as tomotherapy, but it can also be performed by classic accelerators. The design of the tomotherapy device allows accurate irradiation of narrow layers of the patient's body. To irradiate a wider area, successive shifting of the therapeutic table is required, as in CT. Linear accelerators perform VMAT procedures without moving the therapeutic table, but the maximum size of the therapeutic field is limited by the effective size of the MLC collimator (measured in the axis of rotation of the accelerator arm). Verification of both the beam intensity modeling process and its correlation with the machine's rotational motion meant that the implementation of VMAT required the development of complicated methods of quality assurance.

3.7.2.2 Special techniques in external beam radiotherapy

3.7.2.2.1 Radiosurgery and stereotactic treatment

Traditionally, external beam radiotherapy is carried out in several dozen subsequent therapeutic sessions. As mentioned earlier, the total therapeutic dose is split into fractions to enable effective regeneration of healthy cells. This approach is possible in the case of slow-growing tumors. In rare cases, a single-fraction (or only a few fractions) high dose of radiation is needed. This dose fractionation scheme is called hypofractionation, and it is typical for stereotactic radiotherapy. Fast-growing tumors, e.g., brain metastases, require the rapid and precise delivery of a high dose of radiation in a well-defined small area. The goal of treatment is the same as for brachytherapy, but the close proximity of critical structures usually prevents the application of the source into the tumor tissue.

Stereotactic radiotherapy is considered to be an elite technique. It requires great precision both in geometrical control of the patient's position and in dosimetry. The effectiveness of therapy depends on how quickly the metastasis is noticed, which means that very small objects are irradiated. The spatial dimensions of the

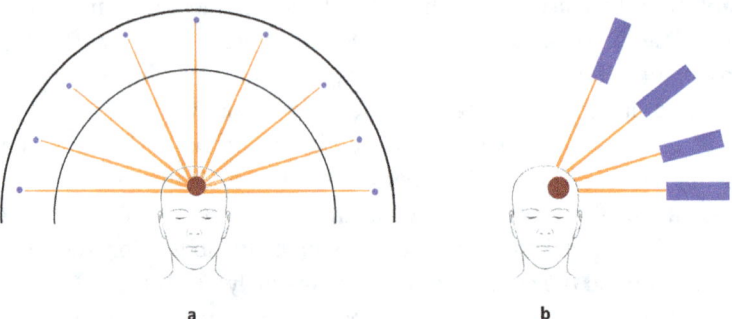

Figure 3.7.2.14: Stereotactic treatment: a – set of Co-60 gamma beams generated by Gamma Knife unit, b – set of 6 MV X-ray beams generated by Cyber Knife accelerator.

area of interest often do not exceed a few millimeters and the target's distance from the OAR is also incredibly small. It means that high-resolution spatial visualization must be used to monitor patient positioning. Stereotactic radiotherapy can be performed using dedicated devices such as Gamma Knife or Cyber Knife, but traditional linear accelerators are also allowed – consider Figure 3.7.2.14.

Gamma knife is a device designed to irradiate brain tumors. Its head has several dozen built-in small cobalt sources located on the surface of the sphere. The special collimation system focuses gamma radiation at the indicated point inside the sphere. The idea is to place the patient on the treatment table in such a way that PTV is at the point where the radiation is focused. Operators have the ability to cover unnecessary sources, which allows reducing the volume of healthy tissues that will be irradiated.

Cyber knife is a small accelerator mounted on an automated arm that allows great freedom of movement. It is intended not only for head tumors radiotherapy. This device generates a beam of X-rays with a nominal accelerating potential of 6 MV. The radiation beam is collimated. The user has the option of using a conical collimator (a set of several collimators with different diameters is available for selection) or an automated collimator with variable diameter. The cone collimator looks like a metal tube mounted in a place where radiation leaves the machine. The idea of an automatic collimator was taken from cameras – it resembles an iris.

A classic linear accelerator usually requires some additional accessories to be able to meet the requirements of stereotactic treatment. First, the patient immobilization systems used in conventional radiotherapy usually do not guarantee sufficient submillimeter precision. In the first period, steel frames were used in stereotactic radiotherapy of brain metastases. The idea was borrowed from neurosurgery: the frame to be screwed to the patient's skull bone. The frame was then rigidly attached to the table for the whole duration of the therapeutic session, which guaranteed stability.

The leaves of the very first MLCs were too wide (10 mm) to be used to shape stereotactic beams. So, vendors began to offer the so-called microcollimators, additional devices mounted on the accelerator head. The collimation of the beam was carried out with the help of conical collimators (an idea also used by Cyber Knife) or with the help of narrow leaves (e.g., 3 mm). Today, vendors offer their customers (clinics) the option of partial configuration of an accelerator. It means that the user can also choose HD MLC, which enables stereotactic irradiation (leaf width 2.5 mm) - Figure 3.7.2.15.

a

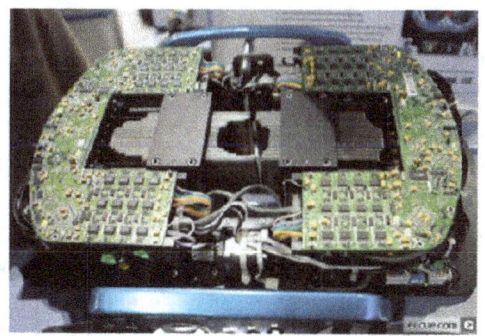
b

Figure 3.7.2.15: Conical collimator (a) and micro-MLC collimator used to stereotactic treatments performed on conventional linear accelerator.

3.7.2.2.2 Total body and total marrow irradiation

Leukemia is a cancer of the hematopoietic system, so cancer cells can be spread throughout the patient's body (blood system, bone marrow). The concept of treatment is simple: remove the diseased marrow and inject the healthy one. Sterilization is carried out by pharmacological method (chemotherapy), but usually it is supported by TBI or TMI.

Irradiation of the whole body requires high concentration, because the patient receives a dose almost three times higher than the lethal one. If such a dose were given within a short time, it would quickly lead to a certain death of the patient. In this particular case, the total therapeutic dose of radiation is evenly distributed over three days, dividing it into six equal long-lasting fractions. This gives a chance to repair the damage that occurs in healthy cells, while cancer cells spreading throughout the body gradually disappear.

In classic TBI therapy, the patient should be as far away from the source as possible. This guarantees low dose rate and uniform intensity for the whole body - consider Figure 3.7.2.16. The patient is not placed on the therapy table like other patients are. If possible, the patient should be placed directly on the floor under the accelerator head, where the dose rate decreases almost four times or at one of the walls of the therapy room – then the intensity decreases more than 16 times. To obtain a homogeneous dose distribution throughout the patient, nonstandard methods of immobilization are required. Due to the incomparably long exposure time compared to typical treatments, the patient must be placed in a comfortable and stable position. This is the task of radiotherapy technicians, a difficult task, because usually young people and children are irradiated, who find it difficult to stay still for a long time.

It is really difficult to get a homogeneous dose if classic technique has been chosen. If the patient were a homogeneous block, e.g., a cylinder, it would be much simpler, but it is not. In those parts of the body where the patient is thinner (neck, calves), additional materials should be placed on the skin to attenuate the radiation. Physicists manually calculate the thickness of such material taking into account the width of the patient. In a similar way, the dose absorbed by the lungs is reduced. Usually the dose must be reduced a little to avoid unnecessary fibrosis that makes it difficult to breathe in the future. The shape of covers and fillers should be designed after placing the patient in the therapeutic position. It is not possible to use a commercial staff, so the accessories are made usually by hand.

TMI can only be performed using the VMAT technique: tomotherapy units are preferred, but modern linear accelerators are also used for such applications. While performing a classic TBI does not require the use of a treatment planning system, planning a TMI without TPS will not be possible. Doctors must first contour all the bones that play here the role of PTV. Only the most sophisticated calculation algorithms and superfast computers are able to find a way to escalate a dose in the bones and lower the dose in soft tissues.

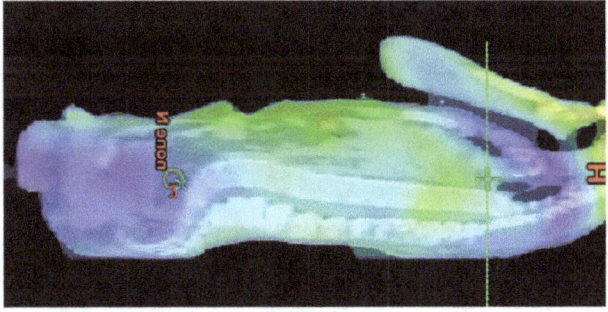

Figure 3.7.2.16: Total Body Irradiation: a – patient position, b – acquired dose distribution.

3.7.2.2.3 Intraoperative radiotherapy

The surgical removal of a tumor often does not give 100% certainty for cancer curing. The human eye does not guarantee the perception of individual cancer cells. There is a risk that little part of the cancer tissue still may remain in the tumor bed. In many cases (e.g., treatment of breast cancer), radiotherapy as an adjunctive treatment is practiced (brachytherapy or external beams) to increase the probability of complete cure. Unfortunately, radiation therapy is usually carried out a few weeks after surgery due to the need to complete the wound healing process.

To deliver the concentrated dose of radiation directly to a tumor during surgery, the Intraoperative Radiotherapy (IORT) has been developed. It is believed that direct irradiation of the tumor bed gives a greater chance for complete sterilization of cancer cells. However, tissue deformation that occurs during surgery means that radiotherapy planning based on presurgery scans does not make sense.

The IORT uses short-range radiation: electrons and X-rays. Smart mobile accelerators have been constructed to work in the operating room: Figures 3.7.2.17 and 3.7.2.18. They must not be heavy and should be able to move closer to the operating table to stand above it. They also do not need an MLC: the beam is collimated usually with the help of conical collimators (applicators).

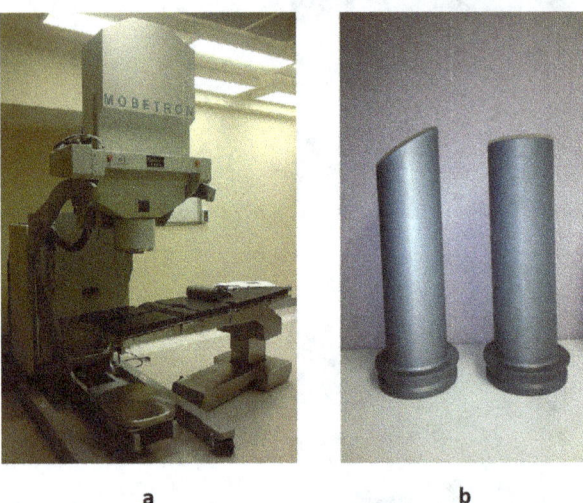

Figure 3.7.2.17: Mobile accelerator for electron IORT (IOeRT): a – accelerator and operating table in the operating room, b – examples of conical aplicators.

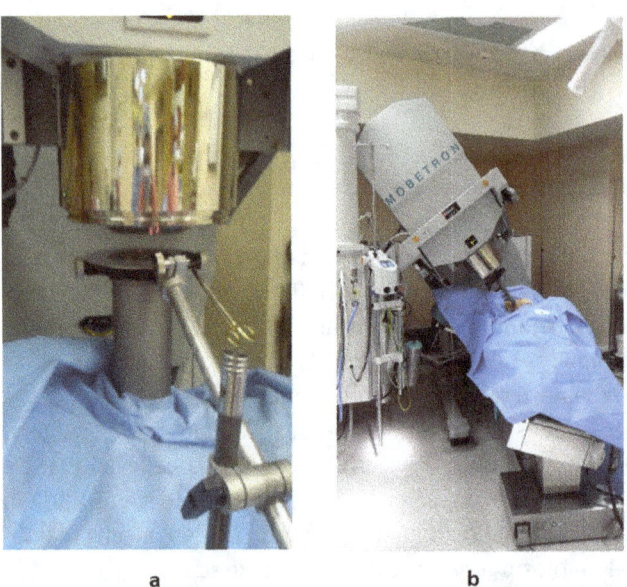

Figure 3.7.2.18: Intraoperative radiotherapy in progress: a – docking (breast tumor irradiation), b – general view.

The radiation falls directly onto the PTV (known and indicated by the surgeon), so no additional patient positioning control systems are needed. The treatment time and dose distribution have to be calculated manually by physicists in the operating room.

IORT provides a unique opportunity to protect healthy tissues that no other technique used in oncology can provide. It is even possible to protect the tissues located between the tumor and the source of radiation (above the tumor, e.g., the skin). Usually in radiation therapy the shields are used to cover tissue to reduce the dose. If necessary, in IORT the surgeon may simply move healthy tissues away from the radiation field. After the radiotherapy, the tissues should be restored to their original shape and place. Figure 3.7.2.19 presents and example of a typical dose distribution obtained for external electron beam in breast cancer [5].

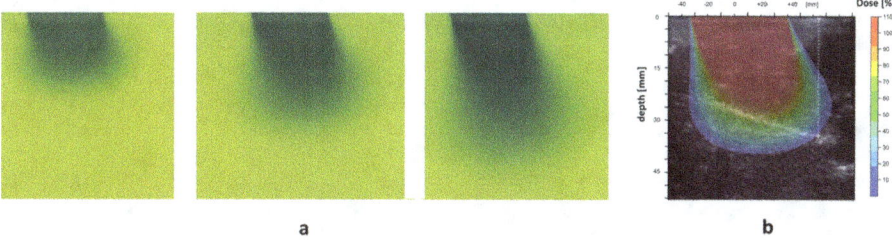

Figure 3.7.2.19: Examples of electron beam dose distribution in IORT: a – dose distribution recorded on self-developing films, b – calculated dose distribution [5].

Author contribution: The author has accepted responsibility for the entire content of this submitted manuscript and approved submission.
Research funding: None declared.
Conflict of interest statement: The author declares no conflicts of interest regarding this article.

References

1. Łobodziec W. *Fundamentals of ionizing radiation physics for radiotherapy and radiological diagnostics.* Rzeszów: Wydawnictwo Uniwersytetu Rzeszowskiego; 2016. – polish language.
2. Prescribing, Recording, and Reporting Photon Beam Radiotherapy (Report 50). International commission on radiation units and measurements. *J Int Comm Radiat Units Meas* 1993;26:NP.
3. Prescribing, Recording, and Reporting Photon Beam Radiotherapy (Report 62). International commission on radiation units and measurements. *J Int Comm Radiat Units Meas* 1999.
4. Khan FM. *The physics of radiation therapy.* Philadelphia: Wolters Kluwer; 2014.
5. Winiecki J, Orzechowska A, Maleszka S, Biedka M, Wiśniewski T, Nowakowski S, et al. Visualization of dose distribution in intraoperative electron beam radiotherapy based on ultrasound images. *J BUON* 2019;24:2570–6. 31983134.

Karolina Balik, Karolina Matulewicz, Paulina Modrakowska, Jolanta Kowalska, Xavier Montane, Bartosz Tylkowski and Anna Bajek

4 Advanced cell culture techniques for cancer research

Abstract: The incessant increase number of cancer cases, motivates scientists to constantly develop and search for new therapies. Along with the dynamic development of anti-cancer drugs and therapies, we are witnessing huge progress in the world of science - the development of personalized medicine. An inseparable element is also a very strong trend in the development of new in vitro animal models for chemotherapeutic research. Cell cultures are commonly undertaken by research models before animal testing. They are the basis for the development of new diagnostic and cancer treatments. It should be emphasized that basic research is a strong foundation for any therapy introduced. This chapter provides an overview of the modern cell culture techniques that are currently developing, which allow the introduction of modern models that reflect the organs and physiological system. Currently available cell culture methods are a key aspect of studying these interactions, however, a method that eliminates the limitations of standard methods is still being sought.

Keywords: cancer research, culture techniques, in vitro culture

4.1 Cell-to-cell interactions

4.1.1 Immunotherapy and cancer research

Over the past decades, we have witnessed enormous advances in the world of science—the development of gene therapy, personalized, and regenerative medicine. Nonetheless, it should be emphasized that basic research is the firm foundation of every introduced therapy. Commonly undertaken research models preceding animal studies are cell cultures. They form the base for the development of new diagnostic methods and treatment of cancer. They are grounded in cellular interactions studies. Cell-to-cell interactions are an extensive intercellular communication system that regulates cellular processes. The analysis of the interplay between the examined cells allows determining their mutual influence on such processes as apoptosis, proliferation, growth, differentiation, movement, or secretion of mediators to the environment. We distinguish indirect and direct types of interactions. An example of direct communication is when a whole tissue is formed by cells. In terms of *in vitro* for cancer research, indirect interactions play another significant role. Cells communicate and interact with one another by secreting mediators, such as cytokine, growth factors, and many other proteins. The adopted culture techniques constitute a key aspect of studying these interactions, however, a method which would eliminate limitations of standard methods is still being searched for.

One of the most substantial research achievements in recent years that relies on basic science is the discovery of surface proteins whose function is to halt T cells' immune response. The proteins are called immune checkpoints and we can list CTLA-4 and PD-1 proteins. As early as in the late 1990s, research on antibodies that block CTLA-4 molecules began, therefore preventing the negative regulation of T cells activity. In 1996, Leach et al. illustrated in an *in vivo* study that blocking T-cell inhibitory signals increases cancer resistance [1]. Then in 2011, based on the results of the first phase III clinical trial, Ipilimumab was launched on the market for the treatment of advanced melanoma [2, 3].

Meanwhile, the ongoing research on PD-1 protein has shown its similarity to CTLA-4 protein. Likewise CTLA-4 molecules, the PD-1 molecules constitute a negative control point for T cells but are based on a different mechanism [4, 5]. Efficacy of the checkpoint therapy against PD-1 was proved in a battle with different types of cancers, such as NSCLC, prostate cancer, kidney cancer, colorectal cancer or melanoma [6–8].

Albeit, like any other therapy, immunotherapy also reveals limitations. Side effects of immunotherapy are mainly concomitant with an excessive immune response, sometimes leading to autoimmune reactions, whereas some types of cancers do not respond to immunotherapy, which is also being investigated [9–11]. Therefore, in addition to the ongoing numerous attempts to use the therapy associated with blocking control points in many cancers, also new proteins as molecular targets are still being sought and tested. Moreover, the mechanisms resulting in immunosuppression in the cancer microenvironment are intensively studied.

4.1.2 Cancer microenvironment

Tumor microenvironment (TME) is defined as all noncancer cells identified near the tumor. These include, among others, cancer-associated fibroblasts (CAFs), stem cells (ASCs), and immune-inflammatory cells. It also consists of extracellular matrix (ECM), blood, and lymphatic vascular network. Additionally, it comprises all molecules synthesized by cells present in the tumor that affect cancer cells [12].

Recent research concerning TME primarily focuses on CAFs and immune-inflammatory cells. Studies to date show, that these cells can both promote and inhibit the development of cancer [12–14]. Notwithstanding decades of experimentation, some of the above-mentioned mechanisms remain unclear. A vast number of studies devoted to that area are based on the analysis of mechanisms that induce resistance to treatment. Kim et al. divided the immune types of triple-negative breast cancers (TNBCs). So far, three subtypes have been distinguished: enriched with macrophages (MESs), with neutrophils (NESs), and called cold subtype—without macrophages or neutrophils. MES and cold types are resistant to checkpoints blockade, while MES has displayed several responses to the same treatment. The research presents, that some cancer cell-derived interleukins can attract both neutrophils and monocytes, whereas the epithelial–mesenchymal transition (EMT) was associated with the upregulation of PD-L1 molecules [15]. Understanding the differences between tumor immune types may be a source of information on new strategies for immunotherapy and heterogeneity in patients.

Research exploring the impact of TME on cancer progression begins with *in vitro* studies, thus it is vitally important to improve the tools available for this type of studies. Until now, the most commonly implemented methods have been either the Transwell system or the use of the conditioned medium (CM). Worth noticing, along with the development of tissue engineering, not only have High-Throughput Strategies (HTs) occurred but also new microfluidic devices providing controlled cell-to-cell studies [16, 17].

4.2 Standard methods of cell-to-cell interactions studies

The previously mentioned standard techniques rely on 2D culture. Cells grown in the 2D system are forced to form a monolayer, they do not adopt natural morphology, and also, the intercellular communication can be altered. Hence, this can directly affect cell migration and differentiation. Nevertheless, cell monolayer analysis is substantially easier and cheaper, and it is also characterized by enhanced repeatability [18, 19].

In general, two methods have traditionally been utilized for cell-to-cell analysis in 2D culture and they are as follows: the Transwell system and the CM medium. The Transwell system is a co-culture devised by the use of commercially available microporous membranes that provide physical cell separation. One type of cells grows on the growth surface of the multiwell plate, while another one on the membrane itself. This in turn, ensures a free flow of culture medium with components synthesized by both cell types between levels at which the cells are located. The other method involves CM, i.e., a medium previously used to grow cells. It is, therefore, a medium enriched with mediators synthesized by this culture which allows for analyzing indirect interactions between the tested cells. Thus, two standard methods can be applied when it comes to studying interactions between cells in a tumor microenvironment based on a traditional 2D culture. One fundamental limitation of this commonly used method for monitoring cell growth is the determination of cell survival, not continuous growth. On this account, the use of microfluidic devices in both 2D and 3D models is considered an innovative approach to cell-to-cell research.

4.3 Microfluidic devices

Microfluidic systems have been used in monitoring cell growth and viability, environmental control in cell aging, and for cell-to-cell research. Commercially available microfluidic systems have begun to appear over the years, but many researchers still opt for designing the devices themselves. These systems are predominantly based on alterations in fluid behavior that can be observed at the microlevel. This is possible via application of channels with either nanometer or micrometer diameters that allow the use of a minute amount of fluid. Microfluidic is derived from microanalytical methods by the use of a capillary form, such as high-performance liquid chromatography (HPLC) or gas chromatography (GC). Confocal and fluorescence microscopy are preferable methods applied for cells behavior monitoring in real-time in microfluidic systems [20].

4.3.1 Microfluidic 2D models

Scrutinizing the interaction of cancer cells with TME cells, i.e., effects on growth, survival, change in gene expression, aging, migration, or secretome is of great importance with regard to expanding our knowledge about cancer progression. Ma et al. developed a device consisting of two layers with three chambers for culture, and a large number of microchannels enabling cells communication and migration. The height of the migration region measured in the device was 5 µm. This contributes to observing the effect of cancer cells on the migration of human embryonic lung fibroblasts as compared with migration toward normal epithelial cells. A migration test was completed by culturing fibroblasts in one chamber and cancer cells or control cells in the other one, while the third chamber held a culture medium without cells. After 48 h of culture, cell viability within this device amounted to 98%, which indicates its utility. The study identified migration and transdifferentiation of fibroblasts cultured in the presence of tumor cells, which was not observed in the presence of control cells [21]. Many examinations report that fibroblasts can be activated in the presence of cancer cells, and after transdifferentiation, they can have a significant impact on the invasion and metastasis of cancer cells [12]. It has been revealed that the devised system allows for the culture of various cell types under physiological conditions and for the analysis of interactions between them. Equally, a microfluidic system for examining cell migration was developed by Agliari et al. The study aimed to analyze the interaction between immune cells and melanoma cells. The device was designed on a similar principle as in the case of Ma et al. It was to contain three chambers for cell growth as well, the first one for mouse metastatic melanoma cells, the second one contained no cells only culture medium, and the last chamber held cells isolated from the spleen of immunodeficient mice without IRF-8 or from immunocompetent mice. Cell migration enabled 10×10 mm^2 microchannels between the chambers. The study points out that splenocytes from immunodeficient mice did not show any response to melanoma cells. In contrast, splenocytes from wild-type mice migrated and formed clusters around tumor cells. The research demonstrated that the application of the microfluidic system reflects the behavior of immune cells *in vivo*, and consequently is a good tool for further examinations regarding mechanisms between cells [22].

Studies have pinpointed that microfluidic systems allow for cell cocultures to be carried out on a controlled microscale. Moreover, they can constitute a tool for screening chemotherapeutics. Liu et al. created a stimulation of the microenvironment in bladder cancer comprising four types of cells. Bladder cancer cells fibroblasts, endothelial cells, and macrophages have been used to stimulate the TME. The device consists of two layers, four chambers for cell culture, matrigel channels, and microchannels. The designed system enabled direct and indirect interactions between cultured cells by soluble extracellular factors and metabolites, as well as cell migration through microchannels. Nevertheless, the major goal of the study was to assess the sensitivity of tumor cells to chemotherapy in the presence of TME cells. The findings of the study show that microfluidic systems can constitute simulations of the tumor microenvironment. A few insights have been shared, such as mobilization of macrophages toward stromal cells, the reticular structure of bladder cancer cells, and different sensitivity of cancer cells to the chemotherapy schemes, which mirrors the physiological behavior of cells *in vivo* [23].

4.3.2 Microfluids 3D models

Together with the development of cell culture, three-dimensional geometry microfluidic systems were established. They have been designed to resemble conditions that are as similar to the natural ones as possible. Three-dimensional structure cultures make it possible to recreate interactions not only between cells but also with other TME components, such as blood and lymphatic vascular network or ECM proteins. Liu et al. developed a microfluidic device for three-dimensional culture for cancer cells and CAFs. The researchers aimed to create a model for studying the cancer invasion mechanisms and for testing anti-invasive drugs. The device consists of multiple culture chambers connected with medium channels. In view of creating a 3D platform, chamber B intended for cell culture was seated higher than chamber A and the channel with the culture medium. The cell suspension was mixed with synthetic ECM to initiate 3D culture. Next, tumor cells were placed in chamber A, while CAFs or the control in chamber B. This platform enabled reflection of *in vivo* conditions. After six days of culture, the spheroid formation was observed in both chambers as well as the cell-cell, and the cell-ECM communications. It should be highlighted, that the scientists have successfully drawn up a model that facilitates the spheroid formation and observation of cells communication in real-time by measuring cells migration using fluorescence microscopy. The experiment confirmed that CAFs promote cancer cell invasion and MMP inhibitors can block the effects of CAFs. It has been found that the designed device can be a platform for studying mechanisms of promoting tumor progression [24]. The invasion of cancer cells is strongly influenced by the interactions between TME and the vascular system, and for this reason, Wong et al. developed a device that allows cell growth along with artificial ECM-based microvessels. Prior studies similarly used fluorescence microscopy to monitor cell growth in real-time. The cylindrical collagen channel lined with endothelial cells has been connected to the gravitational flow system to create artificial microvessels. In the study, varied interactions between breast cancer cells and endothelial cells were observed, most of which are in accordance with the literature. The designed device fostered observations of such events as invasion, metastasis, or angiogenesis. The research team managed to recreate *in vivo* conditions [25].

Standard 3D cultures are most frequently carried out on multi-well plates, require a large number of cells, and in addition, they are time-consuming. Bauer et al. compared a standard 3D culture with a microfluidic 3D culture. For this purpose, a breast cancer model was created in the form of a microfluidic coculture device with 96 arrays single channel with CAFs and breast cancer cells. Yet, the device did not contain microchannels with the flow of nutrients and had to be supplied manually as in a standard one. Compared to the common 3D culture, the designed plate reduced the volume of used reagents and considerably decreased the number of cells necessary for initiation of the culture in three-dimensional geometry. Furthermore, the study shows, that this high-throughput strategy can be successfully employed for studying paracrine cell communication between fibroblasts and cancer cells [26].

Still, one of the greatest challenges of modern medicine is the fight against cancer cells resistance to immunotherapy, targeted therapy, or to standard anticancer treatment methods. The TME is considered to be the first and foremost factor supporting escape from immune surveillance of cancer cells. Due to the above-mentioned, it is crucial to expanding knowledge about the mechanisms that occur between cells. Pavesi et al. presented a device that enables the assessment of adoptive T-cell therapy in a preclinical model. To do so, the previously designed device was used [27]. It consisted of four cell culture channels, channels with culture medium, and microchannels enabling migration and contact between cells. To obtain a 3D culture, the liver cells suspension was mixed with collagen. In this case, confocal microscopy was used to depict prestained cells. The study identified that mRNA-TCR cells show less cytotoxicity than retro V-TCR, as was previously shown by Koh et al. in an *in vivo* study on an animal model [28]. From this, it results that microfluidic systems can be successfully used to test the effectiveness of T cell-based anticancer vaccines with the possibility of constant monitoring of cells viability [29]. Another interesting study of resistance to immunotherapy using microfluidic systems was conducted by Parlato et al. The designed device consists of six chambers interconnected with multiple channels: two culture chambers for interferon-α-conditioned dendritic cells (DCs), two for colorectal cancer cells and two external ones with culture medium. As in other devices, the channels were connected with microchannels enabling migration of cells. The designed device made it feasible to observe and regulate interactions between cancer cells with anticancer agents administered, and interferon-α-conditioned DCs in real-time. Furthermore, the study revealed the underlying factor in these interactions—CXCR4. It has been shown that the designed system can constitute an innovative approach to research on immunotherapy and on resistance of cancer cells to treatment [30].

4.4 Artificial organs—why do we need them?

Along with the dynamic development of drugs and cancer therapies, the development trend of new *in vitro* animal models for chemotherapeutics studies is also clearly visible. The development of medicines is associated with huge costs, and not always bringing the expected results. The report showed that from February 10, 2006, to June 1, 2016 there had been a total of 87 new anticancer drugs approved [31]. According to reports, bringing new molecules to market takes 10–12 years and costs 1.3–2.6 billion USD, moreover <1 out of every 10 drugs entering clinical trials makes it to market [32]. Focusing on discovering new treatment methods, researchers are looking to develop better models for preclinical, *in vitro*, or *in vivo* research. The cell culture, an alternative to *in vivo* organ culture and animal models, is one of the main parts of the development of medicine. Drug development is associated with enormous costs, and not always brings the expected results. Very often, the results of research on new drugs give very promising results *in vitro*, but after starting the *in vivo* tests, it turns out that the complexity of the microenvironment in the human body significantly affects the research results. The result of *in vitro* tests decides whether a costly investment in *in vivo* tests and preclinical tests is likely to be profitable [33]. The long path every compound has to go from *in vitro* testing to clinical testing is extremely expensive and lasts many years (Figure 4.1). The development of accurate and precisely reflecting the microenvironment of organs and cancerous tumors will allow for wise management of research and possible resignation from further steps at an early stage [34].

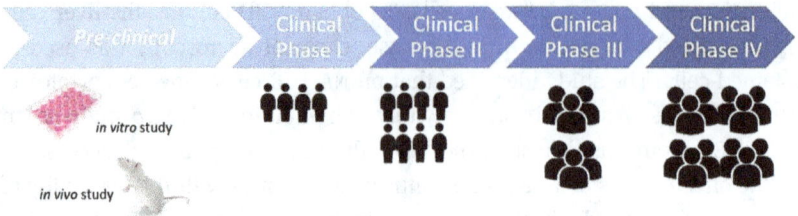

Figure 4.1: Drug development process.

4.4.1 Cell culture analog (CCA) system

The cell culture analog (CCA) system, also known as an animal or human surrogate, is designed to mimic biological and biochemical responses of an animal or human to the effects of a drug or other compounds. An animal surrogate we can call a device that replicates the circulation, metabolism, or adsorption of a chemical substance and its metabolism, reflecting the organs functions. An animal surrogate, in particular a human surrogate, can provide important insight into the toxicity and efficacy of a drug or a chemical when the use of live animals for testing is unjustified [35].

One of the known and described CCA systems is the ADME (Administation-Distribution-Metabolism-Excretion) breeding system, which consists of two parts (Figure 4.2). The first part is intended to reflect the *in vivo* circulation and metabolism of the drug's canalizes. It should include at least the small intestine (main tissue for orally administered drugs), liver (where most drugs are metabolized), and

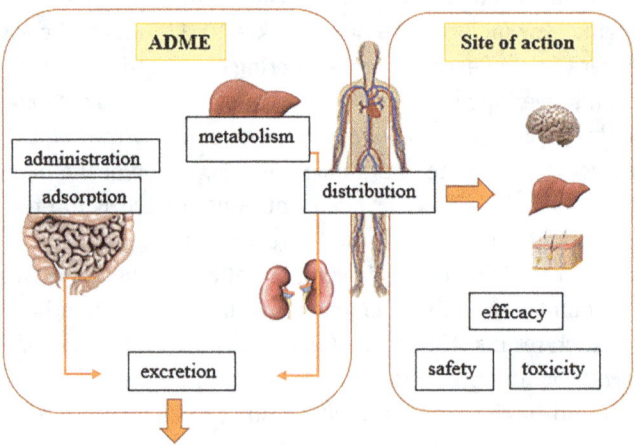

Figure 4.2: Scheme of ADME system.

kidneys (which mediate excretion). The second part consists of organs with which the effectiveness or toxicity of the drug is assessed [36].

Another example of the CCA systems are body-on-chip systems that represent pharmacokinetic models (*PBPKs—physiologically based pharmacokinetic model*). PBPK models are characterized by a detailed reflection because in their assumption they take into account physiological parameters such as the size of organs, the speed of blood flow through the organs, and the activity of the enzyme [32]. The development of such models begins with the planning of the scheme using the drawing (for example see Figure 4.3). Such a drawing only contains organs and reactions that are important for the metabolism of a given drug. The chip is to reflect organs, while the arrows reflect the microfluidic channels. Both the size of the organ chamber and the sizes of the microfluidic channels are adapted to mimic the body conditions. One of the key things in developing schemes is to calculate the size and volume of tissue constructs to represent the organs [37]. To properly construct a PBPK model, it is worth using guides that describe this process in detail [38, 39].

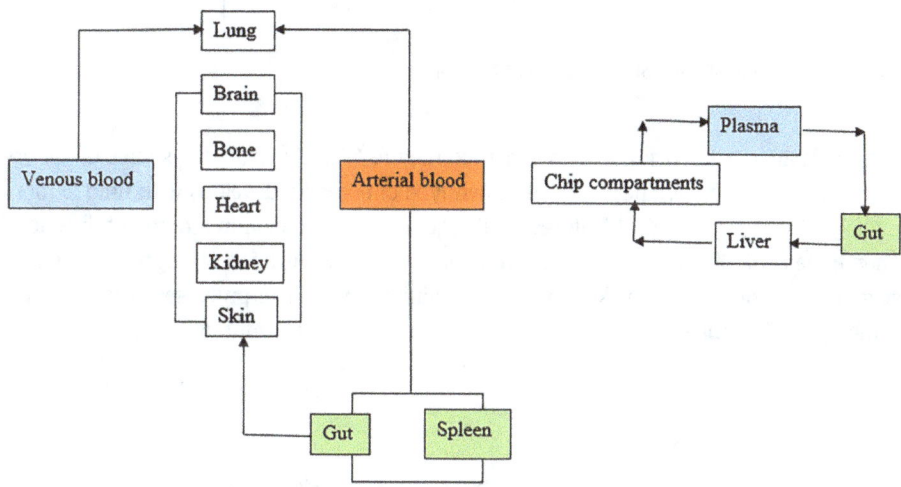

Figure 4.3: Scheme of a physiologically based pharmacokinetic model.

Chip-based body systems can predict side effects of new drugs and show human metabolic pathways. These systems allow the interaction of many types of cells with each other. This makes CCA bioreactors more attractive to predict integrated biological responses to chemicals or pharmaceuticals than traditional *in vitro* convection culture methods, which are often limited in restoring communication between different types of cells building various organs [32].

4.4.2 Organoids system

Designing systems that replace biological pathways *in vivo* involves designing them from the smallest unit, i.e., cells, to organoids (Figure 4.4). Organoids are unique, three-dimensional cell culture systems. Arise through differentiation of stem cells or tissue progenitor cells. Most organoid models represent individual or partial tissue components only, which often causes problems in controlling intercellular interactions [40].

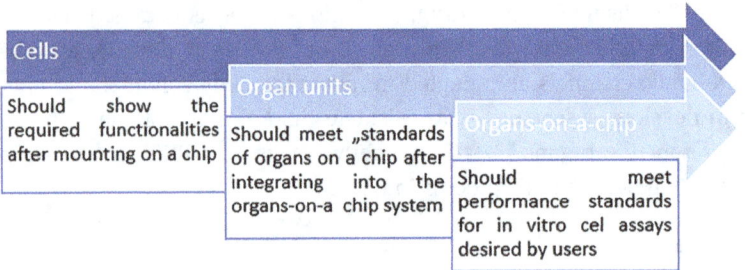

Figure 4.4: Scheme of units of organs on chip systems.

Organoids are created using a simple method (Figure 4.5). Above all, the source of the organoid is its source, i.e., the stem cells. Each type of organoid develops with unique morphological and physiological features, so it requires differentiation in a different direction. Then there is proliferation, early differentiation, and self-segregation of cell types. The late stage of organoids is tissue-like structures, exhibiting organ-like processes. The structure is then placed on the chip using a 3D matrix, e.g., through a bioprinting printer [40].

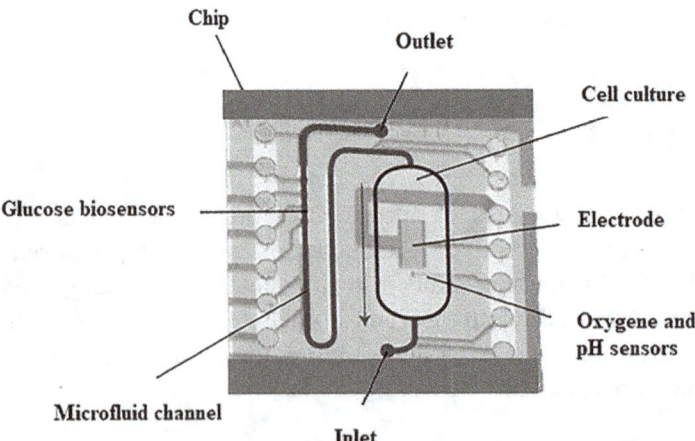

Figure 4.5: Scheme of units of organs on chip systems.

The most frequently created artificial organs are the liver, blood vessels, heart, and kidneys. The blood vessels are created to reflect the transport and circulation of the drug in the body. Studies performed by Wimmer et al. showed that blood vessels can be made from pluripotent stem cells [41]. Blood vessel organoids include pericytes and endothelial cells that self-assemble in a network of capillaries surrounded by a basal membrane. Human blood vessel organoids transplanted to mice form a stable, perfused vascular tree, including arteries, arterioles, and veins. Very promising results about vascularized human brain organoids were presented by Mansour's team [42]. This is a chance to model the development and diseases of the human brain. This model previews the imaging of both the blood vessel network and the neural network of this extremely complexed organ. This combination of human neural organoids and the *in vivo* physiological environment in the animal's brain can facilitate modeling of the disease under physiological conditions. It is also worth mentioning that the growth factors necessary for cell equalization. The results indicate that VEGF-enhanced vascular endothelial (EC) differentiation without reducing neuronal markers in embryonic bodies (EBs), which then successfully transformed into cerebral organoids with vascular structures [43]. Nowadays organoid creation technology provides great *in vitro* models that mimic the structure and function of developing human organs, even an extremely complexed human brain, which allows the organ, its anatomy and pathology, to be tested in normal conditions.

One of the most frequently formed organoids, due to its small contribution to drug metabolism, is the liver. Currently, several three-dimensional cell culture systems are available to form liver organoids. One of them is the transformation of human liver progenitor cells into matrix scaffolds, creating three-dimensional liver organoids. The morphogenesis process was interrupted by inhibition of Notch signaling [44]. An interesting method of forming liver organoids is also the use of hydrogel matrices. The Saheli et al. study shows that the use of three-dimensional liver-derived ECM hydrogel (LEMgel) promotes the physiological function of liver organoids generated by the self-organization of human liver cancer cells together with human mesenchymal and endothelial cells [45]. The generated LEMgel 3D organoids, however, had significantly higher ALB and AAT secretion, urea production, CYP3A4 enzyme activity, and inductance. Undoubtedly, the innovative 3D LEMgel organoid can potentially be used in liver tissue engineering, drug discovery, toxicological research, or artificial liver production.

Kidneys are another organ that is essential in assessing the effects of drugs. Methods for generating renal organoids from human pluripotent stem cells (hPSCs) have been developed by Takasato's team [46]. Units consist of cells with the characteristics of podocytes, proximal tubules, Henle loops, interstitial cells, and distal collapsed tubules in a continuous nephron-like system *in vivo*. The team described how to produce renal organoids using the CRISPR/Cas9 system. Cell scaffolds were obtained by detergent perfusion of animal kidneys, while bioprinting methods led to the creation of three-dimensional vascular constructs with human umbilical vein endothelial cells.

Organoids are opening new possibilities for medicine, and many potential applications of this technology are just beginning to be explored. Organoid technology can, therefore, be used to model the development of the human organ and various human pathologies "in the vessel". What is more, created from cells from patients can predict a personalized response to drugs [47].

4.5 "Omic" techniques in biological research

The development of anaplastic changes in an organism is a complicated process with many complexed steps. Despite constant advances in the field of cancer research, the exact mechanism responsible for tumor formation and further development remains unclear. Since the beginning of the Human Genome Project, the field of genomics started expanding rapidly, giving new possibilities for cancer research and opening doors for technologies aimed in the analysis of DNA, RNA, and many more. Nowadays, "omic" technologies consist of transcriptomics, metabolomics, genomics, proteomics, and radiomics.

Genomics is known as the branch of biological science specialized in the analysis of genome—its functions, structure, and many more. The role of DNA and its abnormalities in carcinogenesis has been explored since 1914 when it was first noticed that during cancer cell division chromosomes spread unusually. The biggest push in the direction of developing and advancing genomic techniques was the beginning of the Human Genome Project [48]. Since every malignancy starts with changes in the DNA sequence, mutations, deletions, insertions, and others are still being sought by researchers. Nowadays, Next-Generation Sequencing greatly improved methods for genetic material sequencing. It is characterized by relatively low cost, high speed, and accuracy. There are many commercially available methods differing in the parameters mentioned above [49]. Despite the presence of genetic alternations in all carcinomas, not all of those changes influence the phenotype of malignancies. To study, among others, functions of genome and development of a disease, it is desired to evaluate transcriptome. That means taking into account the whole range of messenger RNA (mRNA) in a cell or tissue—its total quantity at a given point in time and in specific conditions. The mRNA is a dynamically changing type of information, while the genome is fixed—that is why transcriptome is considered as a link between genetic information and a phenotype of an organism. Nowadays, techniques used to analyze RNA are based on microarrays or sequencing [50].

Another step toward the most accurate phenotype depiction is the analysis of proteome since proteins are the end product of DNA transcription and translation. The general purpose of proteomics is to analyze the function and significance of proteins in a given cell or in the whole organism, their interactions, and pathways.

Since one gene can code multiple different proteins, the analysis of proteome is much more complex than genomics or transcriptomics. A crucial step towards identification of proteins is their effective isolation and separation. Methods to obtain separated, high-quality proteins are most commonly based on chromatography. Among them, gel-based chromatography and liquid-based chromatography (LC) can be distinguished. Nowadays, LC coupled with mass spectrometry (MS) is the most frequently used method for proteome analysis, MS being crucial in protein identification [51].

4.6 Metabolomics

A branch of "omics", that allows researchers to dig deeper into complicated networks and interactions between genome and environment of a given cell or organism is called metabolomics. Metabolomics is focused on the analysis of metabolites—molecules of a size smaller than 1kDa, that are produced as a result of metabolism [52]. For a long time, it was believed that cellular metabolome was fully discovered and there can be no further advancements in the field of metabolite discovery. Nonetheless, when cutting-edge mass spectrometry was applied to analyze metabolome expansively and thoroughly it turned out that unexpected amounts of compounds have been detected, whose masses have never been recognized before. Based on this information, this branch of "omics" was termed metabolomics. Since this revelation, numerous new metabolites have been discovered and the metabolomics itself is of unending interest to researchers.

Every compound found in an organism, except for protein, RNA, and DNA, that can be sampled and studied could be called a metabolite. Those substances affect the state of both physiological and pathological cells. Metabolomics offers numerous advantages over previously described genomics, transcriptomics, and proteomics. As metabolome profiling shows an accurate picture of what happened in a given cell or tissue, important processes that define the authentic phenotype of an organ or disease could be identified and analyzed. Moreover, metabolomics techniques allow detection of small changes in the activity of a gene or its expression because they are much more noticeable in the metabolome—similarly to the butterfly effect. Since metabolites actively take part in the processes that occur in every cell and hence influence its behavior, they can affect proteome and transcriptome. The practical advantage that metabolomics can offer is its relatively low cost compared to the other "omics" techniques described above. Using metabolomics is also beneficial for analysis meant to compare conditions, diseases, or cell types, as well as check for parallels between organisms or species. It is important to differentiate metabolomics form metabonomics, although nowadays they are used interchangeably. First of the aforementioned terms is used in analysis meant to profile a group of metabolites in a given cell or tissue under specific conditions and precise point in time. Those metabolites must create a biochemical profile. Whereas metabonomics, although it also refers to profiling, addresses metabolomics changes in response to some stimuli—either genetic change or chemical substance. It is useful for the prediction of changes in biochemical pathways or the creation of new ones in nonphysiological situations [53]. For the sake of simplicity, in many papers, both concepts are referred to as metabolomics, same as in this text.

Commonly, approaches in metabolomics could be divided into two main trends. The first one is the targeted approach. It requires insight before conducting the analysis. Measurement is focused on a selected group of known metabolites, that have been characterized both chemically and biochemically. Their mass, kinetics, and end products, as well as the general role in biochemical pathways, are known. For targeted analysis, sample preparation can be optimized for a particular chemical compound to eliminate high-abundance molecules as much as possible. Thanks to internal standards—known quantities of substance—it can offer better quantification. This approach can provide a new understanding of well-known metabolites—their association with each other and their contribution to the development of specific changes such as malignancies [54]. The second, untargeted approach, is an analysis that is not focused on any particular metabolite. Instead, it focuses on as many metabolites as possible. To provide a broad analysis of compounds, sample preparation, separation, and detection should be nonselective to not exclude any metabolite. This approach requires a comprehensive analysis of data acquired after detection to properly identify and characterize novel metabolites or explore alternative metabolic pathways [55].

Two leading techniques are used in metabolomics nowadays, both based on spectroscopy. First of them is the Nuclear Magnetic Resonance (NMR) spectroscopy and is characterized by high reproducibility, both between different samples or different laboratories. Using the NMR technique it is possible to identify new, unknown metabolites structures and to applying isotopic labeling. It provides a new angle for analyzing changes in response to used xenobiotics. The second and more popular is MS-based metabolomics. Similarly to NMR, MS could is applied in shotgun metabolomics where chromatography is not involved. Both those methods enable fast detection and quantitation of metabolites, however, in medical research chromatography coupled with MS is of choice, since it offers better coverage. Among those techniques, there are three most used. Capillary electrophoresis (CE) could be used for separation and later detection of molecules that are charged. GC is a separation technique of choice when the main focus of analysis is volatile metabolites or analyte that could be chemically derivatized. Therefore, it would be difficult to analyze larger or heat-labile compounds using GC-MS. The most commonly used is liquid chromatography (LC)—it ensures the greatest coverage of metabolites and is highly sensitive. Yet, like the rest of the techniques described above, it has its disadvantages. Using LC-MS in metabolomics requires careful sample preparation to ensure full enclosure of metabolome [56].

4.7 Techniques for metabolomics in 2D cell culture

Metabolomics finds its use in many areas of biology, being successfully implemented in ecology, microbiology, oncology, and many others. Metabolomics has been effectively applied in the analysis of body fluids such as urine, blood, saliva, or even sweat, as well as in *ex vivo* analysis of tissues. Many papers use metabolomics to emerge new diagnostic or prognostic markers of cancer [57]. Cell culture as a matrix for metabolomics research is steadily gaining popularity as it offers a stable and strictly controlled microenvironment. *In vivo* TMN is a highly complex system, as described previously, and thus, characteristic of specific cancer cells might vary depending on surroundings. This ability to adapt to rapidly changing conditions through freely modeling metabolic pathways is called metabolic reprogramming [58]. One of the best described metabolic changes in TMN is the Warburg effect—excess glucose uptake caused by incorrect regulation of phoinositide 3-kinase signaling. Similarly, the metabolism of glutamine is changed in a way that enables its excess uptake, making cancer cells almost entirely dependent on these metabolites [59]. Understanding the metabolic reprogramming of cancer cells and the role that TMN plays in it can provide new targets for therapy, as well as prognostic and diagnostic markers. *In vitro* research provides greater reproducibility and lower cost than the animal model.

Metabolomics using cancer cell cultures as a matrix, even though gaining popularity, must face certain challenges that are yet to be resolved. Metabolomics is an extremely delicate and precise type of study, while standard 2D cell culture requires maintaining stable conditions for cells to grow. It is necessary to take into account every aspect of cell culture growth conditions—levels of oxygen, medium composition, and others. That's why good experimental design is very important in *in vitro* cell metabolomics—it requires careful sample preparation which takes into account metabolism quenching and metabolite extraction. It is also necessary to choose an appropriate cellular model and an analytical platform depending on needs. Being cautious about every aspect of the analysis is at most importance to ensure the reliable results and proper conclusions.

4.7.1 Experimental design at the level of 2D cell culture

To guarantee a comprehensive analysis of cellular metabolome, research should focus on both intracellular and extracellular metabolites. Metabolites found inside cells (endo-metabolome) and metabolites secreted to or excreted from culture medium (exo-metabolome) make up whole cellular metabolome. Monitoring extracellular metabolome is a relatively easy task, as it does not require integration in cell structure. Analysis of culture media can provide information useful in researching drug effects, identify new markers, or help establishing new cancer primary cultures.

Intracellular metabolome analysis is a bigger challenge in terms of sample preparation as it requires interference in the cell structure. However, it allows picturing a more accurate metabolic state of analyzed cells—their phenotype and actual response to different substances. It is also important to take into account density, amount, and passage number of cells in a sample. Even though the number of cells does not always translate directly into metabolite concentration, it does influence total ion current. It is important to choose an appropriate harvesting method depending on the used cell line and metabolite class. Muschet et al. suggest using fluorometric DNA quantification with Hoechst 33342 on cell homogenates to accurately reflect cell number in a sample, but it is also possible to use standard cell counting or metabolic markers to assess cell count [60]. To eliminate differences between samples while comparing different cell lines, it is desirable to use culture media with the same composition. It is not always possible, as some media could stunt the growth of specific cell lines or even stop it entirely, making analysis doomed by default. Hence, it is extremely important to take care of cell culture normalization at the stage of experiment design.

Sample preparation is a crucial step in performing a metabolomics experiment using cell culture as a matrix. It should be selected based on cell type, type of metabolome (endo- and exo-metabolome), and the end goal of the experiment. Usually, sample preparation involves metabolism quenching, separation of cells and medium, detachment of adherent cells in the case of the intracellular metabolome, it can involve cell homogenization, and lastly metabolome extraction. In case of extracellular metabolome of adherent cells, separation procedure is simple—as merely collecting medium using pipette and centrifuging is enough. The supernatant can be analyzed straight away or it could be dried to later achieve desired concentration in the selected solvent. For cells growing in suspension, fast filtration is usually used to separate cells form medium [61]. As for intracellular or full metabolome analysis, metabolic quenching should be performed as fast as possible, preferably right after sampling, usually using rapid temperature or pH changes. One of the relatively easy procedures uses quenching with −40 °C 60% methanol with 0.85% ammonium bicarbonate. Quenching with methanol only is also used in some protocols [62, 63]. The use of methanol for metabolic quenching should be carefully implemented, as it causes leakage of metabolites from the cells. Instead, it is advised to use buffered methanol. Similarly, caution is required while using liquid nitrogen freezing, but it is gaining popularity as it provides better recovery of metabolites. It is also common to rinse the cells before quenching as it improves metabolites quality, if limited to single washing, does not intervene with cell integrity [64].

After successful metabolic quenching, it is necessary to perform the extraction of metabolites. The ideal extraction method should be effective, not-specific, and should not influence metabolic composition while avoiding losses of metabolites. For an intracellular metabolome analysis, an extraction is usually combined with cell lysis. Dietmair et al. compared 12 extraction methods for suspension mammalian cell studies and proved water: acetonitrile (1:1) solvent to be the best extraction solvent [65]. Lorenz et al. proposed using methanol:chloroform over other solvents for metabolite extraction as it proved to be more efficient for adherent cells [66]. Overall, it is important to consider using multiple different solvents that differ in polarity while keeping low temperature, since it ensures the most complete metabolite extraction. Among all procedures, liquid–liquid extraction is commonly used. An interesting step for extracellular metabolome extraction while warranting quenching at the same time is solid-phase microextraction (SPME). It offers nonexhaustive extraction while inhibiting proteins or enzymes from adhesion, thereby quenching metabolism. It has been successfully used for headspace and standard exo-metabolome analysis on adherent cancer cell lines [67, 68]. This technique allows performing analysis over time, like estimating the effect of a drug. It provides a possibility for additional analysis such as MTT assay on the same sample. When planning an experiment it is necessary to choose quenching and extraction methods depending on the analytical platform of choice, type of used cells, culture media composition, and type of metabolomics (targeted or untargeted).

Further metabolite measurement methods for cell culture metabolomics do not differ from analysis on standard matrixes, using most often LC-MS, GC-MS, or NMR. Data processing is selected based on the chosen approach: qualitative or quantitative.

4.7.2 Experimental design at the level of 3D cell culture

As 3D cell cultures, like spheroids or organoids, could provide a more accurate TMN picture it is also necessary to consider the metabolic analysis and sample preparation in those specific culture conditions. There is a big difference between adherent cell culture and 3D culture metabolome. Since there is no standardized protocol for 3D cell cultures, certain methods of culturing can interfere with some quenching and extraction procedures, which should be taken into account while designing the experiment. It might be difficult to wash spheroids/organoids before quenching, as it might destroy 3D structure or provoke the risk of material loss—washing spheroid away by accident. Mathon et al. compared freeze–thawing and homogenization as extraction methods in 3D culture but did not report any significant differences [69]. Similarly, Rusz et al. compared the extraction of metabolites using boiling ethanol to cold methanol for single tumor spheroid and provided promising sample preparation protocol. They reported that cold 80% methanol combined with single PBS washing delivers the best sample quality [70]. Both of those papers highlight the importance of internal standards in metabolomics analysis. An interesting way of combining platform for easy 3D culture, metabolome extraction, and separation was proposed by Luyao et al. They used microfluidic to culture spheroids, SPE for metabolite extraction, and microchip electrophoresis for separation. This model needs further investigation, but could eventually provide a useful platform for advanced 3D cell culture metabolomics [71]. To this day, there is no commonly used sample preparation protocol for 3D cell cultures. While planning an experiment with the 3D culture, it is necessary to evaluate the culture method (presence and type of scaffold, presence of magnetic beads, etc.) and its possible metabolism influence as well as to choose the best quenching and extraction methods to provide best sample quality.

4.8 The direction of *in vitro* cancer research

Overall, the future of cell culture cancer research seems to lay in 3D culture advanced systems that provide appropriate insight in TNM. Research performed on spheroids or organoids is much more desirable than the one on standard cell culture, since it reflects cell-to-cell interactions, metabolic pathways changes etc.—it reflects an actual disease phenotype. Techniques like microfluidics or organ-on-a-chip make *in vitro* 3D culture studies more applicable, and with proper development, could minimalize animal studies. Application of "omics" on the 3D culture model, especially metabolomics, could provide new prognostic or diagnostic biomarkers as well as discover new cancer treatments.

Author contribution: All the authors have accepted responsibility for the entire content of this submitted manuscript and approved submission.
Research funding: None declared.
Conflict of interest statement: The authors declare no conflicts of interest regarding this article.

References

1. Robert C, Thomas L, Bondarenko I, O'Day S, Weber J, Garbe C, et al. Ipilimumab plus dacarbazine for previously untreated metastatic melanoma. *N Engl J Med* 2011;364:2517–26. https://doi.org/10.1056/nejmoa1104621.
2. Leach DR, Krummel MF, Allison JP. Enhancement of antitumor immunity by CTLA-4 blockade. *Science* 1996;271:1734–6. https://doi.org/10.1126/science.271.5256.1734.
3. Hoos A, Ibrahim R, Korman A, Abdallah K, Berman D, Shahabi, V, et al. Development of ipilimumab: contribution to a new paradigm for cancer immunotherapy. *Semin Oncol* 2010;37:533–46. https://doi.org/10.1053/j.seminoncol.2010.09.015.
4. Freeman GJ, Long AJ, Iwai Y, Bourque K, Chernova T, Nishimura H, et al. Engagement of the Pd-1 immunoinhibitory receptor by a novel B7 family member leads to negative regulation of lymphocyte activation. *J Exp Med* 2000;192:1027–34. https://doi.org/10.1084/jem.192.7.1027.
5. Ishida Y, Agata Y, Shibahara K, Honjo T. Induced expression of PD-1, a novel member of the immunoglobulin gene superfamily, upon programmed cell death. *EMBO J* 1992;11:3887–95. https://doi.org/10.1002/j.1460-2075.1992.tb05481.x.

6. Brahmer J, Reckamp KL, Baas P, Crinò L, Eberhardt WEE, Poddubskaya E, et al. Nivolumab versus docetaxel in advanced squamous-cell non-small-cell lung cancer. *N Engl J Med* 2015;373:123–35. https://doi.org/10.1056/nejmoa1504627.
7. Topalian SL, Hodi FS, Brahmer JR, Gettinger SN, Smith DC, McDermott DF, et al. Safety, activity, and immune-correlates of anti-PD-1 antibody in cancer. *N Engl J Med* 2012;366:2443–54. https://doi.org/10.1056/NEJMoa1200690.
8. Robert C, Long GV, Brady B, Dutriaux C, Maio M, Mortier L, et al. Nivolumab in previously untreated melanoma without BRAF mutation. *N Engl J Med* 2015;372:320–30. https://doi.org/10.1056/nejmoa1412082.
9. Kong YM, Flynn JC. Opportunistic autoimmune disorders potentiated by immune-checkpoint inhibitors anti-CTLA-4 and anti-PD-1. *Front Immunol* 2014;5:206. https://doi.org/10.3389/fimmu.2014.00206.
10. Gettinger SN, Horn L, Gandhi L, Spigel DR, Antonia SJ, Rizvi NA, et al. Overall survival and long-term safety of nivolumab (Anti–Programmed death 1 antibody, BMS-936558, ONO-4538) in patients with previously treated advanced non–small-cell lung cancer. *J Clin Oncol* 2015;33:2004–12. https://doi.org/10.1200/jco.2014.58.3708.
11. McDermott D, Haanen J, Chen TT, Lorigan P, O'Day S. Efficacy and safety of ipilimumab in metastatic melanoma patients surviving more than 2 years following treatment in a phase III trial (MDX010-20). *Ann Oncol* 2013;24:2694–8. https://doi.org/10.1093/annonc/mdt291.
12. Wang M, Zhao J, Zhang L, Wei F, Lian Y, Wu Y, et al. Role of tumor microenvironment in tumorigenesis. *J Canc* 2017;8:761–73. https://doi.org/10.7150/jca.17648.
13. Crinier A, Vivier E, Blery M. Helper-like innate lymphoid cells and cancer immunotherapy. *Semin Immunol* 2019;41:101274. https://doi.org/10.1016/j.smim.2019.04.002.
14. Lazennec G, Lam PY. Recent discoveries concerning the tumor – mesenchymal stem cell interactions. *Biochim Biophys Acta* 2016;1866:290–9. https://doi.org/10.1016/j.bbcan.2016.10.004.
15. Kim IS, Gao Y, Welte T, Wang H, Liu J, Janghorban M, et al. Immuno-subtyping of breast cancer reveals distinct myeloid cell profiles and immunotherapy resistance mechanisms. *Nat Cell Biol* 2019;21:1113–26. https://doi.org/10.1038/s41556-019-0373-7.
16. Seo J, Shin JY, Leijten J, Jeon O, Camci-Unal G, Dikina AD, et al. High-throughput approaches for screening and analysis of cell behaviors. *Biomaterials* 2018;153:85–101. https://doi.org/10.1016/j.biomaterials.2017.06.022.
17. Rothbauer M, Zirath H, Ertl P. Recent advances in microfluidic technologies for cell-to-cell interaction studies. *Lab Chip* 2018;18:249–70. https://doi.org/10.1039/c7lc00815e.
18. Maltman DJ, Przyborski SA. Developments in three-dimensional cell culture technology aimed at improving the accuracy of in vitro analyses. *Biochem Soc Trans* 2010;38:1072–5. https://doi.org/10.1042/bst0381072.
19. Duval K, Grover H, Han LH, Mou Y, Pegoraro AF, Fredberg J, et al. Modeling physiological events in 2D vs. 3D cell culture. *Physiology (Bethesda)* 2017;32:266–77. https://doi.org/10.1152/physiol.00036.2016.
20. Whitesides GM. The origins and the future of microfluidics. *Nature* 2006;442:368–73. https://doi.org/10.1038/nature05058.
21. Ma H, Liu T, Qin J, Lin B. Characterization of the interaction between fibroblasts and tumor cells on a microfluidic co-culture device. *Electrophoresis* 2010;31:1599–605. https://doi.org/10.1002/elps.200900776.
22. Agliari E, Biselli E, De Ninno A, Schiavoni G, Gabriele L, Gerardino A, et al. Cancer-driven dynamics of immune cells in a microfluidic environment. *Sci Rep* 2014;4:6639. doi:https://doi.org/10.1038/srep06639.

23. Liu PF, Cao YW, Zhang SD, Zhao Y, Liu X-g, Shi H-q, et al. A bladder cancer microenvironment simulation system based on a microfluidic co-culture model. *Oncotarget* 2015;6:37695–705. https://doi.org/10.18632/oncotarget.6070.
24. Liu T, Lin B, Qin J. Carcinoma-associated fibroblasts promoted tumor spheroid invasion on a microfluidic 3D co-culture device. *Lab Chip* 2010;10:1671–7. https://doi.org/10.1039/c000022a.
25. Wong AD, Searson PC. Live-cell imaging of invasion and intravasation in an artificial microvessel platform. *Canc Res* 2014;74:4937–45. https://doi.org/10.1158/0008-5472.can-14-1042.
26. Bauer M, Su G, Beebe DJ, Friedl A. 3D microchannel co-culture: method and biological validation. *Integr Biol (Camb)* 2010;2:371–8. https://doi.org/10.1039/c0ib00001a.
27. Zervantonakis, IK, Hughes-Alford, SK, Charest, JL, Condeelis, JS, Gertler, FB, Kamm, RD. Three-dimensional microfluidic model for tumor cell intravasation and endothelial barrier function. *Proc Natl Acad Sci USA* 2012;109:13515–20. https://doi.org/10.1073/pnas.1210182109.
28. Koh S, Shimasaki N, Suwanarusk R, Ho ZZ, Chia A, Banu N, et al. A practical approach to immunotherapy of hepatocellular carcinoma using T cells redirected against hepatitis B virus. *Mol Ther Nucleic Acids* 2013;2:e114. https://doi.org/10.1038/mtna.2013.43.
29. Pavesi A, Tan AT, Koh S, Chia A, Colombo M, Antonecchia E, et al. A 3D microfluidic model for preclinical evaluation of TCR-engineered T cells against solid tumors. *JCI Insight* 2017;2: e89762. https://doi.org/10.1172/jci.insight.89762.
30. Parlato S, De Ninno A, Molfetta R, Toschi E, Salerno D, Mencattini A, et al.. 3D Microfluidic model for evaluating immunotherapy efficacy by tracking dendritic cell behaviour toward tumor cells. *Sci Rep* 2017;7:1093. https://doi.org/10.1038/s41598-017-01013-x.
31. Brown VT, Cho V, Parkey S. Analysis of FDA approvals of targeted anticancer combination regimens. *Am J Health Syst Pharm* 2017;74:1938–42. https://doi.org/10.2146/ajhp170029.
32. Esch MB, King TL, Shuler ML. The role of body-on-a-chip devices in drug and toxicity studies. *Annu Rev Biomed Eng* 2010;13:55–72.
33. Bhatt DL, Mehta C. Adaptive designs for clinical trials. *N Engl J Med* 2016;375:65–74. https://doi.org/10.1056/nejmra1510061.
34. Sant S, Johnston PA. The production of 3D tumor spheroids for cancer drug. *Drug Discov Today Technol* 2017;23:27–36. https://doi.org/10.1016/j.ddtec.2017.03.002.
35. Shuler ML, Ghanem A, Quick D, Wong MC, Miller P. A self-regulating cell culture analog device to mimic animal and human toxicological responses. *Biomed Technol* 1996;52:45–60. https://doi.org/10.1002/(sici)1097-0290(19961005)52:1<45::aid-bit5>3.0.co;2-z.
36. Shida S. Organs-on-a-chip: current applications and consideration points for in vitro ADME-Tox studies. *Drug Metabol Pharmacokinet* 2017;33:49–54.
37. Bronzino JD, Peterson DR. *Tissue engineering and artificial organs*. Boca Raton; 2006. 9780429123054.
38. Bal-Öztürk A, Miccoli B, Avci-Adali M, Mogtader F, Sharifi F, Çeçen B, et al. Current strategies and future perspectives of skin-on-a-chip platforms: innovations, technical challenges and commercial outlook. *Curr Pharmaceut Des* 2018;24:5437–57.
39. Abaci HE, Shuler ML. Human-on-a-chip design strategies and principles for physiologically based pharmacokinetics/pharmacodynamics modeling. *Integr Biol* 2015;7:383–91. https://doi.org/10.1039/c4ib00292j.
40. Kratochvil MJ, Seymour AJ, Li TJ, Pasca SP, Kuo CJ, Heilshorn SC. Engineered materials for organoid systems. *Nature Reviews Materials* 2019;4:606–22. https://doi.org/10.1038/s41578-019-0129-9.

41. Wimmer AR, Leopoldi A, Aichinger M, Wick N, Hantusch B, Novatchkova M, et al. Human blood vessel organoids as a model of diabetic vasculopathy. *Nature* 2019;565:505–10. https://doi.org/10.1038/s41586-018-0858-8.
42. Mansour AAF, Gonçalves JT, Bloyd CW, Li H, Fernandes S, Quang D, et al. An in vivo model of functional and vascularized human brain organoids. *Nat Biotechnol* 2018;36:432–41. https://doi.org/10.1038/nbt.4127.
43. Ham O, Jin YB, Kim J, Lee M-O. Blood vessel formation in cerebral organoids formed from human embryonic stem cells. *Biochem Biophys Res Commun* 2020;521:84–90. https://doi.org/10.1016/j.bbrc.2019.10.079.
44. Vyas D, Baptista PM, Brovold M, Moran E, Gaston B, Booth C, et al. Self-assembled liver organoids recapitulate hepatobiliary organogenesis in vitro. *Hepatology* 2018;67:750–61. https://doi.org/10.1002/hep.29483.
45. Saheli M, Sepantafar M, Pournasr S, Farzaneh Z, Vosough M, Piryaei A, et al. Three-dimensional liver-derived extracellular matrix hydrogel promotes liver organoids function. *Jurnal of Cellular Biochemistry* 2018;119:4320–33. https://doi.org/10.1002/jcb.26622.
46. Takasato M, Er P, Chiu H, Little MH. Generation of kidney organoids from human pluripotent stem cells. *Nat Protoc* 2016;11:1681–92. https://doi.org/10.1038/nprot.2016.098.
47. Clevers H. Modeling development and disease with organoids. *Cell* 2016;165:1586–97. https://doi.org/10.1016/j.cell.2016.05.082.
48. Consortium International. Human Genome Sequencing, Finishing the euchromatic sequence of the human genome. *Nature* 2004;431:931–45.
49. Slatko BE, Gardner AF, Ausubel FM. Overview of Next generation sequencing technologies. *Curr Protoc Mol Biol* 2018;122. https://doi.org/10.1002/cpmb.59.
50. Lowe R, Shirley N, Bleackley M, Dolan S, Shafee T. Transcriptomics technologies. *PLoS Comput Biol* 2017;13:e1005457. https://doi.org/10.1371/journal.pcbi.1005457.
51. Aslam B, Basit M, Nisar MA, Khurshid M, Rasool MH. Proteomics: technologies and their application. *J Chromatogr Sci* 2019;55:182–796.
52. Johnson CH, Ivanisevic J, Siuzdak G. Metabolomics: beyond biomarkers and towards mechanisms. *Nat Rev Mol Cell Biol* 2016;17:451–9. https://doi.org/10.1038/nrm.2016.25.
53. Lindon JC, Holmes E, Nicholson JK. So what's the deal with metabonomics?. *Anal Chem* 2003;75:384A–91A. https://doi.org/10.1021/ac031386+.
54. Dudley E, Yousef M, Wang Y, Griffiths WJ. Targeted metabolomics and mass spectrometry. *Adv Protein Chem Struct Biol* 2010;80:45–83. https://doi.org/10.1016/b978-0-12-381264-3.00002-3.
55. Vinayavekhin N, Saghatelian A. Untargeted metabolomics. *Curr Protoc Mol Biol* 2010:Chapter 30:Unit 30.2.1–24. https://doi.org/10.1002/0471142727.mb3001s90.
56. Segers K, Declerck S, Mangelings D, Heyden YV, Eeckhaut AV. Analytical techniques for metabolomic studies: a review. *Bioanalysis* 2019;11:2297–318. https://doi.org/10.4155/bio-2019-0014.
57. Zhang F, Zhang Y, Zhao W, Deng K, Wang Z, Yang C, et al. Metabolomics for biomarker discovery in the diagnosis, prognosis, survival and recurrence of colorectal cancer: a systematic review. *Oncotarget* 2017;8:35460–72. https://doi.org/10.18632/oncotarget.16727.
58. Phan LM, Yeung S-CJ, Lee M-H. Cancer metabolic reprogramming: importance, main features, and potentials for precise targeted anti-cancer therapies. *Cancer Biol Med* 2014;11:1–19. https://doi.org/10.7497/j.issn.2095-3941.2014.01.001.
59. Wise DR, Thompson CB. Glutamine addiction: a new therapeutic target in cancer. *Trends Biochem Sci* 2010;35:427–33. https://doi.org/10.1016/j.tibs.2010.05.003.

60. Muschet C, Möller GM, Prehn C. Removing the bottlenecks of cell culture metabolomics: fast normalization procedure, correlation of metabolites to cell number, and impact of the cell harvesting method. *Metabolomics* 2016;12:151. https://doi.org/10.1007/s11306-016-1104-8.
61. Hofmann U, Maier K, Niebel A, Vacun G, Reuss M, Mauch K. Identification of metabolic fluxes in hepatic cells from transient 13C-labeling experiments: Part I, Experimental observations. *Biotechnol Bioeng* 2008;100:344–54. https://doi.org/10.1002/bit.21747.
62. Sellick CA, Hansen R, Maqsood AR, Dunn WB, Stephens GM, Goodacre R, et al. Effective quenching processes for physiologically valid metabolite profiling of suspension cultured Mammalian cells. *Anal Chem* 2009;81:174–83. https://doi.org/10.1021/ac8016899.
63. Teng Q, Huang H, Collette TW, Ekman DR, Tan C. A direct cell quenching method for cell-culture based metabolomics. *Metabolomics* 2009;5:199–208. https://doi.org/10.1007/s11306-008-0137-z.
64. Kapoore RV, Coyle R, Staton CA, Brown NJ, Vaidyanathan S. Influence of washing and quenching in profiling the metabolome of adherent mammalian cells: a case study with the metastatic breast cancer cell line MDA-MB-213. *Analyst* 2017;142:2038–49. https://doi.org/10.1039/c7an00207f.
65. Dietmair S, Timmins NE, Gray PP, Nielsen LK, Krömer JO. Towards quantitative metabolomics of mammalian cells – development of a metabolite extraction protocol. *Anal Biochem* 2010;404:155–64. https://doi.org/10.1016/j.ab.2010.04.031.
66. Lorenz MA, Burant CF, Kennedy RT. Reducing time and increasing sensitivity in sample preparation for adherent mammalian cell metabolomics. *Anal Chem* 2010;83:3406.
67. Jaroch K, Boyaci E, Pawliszyn J, Bojko B. The use of solid phase microextraction for metabolomic analysis of non-small cell lung carcinoma cell line (A549) after administration of combretastatin A4. *Sci Rep* 2019;9:402. https://doi.org/10.1038/s41598-018-36481-2.
68. Schallschmidt K, Becker R, Junga C, Rolff J, Fichtner I. Investigation of cell culture volatilomes using solid phase micro extraction: options and pitfalls exemplified with adenocarcinoma cell lines. *J Chromatogr B* 2016;1006:158–66.
69. Mathon C, Bovard D, Dutertre Q, Sendyk S, Bentley M, Hoeng J, et al. Impact of sample preparation upon intracellular metabolite measurements in 3D cell culture systems. *Metabolomics* 2019;15:92. https://doi.org/10.1007/s11306-019-1551-0.
70. Rusz M, Rampler E, Keppler BK, Jakupec MA, Koellensperger G. Single spheroid metabolomics: optimizing sample preparation of three-dimensional multicellular tumor spheroids. *Metabolites* 2019;9:304. https://doi.org/10.3390/metabo9120304.
71. Lin L, Lin J-M. Development of cell metabolite analysis on microfluidic platform. *J Pharm Anal* 2015;5:337–47. https://doi.org/10.1016/j.jpha.2015.09.003.

Adrianna Sobolewska, Aleksandra Dunisławska and Katarzyna Stadnicka

5 Natural substances in cancer—do they work?

Abstract: Owing to anticancer properties of selected natural substances, it is assumed that they have potential to be used in oncological therapy. Here, the recently proven effects of the selected natural polyphenols, resveratrol and curcumin, are described. Secondly, the potential of probiotics and prebiotics in modulation of immunological response and/or enhancing the chemotherapeutic treatments is reported based on the recent clinical trials. Further, the chapter presents current knowledge regarding the targeted supplementation of the patient with probiotic bacteria and known efficacy of probiotics to support immunotherapy. The major clinical trials are listed, aiming to verify whether, and to which extent the manipulation of patient's microbiome can improve the outcome of chemotherapies. In the end, a potential of natural substances and feed ingredients to pose epigenetic changes is highlighted. The chapter provides an insight into the scientific proofs about natural bioactive substances in relation to cancer treatment, leaded by the question – do they really work?

Keywords: cancer, curcumin, microbiome, prebiotics, probiotics, resveratrol

6.1 Introduction

Cancer, which now affects people of all ages, is responsible for one in six deaths according to the World Health Organization, making it the second most common cause of death in the world (*Global Challenges of cancer*, Nature Cancer, 2020). It is a multifaceted global problem that still requires action and has been included in the current list of global challenges. All over the world, due to the change in lifestyle and diet, extension of life time, but above all increasing environmental pollution, living in a fast pace under constant stress, there has been a significant increase in cancer incidence. All this leads to irreversible mutations in the DNA of normal human cells, leading to the initiation of the neoplastic process.

6.2 Microbiome and cancer

Scientists have proven that there is a strong link between the gut microbiota and cancer [1]. Most of the research on the microbiota in relation to cancer focuses on

---

This article has previously been published in the journal Physical Sciences Reviews. Please cite as: Stadnicka, K. Natural substances in cancer—do they work? *Physical Sciences Reviews* [Online] 2021, 6. DOI: 10.1515/psr-2020-0079

https://doi.org/10.1515/9783110662306-005

single species of bacteria and their metabolites that cause cancer. Pathological changes in the organism strongly influence the disturbance of homeostasis and the general condition of the host's organism, including its intestinal microbiota. The complete restoration and rebuilding of normal microbiota takes up to four years after the end of antibiotic therapy [2]. Even a single exposure to an antibiotic causes a reduction of beneficial bacteria in the intestines for several months, directly affecting the intestinal barrier and indirectly affecting the host's organism [3]. The intestinal barrier and microbes permanently interact at the genetic, metabolic and immune levels. The latest literature reports prove that the intestine is the most important immune organ. The initiated immune response in the intestine is responsible for the specific tissue immunity at the systemic level [4]. Conversely, disturbed microbiome homeostasis may be a response to many unfavorable symptoms for human health and well-being. Rebuilding the intestinal microbiota is often supported by supplementation with probiotics or substances of natural origin. Based on the PubMed biomedical database, since 2011, the interest in the subject of intestinal microbiota in relation to cancer has increased 20 times, which may indicate the great interest of researchers, the demand for knowledge, and hence the growing awareness.

The microbiome is a collective genome of all microorganisms inhabiting the host's organism, which shows the genetic makeup of humans. The gut microbiome arouses the greatest interest of scientists. This is due to the fact that the microorganisms present in the intestines are involved in processes that determine every aspect of their host's life. Gastrointestinal microbiota performs the following functions: modulates gene expression, metabolizes drugs, neutralizes carcinogenic compounds, supports the absorption of nutrients, participates in the conversion of carbohydrates for energy, synthesizes hormones, supports the immune system in recognizing pathogens, produces digestive enzymes and supports the balance in the intestines [5]. The composition of the gut microbiome is characteristic of every organism and determines our identity more precisely than our DNA. Thus, it is a clear information about every person, from which information such as: the way of birth, nutrition from the first hours of life, health, place of residence, illnesses or even personal hygiene can be read. In adults, the composition of the gut microbiota is strongly determined by the work performed and exposure to environmental conditions. As a result of a chronic stress, inadequate diet or antibiotic therapy, intestinal dysbiosis, i.e., dysfunction of the intestinal microbiota, may occur [6]. It concerns both the quantity and composition of the microbiota.

The new research clearly points for a crucial role of a microbiome in development of a cancer (referred to as oncobiome) and preventing cancer through modulating innate and adaptive immune response. It was proven on the basis of breast tissue microbiota that women suffering from breast cancer had higher abundance of specific bacteria families: *Bacillus*, *Enterobacteriaceae* and *Staphylococcus* [7]. Using *in vitro* studies on human HeLa cell line, it was shown that some of these bacteria caused DNA damage that may lead to tumorogenesis. If these specific genotoxic bacteria species indeed contribute to oncogenesis, than their elimination from microbiome could potentially significantly reduce the risk of breast cancer development (Table 6.1). In a prospective research, such a preventive modulation could be proposed to breast microbiome, that the potential 'oncogenic'/genotoxic microbiota would be eliminated to shape solely a healthy breast microbiome.

Scientific studies are also focused on providing evidence as to how the patient's microbiome affects the efficacy of molecular drugs. More than 40 drugs have been shown to be metabolized by the gut microbiota (nitroreduction of the radiation sensitizer misonidazole, the hydrolysis of the antimetabolite methotrexate and the deconjugation of the liver-detoxified form of the topoisomerase I inhibitor irinotecan (also known as CPT-11) [15]. Metabolome associated with activity of microbiota affects the dynamics of disease progression. Recently, it has been shown that certain metabolites produced by intestine bacteria in patients with NSCLC and treated with ICIs, where significantly associated with progression of the cancer. SCFA were associated with retardation of disease, but higher activity of ketone and alkane metabolic pathways was found to be associated with progression of cancer [16]. Metabolomic profiling revealed, that metabolism of intestine microbiota plays an important role in response to immunotherapy. Further metabolomic profiling will allow to determine specific microbiome-related markers of early cancer progression and use them in individualized patients care.

6.2.1 Probiotics and synbiotics in immunotherapy

The determination and analysis of patient's microbiome might serve as one of the diagnostic predictors of a response to immunotherapy. Secondly, a targeted supplementation of the patient with probiotic bacteria may improve the efficacy of immunotherapy. Some of the opened clinical trials aim to check, whether, and to which extent the manipulation of patient's microbiome can improve the outcome of chemotherapies. In Table 6.2, the recent and new clinical trials are presented, in which the role of a microbiome and/or supplementation of the microbiome modulating bioactive substances, is considered as an adjuvant element of cancer therapy.

According to FAO/WHO, probiotics are specific strains of microorganisms that, when delivered in the right amount, can have a positive effect on the organism and health of the host. Their effectiveness is strictly dependent on their properties. The key to the effectiveness to probiotic bacteria is the ability to colonize the host's digestive tract. With the probiotics, it is possible to reduce intestinal dysbiosis, improve the intestinal barrier, but also improve the functioning of the immune system and reduce inflammation in the organism [32] The key parameters of a suitable probiotic are: the number of live bacteria, compatibility of strains and activity against pathogenic bacteria. In order for a given strain of bacteria to be classified as probiotic, it must meet a number of criteria such as: safety and nonpathogenicity, survival in the

Table 6.1: Genotoxic microbiota related to oncogenesis.

Cancer type	Microbiota species, toxin	Mode of action	Strategies to reduce or eliminate the risk
Bowel cancer [8]	*Escherichia coli*, colibactin	– Mutations in cells lining the gastrointestinal tract by bacteria metabolites (e.g., colibactin	– A test for detecting this specific pattern of bacteria genotoxic metabolism alongside with bowel cancer screening tests
Stomach cancer	*Heliobacter pylori*	– Local neoplastic pathology through enterotoxins	– Re-evaluation of probiotic strains harboring potential genotoxic signatures
Colorectal cancer [11, 12]	*Escherichia coli* *Fusobacterium* ssp. *Bacteroides fragilis*	– Estrobolome: enteric bacterial genes associated with metabolism of estrogens (postmenopausal, breast cancer)	– Identification of antitumorigenic microbiota:
Gallbladder carcinoma [13]	*Salmonella enterica Typhi*		– Endogenous bacterial strains with strong diagnostic, therapeutic and translational potential (e.g., *Holdemanella biformis* producing short-chain fatty acids that contribute to control protein acetylation and tumor cell proliferation) [9]
HR–positive breast cancer [14]	Estrobolome: enteric bacterial genes associated with metabolism of estrogens (postmenopausal)		– Probiotics and their metabiotics having cytotoxic effect toward cancer cells (*Enterococcus, Lactococcus, Staphylococcus, Lactobacillus*) [10]

Table 6.2: Recent clinical trials on use of probiotics as an adjuvant therapy to mitigate the side effects and adverse events (AE) or accelerate cancer treatment.

Cancer type (clinical trial ID)	Therapeutic drug	Probiotic	Aim	Outcome measure	Outcome	Potential implication/ conclusion	Ref.
NSCLC, advanced (ARCHER 1042, NCT01465802)	Dacomitinib (EGFR tyrosine kinase inhibitor)	VSL#3 polybiotic 9 (Ferring Pharmaceuticals) + topical a.clometasone	Prophylactic intervention (probiotic) on dermatological and GIT adverse events	Plasma drug concentration–time profile; (AUC 0–24 h and AUC0-120 h)	No reduction in the incidence of diarrhea and stomatitis/oral mucositis, dermatologic events, diarrhea, and stomatitis in 88% patients	n/s	[17]
Breast cancer (NCT01723592)	n/s	Probiotic capsules (Astarte HSO Health Care) containing 2.5 × 10⁹ CFU of lyophilized *Lactobacillus crispatus* LbV 88, *Lactobacillus rhamnosus* LbV 96, *Lactobacillus Jensenii* LbV 116, *Lactobacillus gasseri* LbV 150N	Mitigating bacterial vaginosa in treated BC patients	Change in Nugent score (Nugent's criteria evaluate three types of bacteria via Gram stain: *Lactobacillus, Bacteroides/ Gardnerella, and Mobiluncus*)	A positive influence on the vaginal microbiota in 63% women in the intervention group. Shift in Nugent score to normal microbiota (dominated by *Lactobacilli*)	*Lactobacillus* preparation has the potential to improve the vaginal microbiota in women undergoing chemotherapy for BC	[18]

(continued)

Table 6.2 (continued)

Cancer type (clinical trial ID)	Therapeutic drug	Probiotic	Aim	Outcome measure	Outcome	Potential implication/ conclusion	Ref.
Breast cancer (NCT03358511)	n/s therapy, intervention prior to surgery	Primal Defense Ultra® Probiotic (Garden of Life) containing 13 species of beneficial bacteria, including Saccharomyces, Lactobacilli and Bifidobacteria strains	Engineering gut microbiome to target BC prior to surgery	Mean number of cytotoxic T lymphocytes (CD8+ cells) relevant for immune defense against intracellular pathogens, including viruses and bacteria, and for tumor surveillance	Ongoing study	n/s	Reviewed in [19]
Breast cancer ESR receptor positive (NCT03518268)	Aromatase inhibitors (to reduce estrogen levels)	Vivomixx (Mendes) 450 billion viable lyophilized bacteria from eight Lactobacillaceae strains:	Prevention of bone loss in early menopausal women with BC treated with an aromatase inhibitor	Change in number of Collagen type 1 cross-linked C-telopeptide (CTX) Change in Serum type 1 procollagen (N-terminal) P1NP	Ongoing study	n/s	Reviewed in [20]

Cancer patients with enteral nutrition at home (NCT03940768)	n/s	Lactobacillus Plantarum 299 (Sanprobi IBS)	Verify if probiotic improves wellbeing in terms of nutritional status and tolerance to enteral diet	Questionnaires to assess quality of life and tolerance of enteral diet Physical changes in muscle and body mass, physiological examination of protein concentrations and lymphocyte counts	Reduced incidence of diarrhea associated with enteral nutrition	Supplement to reduce side effects of enteral diets in treated cancer patients	[21]
Pediatric malignant neoplasm (NCT03057054)	Hematopoietic cell transplant/ myeloablative chemotherapy	Lactobacillus plantarum strains 299 and 299v	Verify if L. plantarum can prevent GI acute graft versus host disease (aGvHD) in children and adolescents	Intestinal integrity (serum citrulline levels and reduction in mucosal barrier injury (MBI) Incidence of aGvHD blood stream infection and Clostridium difficile-associated diarrhea	Clostridium difficile infections were noted in 20% of the patients Gastrointestinal aGvHD was noted in 22% of patients	Ongoing study Lactobacillus plantarum can safely and feasibly be administered to transplant patients to prevent aGvHD	Nieder et al., 2015

(continued)

Table 6.2 (continued)

Cancer type (clinical trial ID)	Therapeutic drug	Probiotic	Aim	Outcome measure	Outcome	Potential implication/ conclusion	Ref.
Colon cancer (NCT03072641)		ProBion Clinica Synbiotic tablets containing *Bifidobacterium lactis* Bl-04 (ATCC SD5219), *Lactobacillus acidophilus* NCFM (ATCC 700396), and inulin	To reactivate tumor suppressor genes using probiotic	Changes in microbiota composition and DNA methylation levels to assess influence of bioactive on epigenetic changes in colon cancer tissue	Increased microbial diversity CRC-associated genera such as *Fusobacterium* and *Peptostreptococcus* tended to be reduced in the fecal microbiota of probiotic patients	Recommendation on using intestinal mucosa samples rather than fecal to assess microbiota changes in colon cancer. *Peptostreptococcus* was over-represented in both mucosal and fecal samples (a potential CRC marker)	[22]
Gynecological cancer after RT (NCT01549782)	Abdominal and pelvic RT	Prebiotic mixture of fiber (50% of inulin and 50% fructo-oligosaccharide, FOS)	To verify if inulin and FOS could modulate *Lactobacillus* and *Bifidobacterium* can reduce the intestinal injury after RT	Number of bowel movements and stool consistency recorded daily Quantification of *Lactobacillaceae*	Recovery of intestinal microbiota versus placebo (maltodextrin) *Lactobacillus* and *Bifidobacteria* were affected by RT Prebiotics increased *Lactobacillus* and *Bifidobacterium* three weeks after radiotherapy	Prebiotics may reduce intestinal injury (or accelerate recovery) in RT treated gynecological patients	[23]

Colorectal cancer (NCT04131803)	Standard and targeted chemotherapy in metastatic CRC	*Bifidobacterium trifidum* live powder (Bifico, Bifico Pharmaceuticals)	To test if combination of probiotic with current CRC chemotherapy gives synergistic antitumor effect	Tumor size reduction	Ongoing	n/s, apart animal study in mice [24] Potential mechanisms reviewed in [25]
Pediatric Febrile neutropenia condition after chemotherapy (NCT02544685)	n/s	Synbiotic Probio-Fix Inum (Chr. Hansen A/S) containing *Lactobacillus rhamnosus* GG and *Bifidobacterium animalis* Subspecies. *Lactis* BB-12 in combination with inulin and oligofructose	If synbiotic prevention reduces number of febrile neutropenia conditions after chemotherapy	Reduction in the incidence of febrile neutropenia episodes	Ongoing	Previously, probiotic strain *E. faecium* M-74 did not prevent febrile neutropenia [26]

(continued)

Table 6.2 (continued)

Cancer type (clinical trial ID)	Therapeutic drug	Probiotic	Aim	Outcome measure	Outcome	Potential implication/ conclusion	Ref.
Lung cancer (NCT02771470)	Standard chemotherapy	*Clostridium butyricum*	To restore intestinal immunity, mucosal barrier and nutrient absorption during chemotheapry	Composition of microorganisms in stool after probiotic intervention. Frequency and severity of AE during chemotherapy	Chemotherapy-induced diarrhea was lower. The genera producing short-chain fatty acids tended to increase. Lymphocyte/monocyte ratio (LMR) was higher in the probiotic treated patients. *C. butyricum* also weakly influenced the immune and nutrition status	*C. butyricum* may reduce chemotherapy-induced diarrhea in patients with lung cancer, reduce the systemic inflammatory response system and encourage homeostatic maintenance	[27]

Lung cancer (NCT03068663)	n/s	No treatment	To explore microbiota (GI, lungs and upper airways) in NSCLC patients	Difference in diversity of the lungs and upper airways microbiota. Inflammatory status (plasmatic cytokines and interleukins). Immune cells on lung/tumor sample and bronchoalveolar lavage fluid	Specific microbial profiles in bronchoalveolar lavage fluid (*Firmicutes* dominated) and lung tissue microbiota (*Proteobacteria* dominated)	Firmicutes in lower lobes might be a sign of increased pathogenicity and worse prognosis	[28]
Melanoma, renal cancer, lung cancer (NCT04107168)	Nivolumab (Opdivo), Pembrolizumab (Keytruda), Ipilimumab (Yervoy), Durvalumab (Imfinzi), Tremelimumab, Atezolizumab (Tecentriq), Bevacizumab (Avastin)	No treatment	To explore microbiome as a biomarker of efficacy and toxicity in patients receiving immune checkpoint inhibitor therapy	Microbiome signature to predict progression-free survival (PFS) of one year Correlation of microbiome with treatment efficacy	Ongoing	n/s	n/s

(continued)

Table 6.2 (continued)

Cancer type (clinical trial ID)	Therapeutic drug	Probiotic	Aim	Outcome measure	Outcome	Potential implication/conclusion	Ref.
Renal cancer (NCT03829111)	Nivolumab, Ipilimumab	*Clostridium butyricum* CBM588 (MIYARI)	To evaluate effect of CBM588 in treated patients with metastatic renal cell carcinoma	*Bifidobacterium* composition of stool. Change in Shannon index. Circulating regulatory T-cells (Tregs). Circulating myeloid-derived suppressor cells (MDSCs), cytokines and chemokines levels	Higher microbial diversity in patients better responding to immunotherapy. Increased abundance of *Akkermansia muciniphila* in patients responding to checkpoint inhibitors		[29, 30]
Colorectal cancer (NCT01410955)	Irinotecan	Colon Dophilus TM (Monsea), eight Lactobacilli strains	Prevention of Irinotecan induced diarrhea	Prevention of grade 3–4 diarrhea	Probiotics led to a reduction in the incidence of severe diarrhea of grade 3 or 4 (0 vs. 17.4% for placebo)	Probiotics lead to a reduction in the incidence and severity of GI toxicity	[31]

NSCLC, nonsmall-cell lung carcinoma; GI, gastrointestinal; EGFR, epidermal growth factor receptor; BC, breast cancer; AUC, area under the plasma concentration-time curve; ESR receptor, estrogen receptor; aGvHD, acute graft versus host disease; CRC, colorectal cancer; FOS, fructooligosaccharides; RT, radio-therapy.

gastrointestinal tract (at low pH of gastric juice or in the presence of bile), adhesion to the intestinal epithelium, and antibacterial activity. The main key parameters of a suitable probiotic include the appropriate amount of live bacteria and the activity of probiotic strains against local pathogenic bacteria. It is also of great importance to influence the expression of genes or stimulate the production of cytokines [33]. Synbiotic is a term referring to the simultaneous administration of a prebiotic and probiotic in one dose and their synergistic effect. Prebiotics are defined as indigestible nutrients that selectively stimulate the growth and activity of bacteria present in the host's gastrointestinal tract. Probiotics and synbiotics modulate immune system at a molecular and physiological levels. Modulation of cytokines and other inflammation regulatory proteins have been most often types of their function at a molecular level. At the physiological level, the enhancement of barriers against pathogens and mitigating inflammation process in tissues is a crucial effect of probiotics action. It has also been found, that probiotics play an important role in preventing metastasis. Practically, through immunological impact, probiotics and synbiotics have been considered candidate modulators of immunotherapy in cancer, as potential adjuvant agents to immunotherapeutic drugs. The general aim of applying immunotherapy in cancer patients is to enhance or induce action of immune system to fight the tumor back.

At present, the highest efficacy in treatment of different types of cancer in terms of clinical outcome, is provided by monoclonal antibodies, that target immune checkpoints. In particular, two selected antibodies seem to play a crucial role in the current immunotherapy: programmed cell death protein 1 (PD-1) and programmed death-ligand 1 (PD-L1). The expression of PD-L1 molecule on the surface of cancer cells or on the surface of tumor infiltrating immune cells is one of the most important predictors. Still, it has not been explained, why, despite the expression of PD-L1 on the cancer cells, not all the patients respond to the immunotherapy [34].

Understanding the interactions between microbiome (and the diet), tumorigenesis and immune system may be a key to success of immunotherapies. In future, it may be necessary to isolate and fully characterize those microbiota strains, which play a role in clinical response to human therapeutic antibodies. In a perspective, such candidate microbiota strains may be a part of personalized oncodiets, which may improve the clinical response in patients treated with molecular therapies based on PD1/PD-L1 blockers. However, only after ensuring that the medical care systems are opened for the personalized medicine and diagnostics, a full potential of the engineering of microbiome in cancer patients may be achieved.

The amount of probiotics in a single treatment varies from several millions to hundreds of billions cells per dose. It is therefore extremely hard to propose effective doses to see significant effects in cancer patients. Undoubtfully, the microbiome and its modulation has a strong scientific basis to be considered one of fundamental pillars of the organism immune status. But, observing a relatively poor response of cancer patients to existing probiotic formulations may suggest several assumptions:

firstly—the commercial formulations are not dedicated specifically to mitigate the oncological conditions; secondly—the selection of commercial formulations for clinical studies seems accidental. Considering this accidentality of selected probiotics and no clinical proofs for the effects of ready-prepared probiotics to boost immune system of cancer patients, or interacting with immunotherapeutic treatments, allow to suggest that only a oncologically specific studies and optimization of novel, targeted probiotic/synbiotic formulations may be applicable. Such an optimization could, for instance, be based on the metagenomic and proteomic determination of specific oncobiomes (and metabolomes) in patients with different cancers and treated with different chemotherapeutics. Based on specific profiles determined for various cancer types and therapies, a selection of microbiota from responding and recovering patients (ideally until recovery to full health), could serve as candidates to formulate natural, probiotic bioactive formulas as adjuvant supplementation for further clinical trials and implementation.

The probiotics offered as ready preparations, contain various combinations of beneficial bacteria strains producing lactic acid to lower the pH. The diversity of strains as well as quantity of bacteria per dose varies significantly between preparations. For instance, capsules used to test influence of lactic acid bacteria on normalization of microbiota postchemotherapy status in breast cancer patients, consisted of four lyophilized strains: *Lactobacillus crispatus*, *Lactobacillus rhamnosus*, *Lactobacillus jensenii* and *Lactobacillus gasseri*, over 2.5 billions of bacteria per capsule [18]. It was assumed, that colonizing vaginal environment with lactic acid strains would competitively eliminate growth of *Gardnerella vaginalis* and *Mobiluncus* ssp. The probiotic supplementation normalized the microbiota profile in over 60% of treated patients.

When studying mechanisms and strategies to target the oncobiome with probiotics, one must take into account reasons for failure in probiotic functions as adjuvants to chemotherapy. Previously, probiotic strain *Enterococcus faecium* M-74 did not prevent febrile neutropenia is the identified cause was that there are other routes of entrance of infection in neutropenic patients like indwelling catheters and loss of enterococci colonization. Thus, the possible protective effect of the probiotic strain further decreased. Addition of a single probiotic strain could not be sufficient to compensate the modifications induced on the bacterial biota by chemotherapy. Mucositis occurs frequently in pediatric patients with acute leukemia during induction and consolidation chemotherapy. In this situation, attempt to normalize microbiota could be unsuccessful due to disruption of the mucosal barriers [26].

6.3 Food ingredients and cancer

Cancer is increasingly the leading cause of premature death in people aged 30–69 in most countries. Surprisingly, the dominant factor in causing cancer is poor diet, which accounts for as much as 50% of all cancers, ahead of such a factor as smoking. Diet is a major factor that has multiple effects, including alteration of both the transcriptome and the metabolome of the host, and thus it may reduce CRC incidence by as much as 80%—interaction of host genetics with environment is decisive in disease progression [35].

Much scientific research is devoted to the analysis of the relationship between the genetic basis and environmental factors and the impact of this relationship on the development of neoplastic diseases. In recent years, more and more often epigenetic phenomena have been mentioned in this context. The term *"epigenetics"* was first used by Waddington in 1942 to denote heritable changes in gene expression that are not due to changes in DNA sequence [36].

Epigenetic changes can be associated with both physiological and pathological processes, therefore they are currently considered an important patho-mechanism of chronic diseases, including neoplastic diseases [37]. Epigenetic mechanisms can take place throughout the lifetime of each cell division and be modified by environmental factors.

An epigenome is a collection of molecular markers integrated into the genome that suppress or express specific regions of DNA. Molecular markers are added to the genome by appropriate enzymes, e.g., methyltransferases that transfer methyl groups or acetylases that transfer acetyl groups. These markers can determine whether a gene is expressed or not. The epigenome is somewhat flexible and changes dynamically in response to various environmental factors. However, these types of changes do not change the structure of DNA, i.e., they are not a type of genetic mutation that is irreversible. The epigenome is susceptible to intra- and extracellular regulation in a physiological state, for example regulation by cell–cell interactions or by neighboring cells, and regulation by changes in environmental conditions to which molecules such as cytokines, growth factors, and hormones are susceptible. Thus, it can be said that the epigenome expresses the organism's ability to adapt, which is manifested by the expression of selected phenotypic features. It instructs a unique gene expression program specific to a given cell type in normal development and disease, and manages the interactions between distant chromatin segments [38]. The known epigenetic modifications that control and change the transcription of genes concern two main components of the genetic code, namely DNA, and here the essential factor is methylation and changes in the structure and function of chromatin by chemical modification of histones (mainly methylation, acetylation and phosphorylation) [39]. These changes are integrally connected with each other, they give new features to extragene inheritance and create a specific epigenetic code, allowing inheritance or the appearance of certain features independently of the nucleotide notation in the genetic code. Modifications of DNA and histones are the result of the action of various groups of enzymes that directly regulate the inheritance process (they can be inherited and passed on to daughter cells). On the other hand, disturbances in the regulation of the activity of these enzymes can lead to further disturbances, inducing, for example, the occurrence and development of tumors. RNA interference is also indicated as the third possible mechanism of epigenetic modification [40]. The best known epigenetic modification is DNA methylation, which consists in attaching a methyl group to the cytosine. Methylation is a dynamic process as it is after each cell division and for DNA replication, methyl groups may be added or removed. Each cell has its own characteristic methylation pattern which determines the gene expression profile [39] (Figure 6.1).

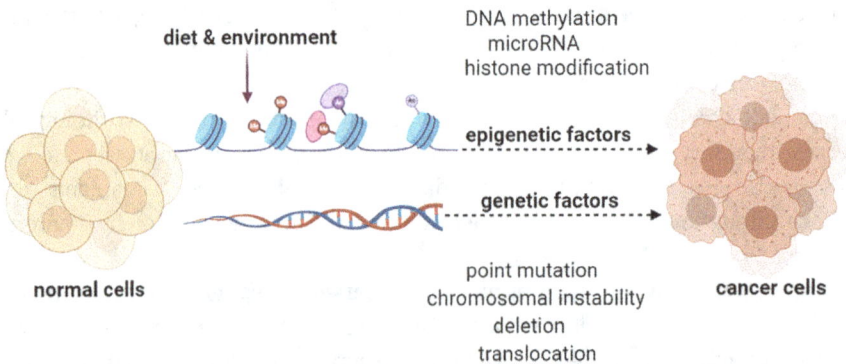

Figure 6.1: Genetic and epigenetic mechanism of carcinogenesis.

Epigenetic phenomena are an extremely interesting subject of research, not only from the cognitive point of view, but also for their possible use in clinical medicine. Cancer cells are characterized by visible changes in the DNA methylation pattern and changes in the activity of enzymes responsible for modifying histone proteins [41]. Abnormal methylation (hypo- or hypermethylation) of certain genes can be a marker of the neoplastic process in the body, it can also be specific for a particular type of cancer, and can even be an indicator of progression. Additionally, the reversibility of epigenetic processes offers an opportunity for new therapeutic interventions. Detailed knowledge on epigenetic modifications of chromatin under the influence of environmental factors may be useful in assessing the risk of cancer occurrence and progression and the effectiveness of treatment.

One of the most important environmental factors modulating the human epigenome is food and its active substances. It has been shown that fruits, vegetables and spices contain many biologically active substances—phytochemicals that exhibit chemopreventive properties and can modify the epigenome. Phytochemicals are bioactive non-nutritive compounds (secondary metabolites) isolated from, among others, from edible parts of fruits and vegetables, medicinal plants, as well as found in fungi, marine organisms and produced by bacteria, with a significant pharmacological effect [42, 43]. Plant secondary metabolites are characterized by biogenetic diversity and diverse structures. This determines their multidirectional biological activity. The literature reports about their broad properties: antibacterial, antifungal, antiviral, enzyme-inhibiting, and also anticancer [44].

The process of cancer formation has many stages, and the purpose of natural chemopreventive compounds is to slow down, block or reverse the process. Polyphenols are among the most studied chemopreventive compounds that modify human epigenome.

Many studies suggest that the chemopreventive potential of dietary polyphenols may be due in part to their ability to modulate epigenetic changes in cancer cells. They have been shown to modify the activity of DNA methyltransferases, acetyltransferases and histone deacetylases [45]. Tumor suppressor gene transcription silencing that is induced by hypermethylation is often an epigenetic defect in many human cancers. Gene hypermethylation reversal, which is achievable by inhibition of DNMT activity in cancer cells, is one of the potential avenues for epigenetic drug development. Several polyphenols in the diet show potential as DNMT inhibitors. They demonstrate the ability to reverse methylation-induced gene silencing and restore expression of suppressor genes [45].

6.3.1 Resveratrol

Resveratrol (3,5,4'-trihydroxystilbene) is a polyphenol with stilbene structure, which exists in the form of two isomers: cis- and trans-. In the natural environment, the dominant form is trans-resveratrol, which shows a higher biological activity compared to cis-resveratrol [46]. This polyphenol is a phytoalexin that is produced by the plant as a result of environmental stress or in response to pathogen attack. Resveratrol was first isolated from the root of a white hellebore plant in 1940 [47]. The best natural sources of resveratrol are knotweed (*Reynoutria japonica*) and dark grape varieties, whose fresh skins contain 50–100 mg of this compound per 1 g of their biomass. This polyphenol is also present in red wine and in fruits such as cranberry, blueberry, mulberry, black currant, strawberries, apples, peanuts and in some herbs [48]. In the human organism, this polyphenol is absorbed in the lumen of the small intestine at the level of about 70%. After absorption, it is almost completely metabolized in hepatocytes with the participation of cytochrome P450 to piceatannol and tetrahydroxystilbene [49]. In humans, it is metabolized in small amounts to the sulfate glucuronic forms of trans-resveratrol [50]. Unfortunately, the half-life of resveratrol is very short, 8–14 min. About 30 min after it appears in the bloodstream, it is converted into sulfite derivatives that circulate in the blood for up to 9 h, after which they are excreted in the urine and feces [51, 52]. However, taking into account the numerous benefits of using resveratrol in the diet, research is still being conducted to increase its bioavailability in humans.

In recent years, many studies have shown that resveratrol can play an important role in the prevention and treatment of cancer [53]. The anticancer properties of resveratrol were first described in 1997. Later studies have shown that it has the ability to block all stages of the neoplastic process, i.e., initiation, promotion and progression [54]. The antioxidant properties of resveratrol also allow inhibition of the initiation stage of carcinogenesis [52]. Resveratrol affects the metabolism of the detoxification of carcinogens in cancer cells by induction of quinone reductase (QR-2), i.e., a phase II enzyme involved in the neutralization of carcinogenic substances and blocks the receptors for aromatic hydrocarbons [52, 53]. In the second phase of carcinogenesis, resveratrol reduces the activity of compounds of the cytochrome P-450 family (CYP1A1, CYP1A2 and CYP1B1), which are involved in the production of oxygen free radicals and cause increased activity of carcinogens. This polyphenol also inhibits cytochrome transcription, thus protecting respiratory epithelial cells against the toxic and carcinogenic effects of various carcinogenic compounds [55]. In the final phase of carcinogenesis, resveratrol silences various cancer cell lines, in part by inhibiting DNA polymerase (the enzyme that catalyzes DNA synthesis) and ribonucleotide reductase (an enzyme needed to synthesize DNA in dividing cells). Moreover, as an antioxidant, this compound inhibits the proliferation of neoplastic cells by arresting the cell cycle in the G1/S phase [55]. In the studies of Shi et al. [56] proved that resveratrol influencing the expression of 2059 genes potentially associated with the development of renal cell carcinoma (line RCC54). It can act as a powerful anticancer agent whose effectiveness depends on the dose used. Its antimutagenic properties, which prevent the formation of potentially mutagenic and carcinogenic oxidative damage to DNA, have been observed in studies in mice that showed a reduction in the incidence of chemically induced skin cancer after prior application of resveratrol to the epidermis [57]. Resveratrol is structurally similar to a synthetic estrogen (diethylstilbestrol), and therefore it can bind to the estrogen receptor and act as a phytoestrogen. As a phytoestrogen, it regulates the expression of genes responsible for the development of estrogen-dependent neoplasms. In a study by Le Corre et al. [58] compared mRNA expression level of each gene in a fibrocystic breast cell line (MCF10a) and three breast tumor cell lines (MCF-7, MDA-MB-231 and HBL100) after treatment with 50 or 30 µmol/l resveratrol, showing that this compound regulates the expression of the BRCA1 suppressor gene.

Resveratrol also induces the expression of the p53 gene responsible for inhibiting neoplastic processes. Activation of p53 induces the transcription of target genes that are involved in cell cycle arrest, apoptosis and DNA repair in some types of cancer [59–63]. In the studies of [60] performed *in vitro*, it was shown that resveratrol exerts a similar level of antiproliferative effects on human melanoma cells with different metastatic potential. The analysis showed, among others a significant increase in the expression of the p53 tumor suppressor gene.

In the third stage of tumorigenesis, this polyphenol is cytotoxic by inhibiting DNA polymerase and ribonucleotide reductase. Resveratrol inhibits the proliferation of cancer cells by arresting the cell cycle in the G1/S phase, which results in apoptosis [55]. *In vitro* studies have shown that resveratrol also increases the effectiveness of chemotherapy by inactivating the NF-κB protein. This factor is created by cancer cells and controls the expression of many genes involved in the inflammatory process. And, as some studies have reported, prolonged inflammation may be the beginning of neoplastic transformation. The NF-κB protein also regulates the expression of more than 150 genes related to survival, proliferation and stress response [64]. The increased activity of the NF-κB protein immunizes cancer cells to chemotherapy and allows them to multiply. By blocking this factor, resveratrol causes chemotherapeutic agents to act at their destination [65]. *In vitro* studies by Kubota et al. [66] showed that resveratrol inhibited by 50% the growth of three lung cancer cell lines (A549, EBC-1, Lu65) at concentrations between 5 and 10 µM. The team of Baatout et al. [67] searching for new substances that could sensitize cancer cells to ionizing radiation, showed in *in vitro* studies that the use of high doses of resveratrol sensitizes cervical cancer cells (HELA human cell line), chronic myeloid leukemia (human cell line K-562) and erythema multiple myeloma (human cell line IM-9). The addition of resveratrol alone at a concentration of 50 (IM-9), 100 (EOL-1) or 200 microM (HELA) to cell cultures induced apoptosis and inhibited cell growth. Simultaneous treatment of cells with resveratrol and X-rays produced a synergistic effect at the highest dose of 200 µM. The tested concentrations of resveratrol used in these studies are much higher than what is available in food. Surh et al. [68] by conducting experiments on the HL60 acute promyelocytic leukemia cell line, showed that resveratrol reduces the viability of these cells and their ability to synthesize DNA. According to the authors, it is probably related to the induction of apoptosis, as the treatment of HL60 cells with this polyphenol resulted in a gradual decrease in the expression of the antiapoptotic protein Bcl-2. Zou et al. investigated the possibility of using resveratrol in the treatment of osteosarcoma, using its influence on the level of β-catenin expression. It has been shown that resveratrol significantly reduces the expression of this protein, thus inhibiting the proliferation of MG-63 osteosarcoma cells. This effect depended on both the incubation time and the polyphenol concentration. These researchers also achieved inhibition of the proliferation of these cells by inhibiting the canonical (classical) Wnt signaling pathway [69]. This pathway plays a significant role in the neoplastic process, particularly in the progression stage. Mutations that permanently activate this signaling pathway lead to the formation of tumors. Deregulation of this pathway results in translocation and accumulation of β-catenin in the nucleus, which contributes to the activation of oncogenes. Experiments performed using flow cytometry showed that treatment of MG 63 cells with resveratrol resulted in an increase in the number of apoptotic cells.

Natural substances, as bioactive compounds, can be taken with food, but their dose is quite low. To achieve a therapeutic effect, they are supplemented in the form of nutraceuticals. In patients with diagnosed colorectal cancer, before resection of the tumor, grape extract was used for 14 days with a relatively low dose of resveratrol: 0.073–0.114 mg, observing in patients inhibition of the expression of genes related to the cancer initiation process in the healthy mucosa, but not in the tissue cancerous. However, after the use of a nutraceutical dose of resveratrol— 500 mg and 1.0 g for eight days in subsequent studies, a dose-dependent decrease in colorectal cancer cell proliferation by 5.6% was observed [70]. In another study, healthy subjects were administered resveratrol in doses ranging from 500 mg to 5.0 g for 29 days. Blood diagnostics revealed a dose-independent reduction in the production of IGF-1 (insulin-like growth factor 1) and IGFBP-3 (insulin-like growth factor−binding protein 3), which correlate with the development of the neoplastic process [71].

Wu et al. [72] treated cells of the human TCC (human transitional cell carcinoma) cell line with resveratrol in order to investigate the safety and efficacy of this polyphenol. Concentrations of 100, 150 and 200 µM were used. The results of the experiments showed that resveratrol at a dose of 150 and 200 µM in these cells led to a significant inhibition of the S-phase cell cycle and apoptosis. This was also accompanied by a reduction in phosphorylation, nuclear translocation, and STAT3 transcription. In addition, a reduction in the expression of survivin, cyclin D1, c-Myc and VEGF genes as well as translocation to the Sirt1 and p53 nucleus were also observed [72]. Due to the lack of documented harmful effects of high doses of resveratrol on liver cells and the lack of significant side effects after oral ingestion, the toxic and lethal dose and acceptable daily intake (ADI) were not determined [73].

6.3.2 Curcumin

Curcumin (diferoylmethane) is an organic chemical compound of plant origin, composed of two feruloyl residues linked by a carbon atom. It belongs to the group of polyphenols. It is obtained from the rhizomes of the Asian perennial turmeric (*Curcuma longa*) called turmeric, in which it makes up 5–10% of the dry weight. Curcumin has low oral bioavailability. It is metabolized in the liver, which inactivates a large part of it [74]. However, the combination of curcumin with piperidine, obtained from the extract of black pepper, increases its bioavailability by as much as 20 times [75]. Curcumin, in addition to such compounds as demethoxycurcumin, bisdemetoxy-curcumin, cyclocurcumin, belongs to the naturally occurring curcuminoids. It accounts for about 77% of all curcuminoids and is also considered the most valuable. Curcumin is a component of the famous spice—turmeric, which is a plant found in the wild in tropical countries and used as a spice and as a food coloring.

Curcumin is a compound with multidirectional action. Its antibacterial, antiviral, antifungal, anti-inflammatory, antioxidant and antitumor activity has been confirmed. Curcumin is classified as one of the blocking and suppressive chemopreventive factors due to the ability to interfere with the cancer process at every stage. This compound inhibits tumor growth at the stage of initiation and inhibits the proliferation of malignant cells at the stages of promotion and progression of the neoplastic process. *In vitro* studies on human and animal cancer cell lines have shown that curcumin exhibits a chemopreventive effect by: inhibiting carcinogen activation, stimulating carcinogen detoxification, suppressing proinflammatory signal cascades, inducing cancer cell death, inhibiting the cell cycle, inhibiting angiogenesis and reducing tumor metastasis capacity [76].

Curcumin shows a chemopreventive effect, closely related to the inhibition of the activity of the transcription factor NF-κB, which controls the expression of genes whose protein products regulate a number of important cellular processes, including: proliferation and cell growth, apoptosis, immune responses, or the cell's response to stress caused by various factors. Such properties of curcumin have been confirmed by the studies of Lin et al. [77] in which it was observed that this polyphenol, by reducing the activity of the NF-κB transcription factor, inhibits tumor growth and angiogenesis in mice transplanted with ovarian cancer cells. Similarly, in the studies by Kunnumakkara et al. [78]; in a mouse model of pancreatic cancer, it was found that curcumin inhibited the formation of blood vessels around the tumor and the proliferation of tumor cells. It also increased the sensitivity of cancer cells to the effects of chemotherapy. The authors suggest that the observed effects are related to inhibition of NF-κB activity. The effect of curcumin may also be the result of its inhibition of the activity of the transcription factor HIF-1, thus reducing the expression of genes dependent on it [79].Very often, in response to the hypoxic state that occurs inside solid tumors, the activation of the transcription factor HIF-1 occurs, inducing the expression of genes whose products are involved in angiogenesis, proliferation, survival, glucose metabolism and migration. As a result of this condition, cancer cells adapt to the unfavorable conditions inside the tumor [80]. In the studies of Aoki et al. [81] showed that the use of curcumin in mice previously implanted with human U87 glioma cells effectively inhibited tumor growth and induced autophagy. Tumor reduction and inhibition of angiogenesis using curcumin were also observed in murine bladder cancer xenograft models [82]. Curcumin, like other plant-derived polyphenols, inhibits the activation of the AP-1 transcription factor, which is involved in the regulation of the expression of genes involved in the processes of apoptosis and cell proliferation. AP-1 can promote cell proliferation by activating the expression of the cyclin D1 gene and inhibit the expression of tumor suppressor genes such as p53, p21 and p16 [83]. Regulation of p53 gene expression by curcumin is mediated by its ability to induce two sets of genes: $p21^{cip1/waf-1}$ oraz $p27^{kip-1}$ [84, 85]. The p53 transcription factor initiates DNA repair processes, stabilizes the cell cycle, or initiates apoptosis due to damage to the genetic material [86].

Studies conducted on human breast epithelial cells of the MCF-7 tumor line showed that curcumin competitively inhibited the activity of CYP1A1, which was induced by 7,12-dimethylbenz[a]anthracene (DMBA) [87]. It was found that curcumin in a dose of 1 μM reduced the activity of CYP1A1 of the microsomal fraction in cells of the MCF-7 line, treated with 1 μM DMBA for 24 h, by as much as 50%. Moreover, curcumin, by reducing the number of DMBA-DNA adducts formed, blocked metabolic activation of DMBA and decreased DMBA-induced cytotoxicity. In studies on

MCF-7 breast cancer epithelial cells in humans and on MCF-10A mammary epithelial cell lines, curcumin has been shown to exert a cytotoxic and apoptotic effect and regulate the expression of certain genes. It was found that it increases the expression of as many as 22 genes threefold and lowers the expression of 17 genes threefold at two concentrations of 25 and 50 µg/ml in the MCF-7 line. Moreover, it was observed that curcumin induced a much higher percentage of apoptosis in cells MCF-7 breast cancer epithelial cells than MCF-10A mammary epithelial cells at all doses [88].

Scientific reports also describe the effect of curcumin on the process of blood vessel formation during the growth of a cancerous tumor, including by regulating the expression of cytokines from the family of vascular endothelial growth factors VEGF. The role of angiogenesis and VEGF in the pathogenesis of neoplastic disease is significant. The presence of VEGF protein and mRNA has been found in cancer of the thyroid gland, bronchus, esophagus, lung, stomach, colon, liver, breast, ovary, uterus, kidney, bladder, brain and bone tumors. The degree of VEGF expression was related to tumor aggressiveness and prognosis in patients. Many studies have shown an inverse correlation between VEGF expression and overall survival in both node-positive and nonmetastatic breast cancer [89–91]. The use of curcumin in breast cancer cells of the T47-D line reduced the secretion of VEGF released by medroxyprogesterone (MPA) [92]. Similar results were observed with MCF-7 and MDA-MB-231 cell lines [93, 94]. In rats, curcumin reduced the proliferation and incidence of MPA and DMBA-induced tumors, and decreased VEGF expression in hyperplastic [95]. Many studies show that increased levels of MMP-2 and MMP-9 metalloproteinases in serum, plasma and other body fluids correlate with neoplastic progression, increased aggressiveness of the disease and the possibility of metastasis in patients with breast cancer [96–99]

There are studies available in the professional scientific literature showing the effect of curcumin on the growth and inhibition of acetylation in cells. Acetylation is one of the mechanisms that regulate a number of signaling pathways and alter some properties of proteins [100]. Balasubramanyam et al. [101] in studies conducted on the HeLa cell line showed that curcumin *in vitro* is a specific inhibitor of p300/CBP acetyltransferase, without affecting the activity of deacetylases. Kang et al. [102] showed that curcumin inhibits the activity of acetyltransferases both *in vitro* and *in vivo*. The inhibition of acetyltransferases by curcumin may be mediated by reactive oxygen species generated by it in concentrations higher than 20 µM. Curcumin has also been shown to induce histone hypoacetylation and caspase-3 dependent death of brain tumor cells [103]. In turn, other studies have shown that curcumin at a dose of 25 µM inhibits the expression of HDAC1, HDAC3, HDAC8 deacetylases, inducing an increase in the level of acetylated form of H4 histone and apoptosis in Raji's B-NHL lymphoma cells [104]. Studies on the same cell line, using the same concentration of curcumin, indicate that it also induces histone H3 acetylation in the p21 gene promoter and enhances the transcription of p21 protein, an inhibitor of cyclin-dependent kinases. It has been shown that curcumin in concentrations above 25 µM induces hyperacetylation of histone H3, with an accompanying increase in p53 protein expression and an increase in its acetylated form. As a consequence, inhibition of cell proliferation of the NB4 tumor line was observed [82]. The available studies show that curcumin may increase the effects of some anticancer drugs. Preclinical studies conducted *in vitro*, assessing the validity of the use of curcumin in the treatment of breast cancer with overexpression of the HER-2 receptor, showed that in five cell lines: MCF-7, BT-474, SK-BR-3-hr, MCF-10A and MDA-MB-231. Curcumin reduced cell growth at all concentrations used (5, 10, 15 µ/mL). Moreover, in combination with herceptin (a HER-2 receptor blocker), curcumin decreased the phosphorylation of the serine-threonine kinase—Akt, mitogen-activated protein kinase—MAPK, the expression of the nuclear factor κB and the HER-2 oncoprotein in BT-474 and SK-BR-hr cells [105]. In BT-474 cells overexpressing HER-2 *in vivo*, curcumin in combination with the cytostatics—taxol reduced the size of the tumor in a manner comparable to herceptin [105]. The dose of curcumin is also an important issue. Clinical trials have shown that curcumin taken orally in a dose of up to 8000 mg daily for three months is not toxic to humans [106].

6.3.3 Quercetin

Quercetin (3,3′,4′,5,7-pentahydroxyflavone) is one of the most important flavonoids belonging to the group of flavonols. It is the most consumed flavonol in the daily diet by humans. Its average daily consumption has been estimated to be around 25–35 mg. Red onions and shallots are the richest in quercetin. Other sources of quercetin include grape, citrus, apples, cabbage, tomatoes, broccoli, soybeans, asparagus, and drinks such as green and black tea and red wine [107]. In food, quercetin occurs mainly in the form associated with sugars, phenolic acids or alcohols. The bioavailability of this compound depends primarily on the diet and the variety of glycoside forms [107, 108]. Quercetin that is not absorbed in the small intestine and that that is combined with sugars other than glucose are transformed in the large intestine by enterobacteria. The absorption of quercetin is facilitated by bromelain, a complex that consists largely of proteolytic enzymes [109].

The resulting quercetin metabolites are detected in urine and serum. The concentration of quercetin metabolites in human blood serum depends primarily on the amount consumed with food and the time of its intake. Studies by Lamson and Brignall [110] and Murakami et al. [111] showed that the concentration of quercetin metabolites in human blood serum reaches the level of 1.5 µM after 2 h of ingestion of about 50 mg of this compound, and their level decreases with time. To maintain high concentrations of quercetin metabolites in blood serum and organs, it must be taken continuously [112]. It is important especially in anticancer therapy, because, according to the available research, quercetin, due to the presence of hydroxyl groups in its chemical structure, is a potential anticancer agent. Hydroxyl groups give it antioxidant properties. The ability to scavenge free radicals by quercetin increases with the increasing number of added hydroxyl groups, but decreases in the presence of sugar substituents [107, 111–114]. In addition to its antioxidant capacity, quercetin also has the ability to modify the course of intracellular pathways that regulate the cell cycle and cell proliferation. However, there are also scientific reports on the mutational activity of this compound and its cytotoxicity in relation to normal cells of the human body. To obtain the antitumor effect of quercetin, the concentration level of this compound in human blood serum should be higher than 10 µM. Based on research by Hollman et al. [115] found that daily consumption of 100 mg of quercetin increased its concentration in blood serum to 0.8 µM. On the other hand, a single dose of 100 mg of quercetin administered intravenously in humans increased the serum concentration of this compound to the level of 12 µM. A single oral dose of 4 g of quercetin did not cause side effects in humans. Same as a single 100 mg intravenous dose of quercetin [116]. Conversely, intravenous administration of 1400 mg quercetin/m^2/week to 10 patients for three weeks, after which the serum concentration of quercetin metabolites increased to 10 µM, resulted in side effects in two patients of nephrotoxicity and a reduction of glomerular flow by approximately 20%. One week after the end of therapy, the abovementioned parameters returned to normal in these patients [117]. In the studies of Ekström et al. [118]; it was observed that daily consumption of min. 12 mg of quercetin in the diet significantly reduces the risk of stomach cancer. When considering quercetin as part of the therapy, the enzymatic polymorphism and the composition of the patient's bacterial microbiota should be taken into account. Consuming large amounts of quercetin in the diet or in the form of a pharmacological supplement leads to a significant increase in one of its metabolites, namely homovanillic acid (HVA). Too high levels of this acid may interfere with the proper diagnosis of neuroblastoma neoplasm, as it is one of the markers of the development of this disease [119]. Quercetin, like the previously described resveratrol and curcumin, regulates the expression of some genes involved in the neoplastic process. In the studies of Jeong et al. [120] and Ramos [121] it was observed that this flavonoid has the ability to activate Chk2 kinase, the function of which is to block the cell cycle. At the same time, the accumulation of the p21 protein then takes place, ultimately resulting in inhibition of the expression of many proteins in the cell cycle. According to the abovementioned authors, quercetin also inhibits the expression of akt1, erb 2, c myc oncogenes. Quercetin, by inhibiting the expression of heat shock proteins (Hsp), mainly Hsp72 and Hsp27, shows proapoptotic activity. Lowering the expression of these proteins, which increase the tumor's invasiveness, makes tumor cells more sensitive to induction of programmed cell death during cytostatic treatment [122–125]. The proapoptotic activity of this flavonoid was observed in MCF-7 breast cancer, A549 lung cancer, BSp73AS and MiaPaCa-2 pancreatic cancer, HepG2 liver cancer, Caco2 colorectal cancer, HL-60 leukemia and HPB thymoma [121, 126–128]. In HepG2 liver cancer cells, this compound additionally induced programmed death by inactivating the transcription factor NF-kappa B [129]. Another group of proteins whose expression is altered by the action of quercetin are tyrosine kinases. They are an integral part of catalytic receptors and are associated with the cell membrane. *In vitro* studies showed that in HT-29 colon cancer cells, quercetin inhibited the expression of the tyrosine kinase receptors—ErbB2 and ErbB3 from the EGFR family (epidermal growth factor receptor). As a result, the process of cell proliferation and apoptosis was inhibited [130]. The effect of quercetin on the regulation of protein expression of the Bcl2 family has been demonstrated in A549 lung cancer cells and HT-29 and SW 480 colon cancer cells. In A549 lung cancer cells, this flavonoid increased the levels of proapoptotic Bad and Bax proteins and the antiapoptotic protein Bcl-xl. In contrast, the expression of antiapoptotic Bcl-2 decreased. On the other hand, in HT-29 and SW 480 colon cancer cells, incubation in the presence of quercetin, at a concentration of 50 µM for 72 h, reduced the level of Bcl-2, without affecting the level of Bax protein expression. An increase in the Bax level was observed in the Jurkat T leukemia line [121, 131].

In addition to studies that have proven the anticancer properties of quercetin, there are also scientific reports proving its carcinogenic activity. However, quite often these observations are contradictory and depend mainly on the dose used. The harmful effects of quercetin are reported mainly by *in vitro* studies. However, it should be remembered that this type of research does not fully reflect the conditions in a living organism.

6.4 Summary

Summarizing the characteristics of anticancer properties of selected natural substances, it can be concluded that they have potential and it is worth leaning toward using them in oncological therapy. However, most of the available scientific data show that they should be used in the so-called combined therapy, based on the use of several compounds acting on tumor cells according to different mechanisms. Simultaneous disruption of several processes or impairment of signaling pathways significantly increases the chance of eliminating cancer cells. The resistance of neoplastic cells to commonly used chemotherapeutic agents is a serious problem in cancer therapy. It can be both primary and acquired during cancer therapy. A significant challenge is therefore to implement into practice natural substances that would be able to overcome the resistance to chemotherapy and induce apoptosis in cancer cells. Future studies are needed to identify putative pathways or molecules that are target of strain-specific gene expression modulation.

Formulations with the best bioactivity and less side effects are another challenge. Probiotics cannot be applied as the only eradication treatment but as supportive treatment. Primarily because probiotics reduce side effects of antibiotics, reduce infections postsurgery and reduce levels of proinflammatory cytokines.

Increasing number of researchers focus on finding substances with antimetastatic properties. For this purpose, it is necessary to find tumor and host factors contributing in the metastasis cascade.

Author contributions: All the authors have accepted responsibility for the entire content of this submitted manuscript and approved submission.
Research funding: None declared.
Conflict of interest statement: The authors declare no conflicts of interest regarding this article.

References

1. Vivarelli S, Salemi R, Candido S, Falzone L, Santagati M, Stefani S, et al., Gut microbiota and cancer: from pathogenesis to therapy. *Cancers (Basel)* 2019;11:38. https://doi.org/10.3390/cancers11010038.
2. Jakobsson HE, Jernberg C, Andersson AF, Sjölund-Karlsson M, Jansson JK, Engstrand L. Short-term antibiotic treatment has differing long-term impacts on the human throat and gut microbiome. *PLoS One* 2010;5:e9836. https://doi.org/10.1371/journal.pone.0009836.
3. Zaura E, Brandt BW, de Mattos MJT, Buijs MJ, Caspers MPM, Rashid MU, et al., Same Exposure but two radically different responses to antibiotics: resilience of the salivary microbiome versus long-term microbial shifts in feces. *mBio* 2015;6:01693–15. https://doi.org/10.1128/mBio.01693-15.
4. Kapp K, Maul J, Hostmann A, Mundt P, Preiss JC, Wenzel A, et al., Modulation of systemic antigen-specific immune responses by oral antigen in humans. *Eur J Immunol* 2010;40:3128–37. https://doi.org/10.1002/eji.201040701.
5. Jandhyala SM, Talukdar R, Subramanyam C, Vuyyuru H, Sasikala M, Nageshwar Reddy D. Role of the normal gut microbiota. *World J Gastroenterol* 2015;21:8787–803. https://doi.org/10.3748/wjg.v21.i29.8787.
6. Carding S, Verbeke K, Vipond DT, Corfe BM, Owen LJ. Dysbiosis of the gut microbiota in disease. *Microb Ecol Health Dis* 2015;26:26191. https://doi.org/10.3402/mehd.v26.26191.
7. Urbaniak C, Gloor GB, Brackstone M, Scott L, Tangney M, Reida G. The microbiota of breast tissue and its association with breast cancer. *Appl Environ Microbiol* 2016;82:5039–48. https://doi.org/10.1128/AEM.01235-16.
8. Pleguezuelos-Manzano C, Puschhof J, Rosendahl Huber A, van Hoeck A, Wood HM, Nomburg J, et al., Mutational signature in colorectal cancer caused by genotoxic pks+ E. coli. *Nature* 2020;580:269–73. https://doi.org/10.1038/s41586-020-2080-8.
9. Zagato E, Pozzi C, Bertocchi A, Schioppa T, Sacchieri F, Guglietta S, et al., Endogenous murine microbiota member Faecalibaculum rodentium and its human homologue protect from intestinal tumour growth. *Nat Microbiol* 2020;5:511–24. https://doi.org/10.1038/s41564-019-0649-5.
10. Sharma M, Chandel D, Shukla G. Antigenotoxicity and cytotoxic potentials of metabiotics extracted from isolated probiotic, Lactobacillus rhamnosus MD 14 on Caco-2 and HT-29 human colon cancer cells. *Nutr Canc* 2020;72:110–9. https://doi.org/10.1080/01635581.2019.1615514.

11. Tilg H, Adolph T, Gerner R, Moschen A. The intestinal microbiota in colorectal cancer. *Canc Cell* 2018;33:954–64.https://doi.org/10.1016/j.ccell.2018.03.004.
12. Zhou Z, Chen J, Yao H, Hu H. Fusobacterium and colorectal cancer. *Front Oncol* 2018;8:371. https://doi.org/10.3389/fonc.2018.00371.
13. Walawalkar Y, Gaind R, Nayak V. Study on Salmonella Typhi occurrence in gallbladder of patients suffering from chronic cholelithiasis—a predisposing factor for carcinoma of gallbladder, *Diagn Microbiol Infect Dis* 2013;77:69–73. https://doi.org/10.1016/j.diagmicrobio.2013.05.014.
14. Kwa M, Plottel CS, Blaser MJ, Adams S. The intestinal microbiome and estrogen receptor-positive female breast cancer. *J Natl Cancer Inst* 2016;108:029. https://doi.org/10.1093/jnci/djw029.
15. Wilson ID, Nicholson JK. Gut microbiome interactions with drug metabolism, efficacy, and toxicity. *Transl Res* 2017;179:204–22. https://doi.org/10.1016/j.trsl.2016.08.002.
16. Botticelli A, Vernocchi P, Marini F, Quagliariello A, Cerbelli B, Reddel S, et al., Gut metabolomics profiling of non-small cell lung cancer (NSCLC) patients under immunotherapy treatment. *J Transl Med* 2020;18:49. https://doi.org/10.1186/s12967-020-02231-0.
17. Kim DW, Garon EB, Jatoi A, Keefe M, Lacoutur ME, Sonis S, et al., Impact of a planned dose interruption of dacomitinib in the treatment of advanced non-small-cell lung cancer (ARCHER 1042). *Lung Canc* 2017;106:76–82. https://doi.org/10.1016/j.lungcan.2017.01.021.
18. Marschalek J, Farr A, Marschalek ML, Domig KJ, Kneifel W, Singer CF, et al., Influence of orally administered probiotic Lactobacillus strains on vaginal microbiota in women with breast cancer during chemotherapy: a randomized placebo-controlled double-blinded pilot study. *Breast Care* 2017;12:335–9. https://doi.org/10.1159/000478994.
19. Laborda-Illanes A, Sanchez-Alcoholado L, Dominguez-Recio ME, Jimenez-Rodriguez B, Lavado R, Comino-Méndez I, et al., Breast and gut microbiota action mechanisms in breast cancer pathogenesis and treatment. *Cancers* 2020;12:2465. https://doi.org/10.3390/cancers12092465.
20. Vivarelli S, Falzone L, Basile MS, Nicolosi D, Genovese C, Libra M, et al., Benefits of using probiotics as adjuvants in anticancer therapy (Review). *World Acad Sci J* 2019;1:125–35.
21. Kaźmierczak-Siedlecka K, Folwarski M, Skonieczna-Żydecka K, Ruszkowski J, Makarewicz W. The use of Lactobacillus plantarum 299v (DSM 9843) in cancer patients receiving home enteral nutrition – study protocol for a randomized, double-blind, and placebo-controlled trial. *Nutr J* 2020;19:98. https://doi.org/10.1186/s12937-020-00598-w.
22. Hibberd AA, Lyra A, Ouwehand AC, Rolny P, Lindegren H, Cedgård L, et al., Intestinal microbiota is altered in patients with colon cancer and modified by probiotic intervention. *BMJ Open Gastroenterol* 2017;4:e000145. https://doi.org/10.1136/bmjgast-2017-000145.
23. Garcia-Peris P, Velasco C, Hernandez M, Lozano MA, Paron L, de la Cuerda C, et al., Effect of inulin and fructo-oligosaccharide on the prevention of acute radiation enteritis in patients with gynecological cancer and impact on quality-of-life: a randomized, double-blind, placebo-controlled trial. *Eur J Clin Nutr* 2016;70:170–4. https://doi.org/10.1038/ejcn.2015.192.
24. Song H, Wang W, Shen B, Jia H, Chen P, Sun Y, et al., Pretreatment with probiotic Bifico ameliorates colitis-associated cancer in mice: transcriptome and gut flora profiling. *Canc Sci* 2018;109:666–77. https://doi.org/10.1111/cas.13497.
25. Sivamaruthi BS, Kesika P, Chaiyasut C. The role of probiotics in colorectal cancer management. *Evid Base Compl Alternative Med* 2020;2020:17. https://doi.org/10.1155/2020/3535982.
26. Mego M, Koncekova R, Mikuskova E, Drgona L, Ebringer L, Demitrovicova L, et al., Prevention of febrile neutropenia in cancer patients by probiotic strain Enterococcus faecium M-74.

Phase II study. *Support Care Canc* 2006;14:285–90. https://doi.org/10.1007/s00520-005-0891-7.

27. Tian Y, Li M, Song W, Jiang R, Li YQ. Effects of probiotics on chemotherapy in patients with lung cancer. *Oncol Lett* 2019;17:2836–48. https://doi.org/10.3892/ol.2019.9906.
28. Bingula R, Filaire E, Molnar I, Delmas E, Berthon JY, Vasson MP, et al., Characterisation of microbiota in saliva, bronchoalveolar lavage fluid, non-malignant, peritumoural and tumour tissue in non-small cell lung cancer patients: a cross-sectional clinical trial. *Respir Res* 2020;21:129. https://doi.org/10.1186/s12931-020-01392-2.
29. Bergerot PG, Dizman N, Ruel N, Frankel PH, Hsu J, Pal SK. *A phase i trial to assess the biologic effect of CBM588 (Clostridium butyricum) in combination with nivolumab plus ipilimumab (nivo/ipi) in patients with metastatic renal cell carcinoma (mRCC)*; J Clin Oncol 2020;38(e6 Suppl). https://doi.org/10.1200/JCO.2020.38.6_suppl.TPS764.
30. Salgia NJ, Bergerot PG, Maia MC, Dizman N, Hsu JA, Gillece JD, et al., Stool microbiome profiling of patients with metastatic renal cell carcinoma receiving anti-PD-1 immune checkpoint inhibitors, *Eur Urol* 2020;78:498–502. https://doi.org/10.1016/j.eururo.2020.07.011.
31. Guthrie L, Kelly L. Bringing microbiome-drug interaction research into the clinic. *EBioMedicine* 2019;44:708–15. https://doi.org/10.1016/j.ebiom.2019.05.009.
32. Wakeman M. A review of the role of probiotics in sport. *Br J Sports Med* 2013;47:e4.25. https://doi.org/10.1136/bjsports-2013-093073.31.
33. Salminen SJ, Gueimonde M, Isolauri E. Probiotics that modify disease risk. *J Nutr* 2005;135:1294–8. https://doi.org/10.1093/jn/135.5.1294.
34. Mocan T, Sparchez Z, Craciun R, Bora CN, Leucuta DC. Programmed cell death protein-1 (PD-1)/programmed death-ligand-1 (PD-L1) axis in hepatocellular carcinoma: prognostic and therapeutic perspectives. *Clin Transl Oncol* 2019;21:702–12. https://doi.org/10.1007/s12094-018-1975-4.
35. Rattray NJW, Charkoftaki G, Rattray Z, Hansen JE, Vasiliou V, Johnson CH. Environmental influences in the etiology of colorectal cancer: the premise of metabolomics. *Curr Pharmacol Rep* 2017;3:114–25. https://doi.org/10.1007/s40495-017-0088-z.
36. Tronick E, Hunter RG. Waddington, dynamic systems, and epigenetics. *Front Behav Neurosci* 2016;10:107. https://doi.org/10.3389/fnbeh.2016.00107.
37. Moosavi A, Ardekani AM. Role of epigenetics in biology and human diseases. *Iran Biomed J* 2016;20:246–58. https://doi.org/10.22045/ibj.2016.01.
38. Kanherkar RR, Bhatia-Dey N, Csoka AB. Epigenetics across the human lifespan. *Front Cell Dev Biol* 2014;2. https://doi.org/10.3389/fcell.2014.00049.
39. Choi SW, Friso S. Epigenetics: a new bridge between nutrition and health. *Adv Nutr* 2010;1:8–16. https://doi.org/10.3945/an.110.1004.
40. Kulczycka A, Bednarek I, Dzierżewicz Z. Modyfikacje epigenetyczne jako potencjalne cele terapii antynowotworowych. Epigenetic modifications as potential targets of anti-cancer therapy. *Ann Acad Med Silesiensis* 2013;67:201–8.
41. Baylin SB, Jones PA. Epigenetic determinants of cancer. *Cold Spring Harb Perspect Biol* 2016;8:a019505. https://doi.org/10.1101/cshperspect.a019505.
42. Nobili S, Lippi D, Witort E, Donnini M, Bausi L, Mini E, et al., Natural compounds for cancer treatment and prevention. *Pharmacol Res* 2009;59:365–78. https://doi.org/10.1016/j.phrs.2009.01.017.
43. Rejhová A, Opattová A, Čumová A, Slíva D, Vodička P. Natural compounds and combination therapy in colorectal cancer treatment. *Eur J Med Chem* 2018;20:582–94. https://doi.org/10.1016/j.ejmech.2017.12.039.

44. Ncube B, Van Staden J. Tilting plant metabolism for improved metabolite biosynthesis and enhanced human benefit. *Molecules* 2015;20:12698–731. https://doi.org/10.3390/molecules200712698.
45. Link A, Balaguer F, Goel A. Cancer chemoprevention by dietary polyphenols: promising role for epigenetics. *Biochem Pharmacol* 2010;80:1771–92. https://doi.org/10.1016/j.bcp.2010.06.036.
46. Shakibaei M, Harikumar KB, Aggarwal BB. Review: resveratrol addiction: to die or not to die. *Mol Nutr Food Res* 2009;53:115–28. https://doi.org/10.1002/mnfr.200800148.
47. Khanna D, Sethi G, Ahn KS, Pandey MK, Kunnumakkara AB, Sung B, et al., Natural products as a gold mine for arthritis treatment. *Curr Opin Pharmacol* 2007;7:344–51. https://doi.org/10.1016/j.coph.2007.03.002.
48. Estrela JM, Ortega A, Mena S, Rodriguez ML, Asensi M. Pterostilbene: biomedical applications. *Crit Rev Clin Lab Sci* 2013;50:65–78. https://doi.org/10.3109/10408363.2013.805182.
49. Piver B, Fer M, Vitrac X, Merillon JM, Dreano Y, Berthou F, et al., Involvement of cytochrome P450 1A2 in the biotransformation of trans-resveratrol in human liver microsomes. *Biochem Pharmacol* 2004;68:773–82. https://doi.org/10.1016/j.bcp.2004.05.008.
50. Kuhnle GG, Paul Edward Spencer J, Hahn U. Absorption and metabolism of resveratrol in the small intestine. Implications for resveratrol in vivo diabetes and epithelial glucose transport view project impact of flavanols on human cognitive function view project; 2000. https://doi.org/10.1006/bbrc.2000.2750.
51. Espín JC, González-Barrio R, Cerdá B, López-Bote C, Rey AI, Tomás-Barberán FA. Iberian pig as a model to clarify obscure points in the bioavailability and metabolism of ellagitannins in humans. *J Agric Food Chem* 2007;55:10476–85. https://doi.org/10.1021/jf0723864.
52. Saiko P, Szakmary A, Jaeger W, Szekeres T. Resveratrol and its analogs: defense against cancer, coronary disease and neurodegenerative maladies or just a fad? *Mutat Res Rev Mutat Res* 2008;658:68–94. https://doi.org/10.1016/j.mrrev.2007.08.004.
53. Mikstacka R, Rimando AM, Ignatowicz E. Antioxidant effect of trans-resveratrol, pterostilbene, quercetin and their combinations in human erythrocytes in vitro. *Plant Foods Hum Nutr* 2010;65:57–63. https://doi.org/10.1007/s11130-010-0154-8.
54. Mikuła-Pietrasik J, Kuczmarska A, Książek K. Biologiczna wielofunkcyjność resweratrolu i jego pochodnych [Biological multifunctionality of resveratrol and its derivatives]. *Postepy Biochem* 2015;61:336–43.
55. Berge G, Øvrebø S, Eilertsen E, Haugen A, Mollerup S. Analysis of resveratrol as a lung cancer chemopreventive agent in A/J mice exposed to benzo[a]pyrene. *Br J Canc* 2004;91:1380–3. https://doi.org/10.1038/sj.bjc.6602125.
56. Shi T, Liou LS Sadhukhan P, Duan ZH, Novick AC, Hissong JG, et al., Effects of Resveratrol on gene expression in renal cell carcinoma. *Canc Biol Ther* 2004;3:882–8. https://doi.org/10.4161/cbt.3.9.1056.
57. Mikulski D, Górniak R, Molski M. A theoretical study of the structure-radical scavenging activity of trans-resveratrol analogues and cis-resveratrol in gas phase and water environment. *Eur J Med Chem* 2010;45:1015–27. https://doi.org/10.1016/j.ejmech.2009.11.044.
58. Le Corre L, Fustier P, Chalabi N, Bignon YJ, Bernard-Gallon D. Effects of resveratrol on the expression of a panel of genes interacting with the BRCA1 oncosuppressor in human breast cell lines. *Clin Chim Acta* 2004;344:115–21. https://doi.org/10.1016/j.cccn.2004.02.024.
59. Bieging KT, Mello SS, Attardi LD. Unravelling mechanisms of p53-mediated tumour suppression. *Nat Rev Canc* 2014;14:359–70. https://doi.org/10.1038/nrc3711.

60. Hsieh TC, Wang Z, Hamby CV, Wu JM. Inhibition of melanoma cell proliferation by resveratrol is correlated with upregulation of quinone reductase 2 and p53. *Biochem Biophys Res Commun* 2005;334:223–30. https://doi.org/10.1016/j.bbrc.2005.06.073.
61. Kai L, Samuel SK, Levenson AS. Resveratrol enhances p53 acetylation and apoptosis in prostate cancer by inhibiting MTA1/NuRD complex. *Int J Canc* 2010;126:1538–48. https://doi.org/10.1002/ijc.24928.
62. Laux MT, Aregullin M, Berry JP, Flanders JA, Rodriguez E. Identification of a p53-dependent pathway in the induction of apoptosis of human breast cancer cells by the natural product, resveratrol. *J Alternative Compl Med* 2004;10:235–9. https://doi.org/10.1089/107555304323062211.
63. Li B, Hou D, Guo H, Zhou H, Zhang S, Xu X, et al., Resveratrol sequentially induces replication and oxidative stresses to drive p53-CXCR2 mediated cellular senescence in cancer cells. *Sci Rep* 2017;7:1–12. https://doi.org/10.1038/s41598-017-00315-4.
64. Aggarwal BB, Bharwaj A, Aggarwal RS, Seeram NP, Shishodia S, Takada Y. Role of resveratrol in prevention and therapy of cancer: preclinical and clinical studies. *Anticancer Res* 2004;24:2783–840.
65. Docherty JJ, Smith JS, Fu MM, Stoner T, Booth T. Effect of topically applied resveratrol on cutaneous herpes simplex virus infections in hairless mice. *Antivir Res* 2004;61:19–26. https://doi.org/10.1016/j.antiviral.2003.07.001.
66. Kubota T, Uemura Y, Kobayashi M, Taguchi H. Combined effects of resveratrol and paclitaxel on lung cancer cells. *Anticancer Res* 2003;23:4039–46.
67. Baatout S, Derradji H, Jacquet P, Ooms D, Michaux A, Mergeay M. Enhanced radiation-induced apoptosis of cancer cell lines after treatment with resveratrol. *Int J Mol Med* 2004;13:895–902. https://doi.org/10.3892/ijmm.13.6.895.
68. Surh YJ, Hurh YJ, Kang JY, Lee E, Kong G, Lee SJ. Resveratrol, an antioxidant present in red wine, induces apoptosis in human promyelocytic leukemia (HL-60) cells. *Canc Lett* 1999;140:1–10. https://doi.org/10.1016/S0304-3835(99)00039-7.
69. Zou Y, Yang J, Jiang D. Resveratrol inhibits canonical Wnt signaling in human MG-63 osteosarcoma cells. *Mol Med Rep* 2015;12:7221–6. https://doi.org/10.3892/mmr.2015.4338.
70. Patel KR, Brown VA, Jones DJL, Britton RG, Hemingway D, Miller AS, et al., Clinical pharmacology of resveratrol and its metabolites in colorectal cancer patients. *Canc Res* 2010;70:7392–9. https://doi.org/10.1158/0008-5472.CAN-10-2027.
71. Chachay VS, Kirkpatrick CMJ, Hickman IJ, Ferguson M, Prins JB, Martin JH. Resveratrol – pills to replace a healthy diet? *Br J Clin Pharmacol* 2011;72:27–38. https://doi.org/10.1111/j.1365-2125.2011.03966.x.
72. Wu ML, Li H, Yu LJ, Chen XY, Kong QY, Song X, et al., Short-term resveratrol exposure causes in vitro and in vivo growth inhibition and apoptosis of bladder cancer cells. *PLoS One* 2014;9:e89806. https://doi.org/10.1371/journal.pone.0089806.
73. Johnson JJ, Nihal M, Siddiqui IA, Scarlett CO, Bailey HH, Mukhtar H, et al., Enhancing the bioavailability of resveratrol by combining it with piperine. *Mol Nutr Food Res* 2011;55:1169–76. https://doi.org/10.1002/mnfr.201100117.
74. Shehzad A, Wahid F, Lee YS. Curcumin in cancer chemoprevention: molecular targets, pharmacokinetics, bioavailability, and clinical trials. *Arch Pharm (Weinheim)* 2010;343:489–99. https://doi.org/10.1002/ardp.200900319.
75. Shoba G, Joy D, Joseph T, Majeed M, Rajendran R, Srinivas PSSR. Influence of piperine on the pharmacokinetics of curcumin in animals and human volunteers. *Planta Med* 1998;64:353–6. https://doi.org/10.1055/s-2006-957450.
76. Szczepański MA, Grzanka A. Chemoprewencyjne i przeciwnowotworowe właściwości kurkuminy, nowotwory. *J Oncol* 2009;59:377.

77. Lin YG, Kunnumakkara AB, Nair A, Merritt WM, Han LY, Armaiz-Pena GN, et al., Curcumin inhibits tumor growth and angiogenesis in ovarian carcinoma by targeting the nuclear factor-κB pathway. *Clin Canc Res* 2007;13:3423–30. https://doi.org/10.1158/1078-0432.CCR-06-3072.
78. Kunnumakkara AB, Bordoloi D, Padmavathi G, Monisha J, Roy NK, Prasad S, et al., Curcumin, the golden nutraceutical: multitargeting for multiple chronic diseases. *Br J Pharmacol* 2017;174:1325–48. https://doi.org/10.1111/bph.13621.
79. Choi H, Chun YS, Kim SW, Kim MS, Park JW. Curcumin inhibits hypoxia-inducible factor-1 by degrading aryl hydrocarbon receptor nuclear translocator: a mechanism of tumor growth inhibition. *Mol Pharmacol* 2006;70:1664–71. https://doi.org/10.1124/mol.106.025817.
80. Patiar S, Harris AL. Role of hypoxia-inducible factor-1α as a cancer therapy target. *Endocr Relat Cancer* 2006;13:61–75. https://doi.org/10.1677/erc.1.01290.
81. Aoki H, Takada Y, Kondo S, Sawaya R, Aggarwal BB, Kondo Y. Evidence that curcumin suppresses the growth of malignant gliomas in vitro and in vivo through induction of autophagy: role of Akt and extracellular signal-regulated kinase signaling pathways. *Mol Pharmacol* 2007;72:29–39. https://doi.org/10.1124/mol.106.033167.
82. Li L, Braiteh FS, Kurzrock R. Liposome-encapsulated curcumin: in vitro and in vivo effects on proliferation, apoptosis, signaling, and angiogenesis. *Cancer* 2005;104:1322–31. https://doi.org/10.1002/cncr.21300.
83. Shishodia S, Singh T, Chaturvedi MM. Modulation of transcription factors by curcumin. *Adv Exp Med Biol* 2007;595:127–48. https://doi.org/10.1007/978-0-387-46401-5_4.
84. Ramachandran C, You W. Differential sensitivity of human mammary epithelial and breast carcinoma cell lines to curcumin. *Breast Canc Res Treat* 1999;54:269–78. https://doi.org/10.1023/A:1006170224414.
85. Sa G, Das T. Anti-cancer effects of curcumin: cycle of life and death. *Cell Div* 2008;3:14. https://doi.org/10.1186/1747-1028-3-14.
86. Harris SL, Levine AJ. The p53 pathway: positive and negative feedback loops. *Oncogene* 2005;18:2899–908. https://doi.org/10.1038/sj.onc.1208615.
87. Ciolino HP, Daschner PJ, Wang TTY, Yeh GC. Effect of curcumin on the aryl hydrocarbon receptor and cytochrome P450 1A1 in MCF-7 human breast carcinoma cells. *Biochem Pharmacol* 1998;56:197–206. https://doi.org/10.1016/S0006-2952(98)00143-9.
88. Ramachandran C, Rodriguez S, Ramachandran R, Nair PKR, Fonseca H, Khatib Z, et al., Expression profiles of apoptotic genes induced by curcumin in human breast cancer and mammary epithelial cell lines. *Anticancer Res* 2005;25:3293–302.
89. Gasparini G, Toi M, Gion M, Verderio P, Dittadi R, Hanatani M, et al., Prognostic significance of vascular endothelial growth factor protein in node-negative breast carcinoma. *J Natl Cancer Inst* 1997;89:139–47. https://doi.org/10.1093/jnci/89.2.139.
90. Gasparini G, Toi M, Miceli R, Vermeulen PB, Dittadi R, Biganzoli E, et al., Clinical relevance of vascular endothelial growth factor and thymidine phosphorylase in patients with node-positive breast cancer treated with either adjuvant chemotherapy or hormone therapy. *Canc J Sci Am* 1999;5:101–11.
91. Yoshiji H, Gomez D, Shibuya U, Thorgeirsson UP. Expression of vascular endothelial growth factor, its receptor, and other angiogenic factors in human breast cancer. *Canc Res* 1996;56:2013–6.
92. Carroll CE, Ellersieck MR, Hyder SM. Curcumin inhibits MPA-induced secretion of VEGF from T47-D human breast cancer cells. *Menopause* 2008;15:570–4. https://doi.org/10.1097/gme.0b013e31814fae5d.
93. Schindler R, Mentlein R. Flavonoids and vitamin E reduce the release of the angiogenic peptide vascular endothelial growth factor from human tumor cells. *J Nutr* 2006;136:1477–82. https://doi.org/10.1093/jn/136.6.1477.

94. Shao ZM, Shen ZZ, Liu CH, Sartippour MR, Liang GV, Herbert D, et al., Curcumin exerts multiple suppressive effects on human breast carcinoma cells. *Int J Canc* 2002;98:234–40. https://doi.org/10.1002/ijc.10183.
95. Carroll CE, Benakanakere I, Besch-Williford C, Ellersieck MR, Hyder SM. Curcumin delays development of medroxyprogesterone acetate-accelerated 7, 12-dimethylbenz[a]anthracene-induced mammary tumors. *Menopause* 2010;17:178–84. https://doi.org/10.1097/gme.0b013e3181afcce5.
96. Djonov V, Cresto N, Aebersold DM, Burri PH, Altermatt HJ, Hristic M, et al., Tumor cell specific expression of MMP-2 correlates with tumor vascularisation in breast cancer. *Int J Oncol* 2002;21:25–30.
97. Leppa S, Saarto T, Vehmanen L, Blomqvist C, Elomaa I. A high serum matrix metalloproteinase-2 level is associated with an ad verse prognosis in node-positive breast carcinoma. *Clin Canc Res* 2004;10:1057–63. https://doi.org/10.1158/1078-0432.ccr-03-0047.
98. Sheen-Chen SM, Chen HS, Eng HL, Sheen CC, Chen WJ. Serum levels of matrix metalloproteinase 2 in patients with breast cancer. *Canc Lett* 2001;173:79–82. https://doi.org/10.1016/s0304-3835(01)00657-7.
99. Talvensaari-Mattila A, Turpeenniemi-Hujanen T. Preoperative serum MMP-9 immunoreactive protein is a prognostic indicator for relapse-free survival in breast carcinoma. *Canc Lett* 2005;217:237–42. https://doi.org/10.1016/j.canlet.2004.06.056.
100. Kouzarides T. Acetylation: a regulatory modification to rival phosphorylation. *EMBO J* 2000;19:1176–9. https://doi.org/10.1093/emboj/19.6.1176.
101. Balasubramanyam K, Altaf M, Varier RA, Swaminathan V, Ravindran A, Sadhale PP, et al., Polyisoprenylated benzophenone, garcinol, a natural histone acetyltransferase inhibitor, represses chromatin transcription and alters global gene expression. *J Biol Chem* 2004;279:33716–26. https://doi.org/10.1074/jbc.M402839200.
102. Kang J, Chen J, Shi Y, Jia J, Zhang Y. Curcumin-induced histone hypoacetylation: the role of reactive oxygen species. *Biochem Pharmacol* 2005;69:1205–13. https://doi.org/10.1016/j.bcp.2005.01.014.
103. Kang SK, Cha SH, Jeon HG. Curcumin-induced histone hypoacetylation enhances caspase-3-dependent glioma cell death and neurogenesis of neural progenitor cells. *Stem Cell Dev* 2006;15:165–74. https://doi.org/10.1089/scd.2006.15.165.
104. Liu HL, Chen Y, Cui GH, Zhou JF. Curcumin, a potent anti-tumor reagent, is a novel histone deacetylase inhibitor regulating B-NHL cell line Raji proliferation. *Acta Pharmacol Sin* 2005;26:603–9. https://doi.org/10.1111/j.1745-7254.2005.00081.x.
105. Chen DR, Lai HW, Chien SY, Kuo SJ, Tseng LM, Lin HY, et al., The potential utility of curcumin in the treatment of HER-2-overexpressed breast cancer: an in vitro and in vivo comparison study with herceptin. Evidence-based complement. *Altern Med* 2012; Article ID 486568. https://doi.org/10.1155/2012/486568.
106. Cheng AL, Hsu CH, Lin JK, Hsu MM, Ho YF, Shen TS, et al., Phase I clinical trial of curcumin, a chemopreventive agent, in patients with high-risk or pre-malignant lesions. *Anticancer Res* 2001;21:2895–900.
107. Nemeth K, Piskula MK. Food content, processing, absorption and metabolism of onion flavonoids. *Crit Rev Food Sci Nutr* 2007;47:397–409. https://doi.org/10.1080/10408390600846291.
108. Scholz S, Williamson G. Interactions affecting the bioavailability of dietary polyphenols in vivo. *Int J Vitam Nutr Res* 2007;77:224–35. https://doi.org/10.1024/0300-9831.77.3.224.
109. Valentová K, Káňová K, Di Meo F, Pelantová H, Chambers CS, Rydlová L, et al., Chemoenzymatic preparation and biophysical properties of sulfated Quercetin metabolites. *Int J Mol Sci* 2017;18:2231. https://doi.org/10.3390/ijms18112231.

110. Lamson DW, Brignall MS. Antioxidants and cancer, part 3: quercetin. *Alternative Med Rev* 2000;5:196–208.
111. Murakami A, Ashida H, Terao J. Multitargeted cancer prevention by quercetin. *Cancer Lett* 2008;269:315–25. https://doi.org/10.1016/j.canlet.2008.03.046.
112. De Boer VCJ, Dihal AA, Van Der Woude H, Arts ICW, Wolffram S, Alink GM, et al., Tissue distribution of quercetin in rats and pigs. *J Nutr* 2005;135:1718–25. https://doi.org/10.1093/jn/135.7.1718.
113. Gerhauser C. Cancer chemopreventive potential of apples, apple juice, and apple components. *Planta Med* 2008;74:1608–24. https://doi.org/10.1055/s-0028-1088300.
114. Terao J. Dietary flavonoids as antioxidants. *Forum Nutr* 2009;61:87–94. https://doi.org/10.1159/000212741.
115. Hollman PCH, Van Trijp JMP, Mengelers MJB, De Vries JHM, Katan MB. Bioavailability of the dietary antioxidant flavonol quercetin in man. *Cancer Lett* 1997;114:139–40. https://doi.org/10.1016/S0304-3835(97)04644-2.
116. Gugler R, Leschik M, Dengler HJ. Disposition of quercetin in man after single oral and intravenous doses. *Eur J Clin Pharmacol* 1975;9:229–34. https://doi.org/10.1007/BF00614022.
117. Ferry DR, Smith A, Malkhandi J, Fyfe DW, Takats PG, Anderson D. Anderson D. *Phase I clinical trial of the flavonoid quercetin: pharmacokinetics and evidence for in vivo tyrosine kinase inhibition*. Clin Cancer Res 1996;2:659–68. PubMed: 9816216.
118. Ekström AM, Serafini M, Nyrén O, Wolk A, Bosetti C, Bellocco R. Dietary quercetin intake and risk of gastric cancer: results from a population-based study in Sweden. *Ann Oncol* 2011;22:438–43. https://doi.org/10.1093/annonc/mdq390.
119. Weldin J, Jack R, Dugaw K, Kapur RP. Quercetin, an over-the-counter supplement, causes neuroblastoma-like elevation of plasma homovanillic acid. *Pediatr Dev Pathol* 2003;6:547–51. https://doi.org/10.1007/s10024-003-5061-7.
120. Jeong JH, An JY, Kwon YT, Rhee JG, Lee YJ. Effects of low dose quercetin: cancer cell-specific inhibition of cell cycle progression. *J Cell Biochem* 2009;106:73–82. https://doi.org/10.1002/jcb.21977.
121. Ramos S. Effects of dietary flavonoids on apoptotic pathways related to cancer chemoprevention. *J Nutr Biochem* 2007;18:427–42. https://doi.org/10.1016/j.jnutbio.2006.11.004.
122. Jakubowicz-Gil J, Paduch R, Piersiak T, Głowniak K, Gawron A, Kandefer-Szerszeń M. The effect of quercetin on pro-apoptotic activity of cisplatin in HeLa cells. *Biochem Pharmacol* 2005;69:1343–50. https://doi.org/10.1016/j.bcp.2005.01.022.
123. Calderwood SK, Khaleque MA, Sawyer DB, Ciocca DR. Heat shock proteins in cancer: chaperones of tumorigenesis. *Trends Biochem Sci* 2006;31:164–72. https://doi.org/10.1016/j.tibs.2006.01.006.
124. Garrido C, Brunet M, Didelot C, Zermati Y, Schmitt E, Kroemer G. Heat shock proteins 27 and 72: anti-apoptotic proteins with tumorigenic properties. *Cell Cycle* 2006;5:2592–601. https://doi.org/10.4161/cc.5.22.3448.
125. Jakubowicz-Gil J, Rzeski W, Zdzisińska B, Piersiak T, Weiksza K, Głowniak K, et al., Different sensitivity of neurons and neuroblastoma cells to quercetin treatment. *Acta Neurobiol Exp* 2008;68:463–76.
126. Granado-Serrano AB, Martín MA, Bravo L, Goya L, Ramos S. Quercetin induces apoptosis via caspase activation, regulation of Bcl-2 and inhibition of PI-3-kinase/Akt and ERK pathways in human hepatoma cell line (HepG2). *J Nutr* 2006;136:2715–21. https://doi.org/10.1093/jn/136.11.2715.

127. Chou CC, Yang JS, Lu HF, Ip SS, Lo C, Wu CC, et al., Quercetin-mediated cell cycle arrest and apoptosis involving activation of caspase cascade through the mitochondrial pathway in human breast cancer MCF-7 cells. *Arch Pharm Res* 2010;33:1181–91. https://doi.org/10.1007/s12272-010-0808-y.
128. Philchenkov A, Zavelevich M, Savinska L, Blokhin D. Jurkat/A4 cells with multidrug resistance exhibit reduced sensitivity to quercetin. *Exp Oncol* 2010;32:76–80.
129. Granado-Serrano AB, Martín MA, Bravo L, Goya L, Ramos S. Quercetin modulates NF-κB and AP-1/JNK pathways to induce cell death in human hepatoma cells. *Nutr Canc* 2010;62:390–401. https://doi.org/10.1080/01635580903441196.
130. Kim WK, Bang MH, Kim ES, Kang NE, Jung KC, Cho HJ, et al., Quercetin decreases the expression of ErbB2 and ErbB3 proteins in HT-29 human colon cancer cells. *J Nutr Biochem* 2005;16:155–62. https://doi.org/10.1016/j.jnutbio.2004.10.010.
131. Chen D, Daniel KG, Chen MS, Kuhn DJ, Landis-Piwowar KR, Dou QP. Dietary flavonoids as proteasome inhibitors and apoptosis inducers in human leukemia cells. *Biochem Pharmacol* 2005;69:1421–32. https://doi.org/10.1016/j.bcp.2005.02.022.

Anna Helmin-Basa, Lidia Gackowska, Sara Balcerowska, Marcelina Ornawka, Natalia Naruszewicz and Małgorzata Wiese-Szadkowska

6 The application of the natural killer cells, macrophages and dendritic cells in treating various types of cancer

Abstract: Innate immune cells such as natural killer (NK) cells, macrophages and dendritic cells (DCs) are involved in the surveillance and clearance of tumor. Intensive research has exposed the mechanisms of recognition and elimination of tumor cells by these immune cells as well as how cancers evade immune response. Hence, harnessing the immune cells has proven to be an effective therapy in treating a variety of cancers. Strategies aimed to harness and augment effector function of these cells for cancer therapy have been the subject of intense researches over the decades. Different immunotherapeutic possibilities are currently being investigated for anti-tumor activity. Pharmacological agents known to influence immune cell migration and function include therapeutic antibodies, modified antibody molecules, toll-like receptor agonists, nucleic acids, chemokine inhibitors, fusion proteins, immunomodulatory drugs, vaccines, adoptive cell transfer and oncolytic virus–based therapy. In this review, we will focus on the preclinical and clinical applications of NK cell, macrophage and DC immunotherapy in cancer treatment.

Keywords: cancer immunotherapy, dendritic cells, macrophages, NK cells

6.1 Introduction

Innate immune cells such as natural killer (NK) cells, macrophages and dendritic cells (DCs) are also involved in the surveillance and clearance of cancer. We know a lot about the mechanisms of recognition and elimination of tumor cells by these cells as well as how cancer evades immune response. The role and function of these cells are very attractive tools for the design of immunotherapeutic approaches, namely boosting of the immune system during tumor treatment. Hence, harnessing these cells has proven to be an effective therapy in treating a variety of cancers.

This article has previously been published in the journal Physical Sciences Reviews. Please cite as: Helmin-Basa, A., Gackowska, L., Balcerowska, S., Ornawka, M., Naruszewicz, N. Wiese-Szadkowska, M.The application of the natural killer cells, macrophages and dendritic cells in treating various types of cancer *Physical Sciences Reviews* [Online] 2021, 6. DOI: 10.1515/psr-2019-0058

NK cells develop in the bone marrow and circulate in the blood. They are large granular lymphocytes (LGLs) and are known to be the most efficient anti-tumor effector cells.

NK cells have the ability to kill target cells without the need of prior sensitization or restriction of the major histocompatibility complex (MHC) molecule. These cells express a series of activating and inhibitory receptors that recognize stress proteins, viral proteins, the constant fragment (Fc) of immunoglobulin G (IgG) and MHC class I molecules. Activation of NK cells depends on the balance of signals generated from the activating and inhibitory receptors. When the activating signals override inhibitory ones, the cells become activated.

NK cells have the ability to recognize and kill cells with downregulated expression of inhibitory ligands (e.g. MHC class I molecule) along with high levels of activatory ligands (e.g. stress-induced ligands, antigen-specific IgG). Since NK cells express inhibitory receptors that bind to MHC class I, cells with downregulated MHC class I become targets to NK cells as the inhibitory binding does not occur. Tumor cells and virally infected cells often express downregulated MHC class I molecules that act as inhibitory ligands but obtain the expression of high levels of stress-induced ligands (e.g. MHC class I polypeptide-related sequence A/B [MICA/B]), which acts as activatory ligands. In this way, activatory receptors (e.g. NKG2D) are activated but inhibitory receptors are not.

NK cells can recognize and eliminate target cells directly or indirectly in a process called antibody-dependent cell cytotoxicity (ADCC), when targets are coated by IgG. These immune cells can kill targets using a variety of mechanisms. Primarily, they release perforin and granzymes from secretory granules. Perforin forms pores in the target cell membrane, permitting granzymes to penetrate into the cytoplasm and induce apoptosis. NK cells can also mediate killing of target cells by dead receptors (Fas, tumor necrosis factor [TNF]–related apoptosis inducing ligand [TRAIL]). Finally, NK cells produce pro-inflammatory cytokines, such as interferon γ (IFN-γ), TNF-α and other cytokines that are responsible for T-cell recruitment to the site of malignancy and impact the function and maturation of DCs. IFN-γ also promotes cytotoxic T lymphocyte (CTL) differentiation and a T helper cell type 1 (Th1) response.

Monocytes and macrophages are part of the myeloid family, a group of hematopoietic cells included in the mononuclear phagocyte system (MPS). They are essential for correct innate immune functions, support of adaptive immunity and tissue homeostasis. The ability to recognize, engulf and destroy pathogens and present antigens (Ags) makes monocytes and macrophages some of the key cells in innate immunity [1].

Monocytes are a population of mononuclear leukocytes that develop in the bone marrow from monoblasts and then are released into the bloodstream. The fate of monocytes strictly depends on their microenvironment. The main factors that stimulate monocyte differentiation into macrophages are macrophage colony–stimulating factor (M-CSF), granulocyte-macrophage colony–stimulating factor (GM-CSF) and interleukin 3 (IL-3) [2].

Based on the phenotype (surface receptor expression) and function, human monocytes can be divided into two main subsets: (a) classical monocytes ($CD14^{high}$/$CD16^-$/$CD62L^+$/$CX3CR1^{low}$/$CCR2^+$/$VEGFR1^{high}$ cells) and (b) non-classical monocytes ($CD14^{low}$/$CD16^+$/$CD62L^-$/$CX3CR1^{high}$/$CCR2^-$/$VEGFR1^{low}$ cells). At present, an intermediate population between classic and non-classic monocytes can also be distinguished (termed intermediate monocytes), which is characterized primarily by high expression of the MHC class II and CD14 receptor, and a low CD16 receptor expression ($CD14^{high}$/$CD16^{low}$). Classical monocytes take part mainly in the inflammatory responses induced by pathogens or tissue damage; non-classical monocytes decrease inflammatory properties, and they are involved in the repair of damaged vessels and the process of angiogenesis. The intermediate monocytes are found at low frequencies in the bloodstream, with a tendency to increase in population as a response to cytokine treatment and during inflammation; however, their immunotherapeutic properties are still to be fully not characterized [2]. The phenotype of monocytes determines their function and their further fate. Classical monocytes migrate mainly towards inflammation and tissue remodeling sites. Non-classical monocytes are mainly responsible for protecting the integrity of vascular endothelium and recruiting neutrophils during self-repair. Once inside the tissue, the classical monocytes differentiate into the tissue, differentiate into M1 pro-inflammatory/anti-tumor, while non-classical monocytes give rise to M2 anti-inflammatory/pro-tumor. Polarization toward M1 is observed in the early stages of the inflammatory response in bacterial, viral and fungal infections and in the case of tissue damage. M1, activated by lipopolysaccharides (LPS) and pro-inflammatory cytokines, such as IFN-γ, present the ability to kill tumor cells, inhibit angiogenesis and promote adaptive immune responses. M1 present their antigens mainly to Th1 and Th cells–producing IL-17 (Th17). They produce cytokines including TNF-α, IL-1, IL-6, IL-12 and IL-23 and chemokines such as chemokine (C–C motif) ligand 5 (CCL5), chemokine (C–C motif) ligand 8 (CCL8), chemokine (C–X–C motif) ligand 2 (CXCL2) and chemokine (C–X–C motif) 4 (CXCL4). On the other hand, M2 macrophages can be induced by anti-inflammatory cytokines, such as IL-4 or IL-13, and promote tumor initiation, progression and survival. They also inhibit immune-stimulatory signals and are devoid of cytotoxic activity. M2 macrophages are activated in the processes of silencing inflammation, fibrosis and formation of atherosclerotic plaque and during parasitic infections. Macrophages of this population activate mainly Th cells type 2 (Th2) and regulatory T cells (Tregs), and they secrete cytokines including IL-10, tumor grow factor (TGF-β), chemokine (C–C motif) ligand 16 (CCL16), chemokine (C–C motif) ligand 18 (CCL18) and chemokine (C–C motif) ligand 22 (CCL22) [3]. However, uncontrolled and long-time activation of inflammatory M1 macrophages could shift toward an M2 macrophage polarization.

DCs have a potent Ag presenting function. Moreover, due to their function, these cells constitute a "link" between the non-specific and specific immune responses. DCs differentiate from myeloid or lymphoid bone marrow–derived progenitors, and hence they are being referred to as conventional DCs (cDCs) or plasmacytoid DCs (pDCs). cDCs provide specific stimulatory function, whereas pDCs exhibit tolerogenic properties. However, both forms are highly specialized professional Ag presenting cells (APCs) with functional plasticity that can express immunostimulating and immunosupressive potential [4–6]. DCs exist in three consecutive stages of DC maturation: immature DCs (iDCs), semimature DCs (smDCs) and mature DCs (mDCs). The immature or semimature state populates non-lymphoid tissues, whereas the inversely mature DCs migrate to lymphoid tissue, activate T cells and induce an immune response. Cells representing these three phenotypes can be distinguished via cytometric analysis of surface receptor expressions (HLA-DR, CD80, CD86, CD83 and CD40) and the cytokine profile (IL-10, IL-12p70, IL-23 and TNF-α) [5–7]. Binding of foreign Ags, such as bacterial Ags, activates iDCs and transforms them into smDCs or mDCs. The phenotype of smDCs does not differ from that of mDCs; however, their functions differ. smDCs have reduced ability to produce pro-inflammatory cytokines and can produce moderate level of IL-10. In contrast, mDCs activate T-cell response and synthesize an array of cytokines, including IL-12p70, IL-12p40, IL-6 and TNF-α [5–8]. These activated T cells, mainly CTLs, are ultimately responsible for killing cancer cells and eradicating the tumor. Therefore, DCs play a major role in cancer immunosurvivalence and are a critical factor in anti-tumor immunity. The newest research concluded that functioning of tumor-infiltrating DCs (TIDCs) may be associated with the delay of tumor progression and lymph node metastasis [9, 10].

Strategies to engage NK cells, macrophages and DCs in cancer therapy are constantly developing. Recent studies have demonstrated that therapeutic monoclonal antibodies, modified antibody molecules, toll-like receptor (TLR) agonists, nucleic acids, chemokine inhibitors, fusion proteins, immunomodulatory drugs, vaccines, adoptive cell transfer and oncolitic virus based therapy influence immune cell presence and function. Immunotherapy is more specific than chemotherapy and radiation therapy. Currently, it is being used to complete conventional cancer treatment. Intensive research involving the potential of the immune system may allow scientists to use it as an alternative to traditional, invasive treatment options available for cancer patients.

Immunotherapy is effective in both hematologic and in solid tumors. In this review, we will focus on the preclinical and clinical applications of NK cell, macrophage and DC immunotherapy in cancers.

6.2 Innate immune cells and the tumor microenvironment

Tumors develop different strategies to escape the immune response: the loss or downregulation of MHC class I molecules, the loss of co-stimulatory molecules on their cell surface, the expression of co-inhibitory receptors, the secretion of immunosuppresive cytokines and production of tumor-derived factors [11] and the formation of tissue barriers around tumors [12]. Factors from both tumors and tumor-associated cells (IL-6, M-CSF, vascular endothelial growth factor [VEGF], *cyclooxygenase-1* [COX-1], *cyclooxygenase-2* [COX-2], prostaglandin E2 [PGE2], IL-10, TGF-β, indoleamine 2,3-dioxygenase [IDO] and arginase I), generally referred to as the tumor microenvironment (TME), have influence on immune cells presence and their differentiation and migration (Figure 6.1).

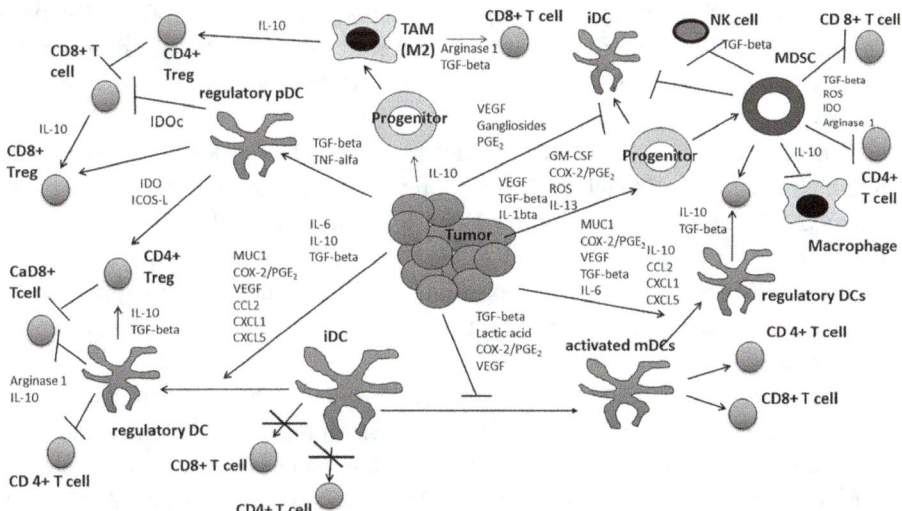

Figure 6.1: Leucocytes function and the tumor microenvironment. Tumors secrete a variety of factors that inhibit immune response. The specific microenvironment altered the function of cells from non-specific and adaptive immune system. Modified according to K. M. Hargado, Frontiers in Immunology, 2013.

When challenged by tumors, NK cells transmigrate toward the vessel endothelium to the tumor tissue. However, extracellular hypoxia, low glucose concentration and acidic pH within the tumor tissue are enemies to correct functioning of immune effectors [13]. Thus, the intratumoral NK cells are more immature, incompletely activated, functionally defective or anergic than peripheral NK cells. Additionally, the low frequency of NK cell accumulation is observed at tumor sites because of fibrotic shield surrounding cancer tissue [12]. Once the NK cells penetrate the tumor, they tend to demonstrate more reduced CD16a (receptor for IgG Fc) expression and have lower proliferative and cytolytic activity and cytokine production than blood NK cells [14]. There are some known mechanisms employed by tumors that interfere with NK cell effector function. One of them is interaction of NK cells with immune suppressive cells, including myeloid-derived suppressor cells (MDSCs), tumor-associated macrophages (TAMs) and Tregs. MDSCs in mice with tumors inhibit NK cell development and downregulate NKG2D expression on NK cells [15]. Tregs suppress NK cells by deprivation of IL-2, induction of NK cell apoptosis and generation of IDO-expressing DCs and adenosine that interferes with the signaling of NK cells [16–18].

Another mechanism of tumor defense is secretion of immunosupressive cytokines (e.g. IL-8, TGF-β, IL-1β) and tumor-derived soluble MIC ligands that attenuate NK cell function *via* reduction of activating signals or increase in inhibitory signals delivered to NK cells.

TGF-β produced by TAMs downregulates the surface expression of NKG2D, resulting in reduced NK cytotoxicity [19]. TGF-β is also able to alter ADCC, IFN-γ production, metabolism and mitochondrial function in NK cells [20–24]. Additionally, TGF-β is capable of converting NK cells into non-cytotoxic group 1 innate lymphoid cells [24, 25].

Recent studies show that IL-1β produced by DCs induces NK cell apoptosis [26] but activin-A inhibits NK cell IFN-γ secretion without interfering with NK cell cytotoxicity [27]. The production of soluble MICA/B (NKGD ligand), B7–H6 (natural cytotoxicity receptor [NCR] ligand) and CD155 (DNAM-1 ligand) also contributes to tumor immune escape [28–30]. For example, MICA/B ligands impair expression of NKG2D, whereas B7–H6 downregulates the expression of NKp30-activating receptor in NK cells.

Current research studies are uncertain as to which of the two main circulating monocytes subpopulation (classical on non-classical) represent the main source of TAMs. Mohavedi et al. have demonstrated that the major precursors of these cells are classical monocytes. Contrasting results have been demonstrated by MacDonald et al. who showed that TAMs mainly differentiate from non-classical monocytes. These conflicting findings suggest that the phenotype of monocytes (precursors of TAMs) and their activation status and functions strictly depend on the type of tumor, stage, size, tumor location and TME [2], [31], [32]. M1 macrophages occur in the TME already in the early stage of tumor development, and they differentiate from monocytes under the influence M-CSF, chemokine (C–C motif) ligand (CCL2) and CXCL12 chemokines, secreted by cancer cells and by surrounding T cells (Th1) and NK cells [33]. M1 macrophages secrete anti-tumor cytokines such as IL-12 and IFN-γ synergistically with NK cells. They also secrete the TNF-α and nitric oxide (NO). As a result, it releases tumor-associated antigens (TAA) that macrophages present to T lymphocytes [34]. Then, in the tumor, and especially in fragments with insufficient oxygen supply, M2 macrophages appear, which have a tumor-promoting effect. M2 macrophages secrete IL-10, TGF-β and CCL22, stimulate immunosuppression and activate of Tregs (Figure 6.1) [33]. M2 macrophages participate in the induction of angiogenesis through the secretion of VEGF, IL-8, IL-23 and platelet-derived growth factor. In addition, they secrete proteins with lytic properties such as metalloproteinases (MMP-2, MMP-7 and MMP-9) [34]. Additionally, it was confirmed that M2 macrophages promote tumor initiation, progression and survival. M2 cells also inhibit anti-tumor immune response, and M2 macrophage infiltration in tumors has been correlated with poor prognosis [35]. TAMs affect most aspects of tumor cell biology and drive pathological phenomena including tumor cell proliferation, tumor angiogenesis, metastasis, immunosuppression and drug resistance [36]. Furthermore, numerous findings have revealed that M2 macrophages are also responsible for resistance to classical anti-tumor treatments (chemotherapy or radiotherapy), and they also limit the efficacy of new immunotherapies (anti-PD-1) [37]. Therefore, attention was paid to the possibility of using TAMs in cancer therapy.

DCs activated by tumor Ags induce tumor-specific CTL that can lead tumor rejection (elimination phase). Several factors from both tumors and tumor-associated cells within the TME are capable of altering DC development and function (immunoescape phase). This is the main reason of tumor-associated immunoescape. Constant exposure to tumor's Ags and the TME factors blocks DC differentiation and can induce immunosuppressive cells (e. g. MDSCs, M2; Figure 6.1).

A number of factors associated with TME such as IL-6, M-CSF, VEGF, COX-1, COX-2 and PGE$_2$ can completely block DC development from CD34$^+$ progenitors [38–42]. In addition, tumor cells' presence affects the differentiation program of DC precursors. This alteration promotes an accumulation of MDSCs. These cells are associated with various cancer types. The induction of MDSCs is driven by tumor-related factors including VEGF, TGF-β, IL-13, GM-CSF, PGE2 and reactive oxygen species (ROS). MDSCs induce T-cell tolerance and Tregs. MDSCs have been shown to release IDO and arginase I which in turn induces the T cells with tolerogenic potential that affect DC maturation [43–45]. Moreover, tumors can also drive the differentiation of DC precursors into other immunosuppressive cells, where the most common ones are M2 TAMs [46–50].

DCs in TME may overexpress genes that affect their regulatory character. For example, increasing the expression of transcription factor STAT3 is associated with prevention of TIDC maturation. Additionally, a number of studies demonstrate the interference of DCs with immunosuppressive factors including IL-10, TGF-β, IDO, arginase 1 and PGE2 in TME. Consequently, we observe the accumulation of regulatory DCs (regDCs) and iDCs. These cells are not effective in Ag presentation and T-cells activation [38–40], [43], [44].

DCs in TME may show increased expression of programmed cell death-ligand 1 (PD-L1). The interaction of PD-L1 with programmed cell death protein-1 (PD-1) on the surface of T cell–infiltrating tumor tissue induces their apoptosis. Such a mechanism of inhibiting the anti-tumor response is observed in ovarian cancers [51, 52], [53]. Blocking the PD-1 molecule by anti–PD-1 antibodies resulted in a decrease in tumor mass and an increase in T-cell immune response [52].

Tumor presence has an effect on the immune cells: it is capable of altering and inhibiting immune cell activation, phenotype and function. Therefore, maintaining NK cell, macrophage and DC anti-tumoral activity is crucial in restoring balance within the immune system.

6.3 NK cell–based immunotherapy in treating cancers

Enhancement of the cytolitic effector function of NK cells to eliminate tumor cells may be beneficial because NK cells are able to eliminate tumor cells in an MHC restriction–independent manner, NK cells are activated and respond to variety of different stimuli (stress proteins, viral proteins, downregulated MHC class I, mismatched MHC class I) and NK cells do not induce immunopathologic disease such as graft versus host disease [54]. The enhancement of cytotoxic potential of NK may be an attractive therapeutic strategy for treatment of a variety of tumors.

6.3.1 Novel strategies for restoring NK cell anti-tumoral activity

There are two parallel strategies for restoring NK cells' anti-tumor responsiveness: increasing factors that activate NK cell function or decreasing inhibitory factors [55]. Blocking the NK cell inhibitory receptors with antibodies or haploidentical NK transplantation reduces the inhibitory signals delivered to NK cells. In contrast, blocking the soluble NK-activating ligands restores NK cell–activating signals. Strategies including the use of therapeutic antibodies, bispecific proteins, chimeric antigen receptor (CAR)–expressing NK cells and genetically modified NK cells enhance the specificity of NK cells.

NK cell adaptive immunotherapy provides activated NK cells to tumor side or replaces functionally defective NK cells in patients with cancer. Cytokines are used to enhance the survival of NK cells in tumor. In this review, we will review recent studies in this field.

6.3.2 Anti-checkpoint receptor monoclonal antibodies

Checkpoint receptors of the NK cells are a group of inhibitory receptors that reduce the effector function of these cells. These checkpoints induce exhausted phenotype of immune cells. NK cells express many checkpoint receptors. It is well known that cancer exploits them to reduce anti-tumor immunity. Thus, some of the checkpoint receptors have been targeted by cancer immunotherapy. Blocking the NK cell checkpoint receptors with monoclonal antibodies minimizes the NK cell inhibitory signals and auguments NK cell function.

Inhibitory killer cell immunoglobulin–like receptor–**blocking antibody.** Inhibitory killer cell immunoglobulin–like receptors (KIRs) exhibit a long form with an immunoreceptor tyrosine–based inhibition motif (ITIM) in the cytoplasm and engage to human leukocyte antigen (HLA) class I groups HLA-C1, C2 and Bw4-inhibiting NK cell activation [56]. The inhibitory KIR recognition of HLA class I ligands on healthy cells prevents these cells from NK cell killing. Tumor cells that preserve or upregulate HLA class I can also inhibit NK cell killing. Thus, blocking the interaction between KIRs and HLA class I, using anti-inhibitory KIR monoclonal antibodies, is a good method to enhance NK cell–mediated killing. Lirilumab (also known as IPH2102) is a human IgG4 monoclonal antibody that engages with high affinity to three major KIRs such as KIR2DL-1, 2 and 3 and blocks their interaction with HLA-C1 and C2 [57, 58]. Moreover, research continues to develop new anti-KIR monoclonal antibodies. Recently, anti-KIR3DL IPH4102 was generated.

In vitro, lirilumab augments NK cell–mediated cytotoxicity toward lymphoma, leukemia and multiple myeloma (MM) [58]. It was effective in phase I trials only in patients with relapsed/refractory MM [59], but a phase II trial was halted because of lack of efficacy [60]. Interestingly, the effect of this therapy depended on the frequency of KIR2DL-positive NK cells [61]. Lirilumab treatment did not show efficiency in acute myeloid leukemia (AML) patients in phase I trials [62].

Lirilumab can be administrated as a monotherapy or as combination therapy with multiple agents. The therapeutic effect of lirilumab was more effective in combination with immunomodulator lenalidomide in phase I trials in MM patients [61]. In combination with rituximab (anti-CD20 antibody), lirilumab treatment has improved anti-lymphoma activity of NK cells [63]. Additionally, lirilumab enhances the effect of daratumumab (anti-CD38 antibody)-mediated ADCC in the killing of MM cells [64] and nivolumab (anti-pd-1 antibody)-mediated killing of squamous cell carcinoma of the head and neck. KIR blockade therapy in combination with KIR-mismatched NK cell transplantation also has improved NK cell responsiveness.

CD94/NKG2A-blocking antibody. CD94/NKG2A is heterodimeric inhibitory receptor of NK and T cells that recognizes peptide-bound HLA class I histocompatibility antigen (HLA-E) in competition with the activating receptor NKG2C. It belongs to the C-type lectin family. CD94/NKG2A is a well-known inhibitor of NK cell anti-tumor activity in different types of tumors [65, 66]. In tumors, HLA-E is increased to evade immune cell recognition [67, 68]. The NKG2A-blocking antibody monalizumab (IPH2201) has been generated. *In vitro*, it enhances NK cell and T-cell anti-tumor immunity in a variety of tumor models [69], [70]. Several clinical trials are also ongoing. Monalizumab can be administrated as a monotherapy (in patients with hematological malignancies) or as combination therapy with cetuximab (anti–epidermal growth factor receptor [EGFR]), afatinib (tyrosine kinase inhibitor) and palbociclib (selective *inhibitors of* cyclin-dependent kinase 4 and 6) in patients with head and neck neoplasmas and durvalumab (anti-PD-L1) in colorectal cancer [71]. Results of the combination trials report safety profiles and acceptable efficacy [69], [71].

Recently, NKG2A expression blockers have also been generated and used in preclinical studies. It consists of a scFv against NKG2A and an ER/Golgi retention peptide and prevents the transport of NKG2A to the cell surface [72]. Interestingly, blockers of NKG2A surface cell expression *in vitro* enhance NK cell cytotoxicity more than anti-NKG2A-blocking antibody [72].

LIR-blocking antibody. The leukocyte immunoglobulin–like receptor (LIR), also known as immunoglobulin-like transcript (ILT), is an NK cell–inhibitory receptor that recognizes HLA-G on fetal cells and several tumor cells [73]. However, blocking the NK cell receptor LIR-1 did not enhance NK cytotoxicity toward MM cells [74]. Perhaps combining therapies will be more efficient, yet many studies need to be conducted in order to test this hypothesis.

TIM-3-blocking antibody. T cell immunoglobulin and mucin-containing domain (TIM)-3 is an inhibitory receptor that is constitutively expressed on NK cells and is upregulated by a variety of activation stimuli (e.g. cytokine) [75, 76] and in tumor patients [77–80]. It binds to galectin-9, phosphatidylserine, high-mobility group box 1 protein (HMGB1) and carcinoembryonic antigen–related cell adhesion molecule-1 (CEACAM1) [81–83]. It is one of the markers of NK cells that produces IFN-γ and releases cytotoxic granules [75, 76] as well as NK cell exhaustion [84]. NK cells in melanoma patients [85] and patients with advanced lung adenocarcinoma [85] express higher level of TIM3 than NK cells from healthy donors. Interestingly, the expression of TIM3 on NK cells correlated with reduced expression of activating receptors (NKG2D, NKp46 and DNAX accessory molecule-1 [DNAM-1]), enhanced expression of the inhibitory receptors (KIR3DL1 and KIR2DL3) and decreased patient survival [85]. Additionally, TIM3-positive NK cells are unresponsive to stimulation. There is currently a phase I trial that investigates anti–TIM-3 blocking antibody as a monotherapy, as well as in combination with anti-PD-1 antibody to treat patients with a variety of advanced solid tumors (NCT02817633).

TIGIT-blocking antibody. T-cell immunoreceptor with immunoglobulin and immunoreceptor tyrosine-based inhibitory motif domains (TIGIT) is an inhibitory receptor that interacts with CD155 (poliovirus receptor [PVR]) and CD112 (poliovirus receptor-related 2 [PVRL2], also known as nectin 2) [86–88]. TIGIT and its ligands are upregulated in a variety of different cancers. It inhibits immunosurveilance by direct inhibition of activating receptor CD226 (DNAX accessory molecule-1, DNAM-1) and competition with DNAM-1 and CD96 (inhibitory receptor) for the same ligands expressed on tumor cells. Antibody-mediated blocking of the TIGIT increases NK cell–mediated cytolytic activity (including ADCC) against tumor cells [87]. However, most patients still do not respond to anti-TIGIT therapy. Currently, there are multiple phase I trials investigating the anti-TIGIT monoclonal antibody as a monotherapy and in combination with anti-PD-1/PD-L1 antibodies (NCT02913313, NCT03628677, NCT02794571) in advanced solid tumors and metastasis. Additionally, anti-TIGIT therapy improves the anti-tumor effect of anti-HER2 monoclonal antibody trastuzumab [89].

CD96-blocking antibody. CD96 is an inhibitory receptor that has a similar effect to TIGIT. It competes with TIGIT and DNAM-1 for PVR binding. However, its affinity is weaker than that of TIGIT but higher than that of DNAM-1 [90]. The blocking the CD96 appears to be a promising strategy in using therapeutic antibodies in cancer patients. Previous research using blocked CD96 in murine tumor models has promoted NK cell anti-metastatic activity [91].

CTLA-4–blocking antibody. CTLA-4 is the first checkpoint receptor expressed by tumor-infiltrating NK cells in murine models of lymphoma and melanoma and is upregulated by exogenous IL-2 and upon bindings to its ligands B7-1 and B7-2 on APCs. CTLA-4 engagement inhibits production of IFN-γ by NK cells [92] and suppresses effector function of NK cells. Anti–CTLA-4 antibody (ipilimumab) can enhance NK-mediated killing of melanoma cells through ADCC. *In vivo*, this checkpoint inhibitor increases the NK cell's production of TNF-α [93]. Also, anti–CTLA-4 therapy may act indirectly on NK cells by targeting Tregs. Tregs control NK cell's activity by limiting the amounts of IL-2. Thus, blocking CTLA-4 may synergisticallly enhance NK cell's cytotoxicity by targeting CTLA-4 on both NK cells and Tregs.

PD-1-blocking antibody. PD-1 is the second checkpoint receptor expressed by resting and activated NK cells. It has two ligands: PD-L1 and PD-L2. PD-1 is induced upon prolonged stimulation by NK cell–activating receptors or by MHC I–deficient tumor cells and is upregulated by IL-2 [94, 95]. PD-1$^+$ NK cells express both activation and apoptosis markers. PD-1 mediates functional exhaustion of activated NK cells, including reduced proliferation, cytokine production and reduced cytolitic activity [95–97].

Anti–PD-1 antibody (pembrolizumab) can enhance NK cell–mediated cytotoxicity against PD-L1$^+$ myeloma cells [94]. This antibody enhances the interaction between NK cells and PD-L1$^+$ myeloma cells and stimulates the production of granzyme B and IFN-γ by NK cells. Pembrolizumab is safe and demonstrates clinical efficacy in patients with advanced hematologic malignancies [98]. It can be used as a monotherapy for classic Hodkins's lymphoma (HL) [99], as a combination therapy with lenalidomide for the MM and in combination with rituximab (anti-CD20 antibody), which activates NK cell–mediated ADCC in follicular lymphoma [100].

LAG3-blocking antibody. Lymphocyte-activation gene 3 (LAG3) is an inhibitory receptor expressed on NK and T cells. It binds to MHC class II, lymph node sinusoidal endothelial cell C-type lectin (LSECtin) [101] and fibrinogen-like protein 1 (FGL1) [102] that are expressed on tumor cells. LAG3 function in NK cells is still unclear. NK cells from LAG3-deficient mice show impaired cytotoxicity toward cancer cells [103]. Blocking the LAG3 did not impact NK cells' cytotoxicity [104]. Antibody-targeting LAG3 are currently in clinical studies [105].

In conclusion, activation of NK cells depends on the balance of activating and inhibitory signals. Thus, blocking the inhibitory signals alone will not restore NK cell activity to kill tumor cells. It will only be successful if combined activating signals occur.

6.3.3 Other antibodies

Soluble MHC class I polypeptide–related sequence A (sMICA)-blocking antibody. A well-known mechanism of immune escape from NK cells is the shedding of NKG2D ligands such as MICA/MICB from tumor cells [106]. Soluble MICA/MICB blocks activating receptor NKG2D [28]. Thus, antibody (IPH4301) anti-soluble MICA/B ligands overcomes the suppression of receptor

NKG2D by soluble MICA/MICB. Additional effect of this therapy is the induction of ADCC effects on MICA/MICB expressed by tumor cells.

CD39- and CD73-blocking antibodies. CD39 and CD73 are membrane-bound extracellular enzymes that are responsible for catalization of transformation of ATP to AMP, which is upstream of indoleamine 2,3-dioxygenase (IDO) and adenosine (ADO). IDO and ADO are inhibitors of NK cell–mediated cytotoxicity [107]. IDO suppress the expression of activating receptors NKG2D and NKp46 and inhibit NK cell activity [18]. ADO has also been found to reduce lytic function of activated NK cells and their anti-tumor activity [108]. CD39 and CD73 are expressed on Tregs, NK cells and some tumor cells [109]. It is known that CD73 expression on tumor-infiltrating NK cells is associated with suppression of NK cell function. CD39- and CD73-blocking antibodies restore NK cell activity in ovarian cancer [110].

Anti-TAA antibody. TAA-targeting antibodies bind TAA via the Fab fragment and induce immune response *via* the CD16a receptor (FcγRIIIa). The therapeutic effects of these antibodies depend on ADCC effect, complement-dependent cytotoxicity and the opsonization. ADCC is mediated by the activating CD16a receptor that is expressed on NK cells. Anti-TAA antibodies enhance NK cell–mediated ADCC against tumor cells. Additionally, they target TAA including HER2 (trastuzumab), EGFR (cetuximab) and CD20 (rituximab). Preclinical research has shown that some of the anti-TAA antibodies (rituximab, obinutuzumab, trastuzumab, cetuximab) induce the ADCC effect but others (elotozumab) induce both ADCC of cancer cells and activation of NK cells in patients with MM. Human studies reveal that the therapeutic efficacy of targeting antibodies correlate with NK activity. NK cell activation mediated by CD16a also depends on polymorphism in CD16a (that determines the affinity for IgG antibodies) and additional NK cell receptors (either activating or inhibitory). It is known that activating receptors enhance the efficiency of ADCC, while inhibitory receptors may limit the efficacy of targeting antibodies. NK cell–mediated ADCC also depends on the binding affinity of therapeutic antibodies to CD16a [111].

Anti-TAA antibodies in combination with NK cell infusion (*ex vivo* expanded and activated NK cells) improve ADCC effect specially in patients with poor NK cells. Third-generation monoclonal antibodies that bind with high affinity to CD16a receptor enhance ADCC effect. However, chronic stimulation of NK by therapeutic antibodies induces NK cell hyporesponsiveness, which is responsible for clinical resistance.

6.3.4 Agonist antibodies

Some agonist antibodies are capable of restoring NK cell anti-tumor activity. Anti-CD137 (also known as 4-1 BB) agonist antibody augments tumor-specific cytotoxicity, 4-1 BBL and IL-15 on K562 cell lines and NK cell expansion [112, 113]. Stimulants of NK-activating receptors such as CD16a and NKG2D are also applicable for restoration of NK cell activity [114]. CD16a and NKG2D are present not only on NK cells, so the application of CD16a and NKG2D agonists may cause unknown severe systemic toxicity.

6.3.5 Modified antibody molecules

Modified antibody molecules in the form of single-chain variable fragments (scFvs) are also used as an immunotherapy technique aiming to treat cancer patients. These modified antibody molecules display two binding fragments: one specific to NK cell's activating receptor and the other specific to a TAA. These antibody molecules known as bispecific killer engagers (BiKEs) promote an immune synapse between NK cells and tumor cells and induce ADCC-mediated lysis of target cells. They are small in size, so they have enhanced biodistribution in comparison to monoclonal antibodies [115].

One example utilized anti-CD16 scFV spliced to anti-EpCAM scFv [116]. Other BiKEs have been generated to engage CD16 with CD19/CD20 (B cell non-Hodgkin's lymphoma), CD30 (Hodgkin's lymphoma, HL), CD33 (myelodysplastic syndrome, MDS and AML), CD133 (colorectal cancer), HER2 (breast cancer), EGFR (EGFR-expressing cells) and CEA (colon carcinoma) [116–123]. Some of them have double target. As an example, CD16xCD33 BiKE (1633 BiKE) links to both $CD33^+$ MDS cells and $CD33^+$ MDSCs and activates NK cells against primary MDS and disrupts MDSCs-induced suppression [120]. Additionally, it targets CD33 on AML and enhances NK cell cytotoxicity and cytokine production against AML cells [124].

Some of the BiKEs have already been applied in clinical trials (NCT01221571, NCT02321592 and NCT02665650) to treat patients with HL. CD16xCD30 BiKEs showed tolerance and good safety in phase I trials [125] and rescues of impaired NK cell function in patients with HL [117] but not in patients with downregulated expression of CD16a.

BiKEs were extended to trispecific killer engagers (TriKEs) and tetraspecific killer engagers (TetraKEs). In addition to binding to TAA and CD16, TriKEs also link to IL-15 receptors on NK cells. The incorporation of modified IL-15 into the BiKEs improves the activation, proliferation and survival of NK cells. One of the TriKEs is 161533 TriKEs, which enhances NK cell cytotoxicity against AML tumor cells in patients post allo-hematopoietic stem cell transplantation (HSCT). There are phase I and II trials to study the safety and effectiveness of 161533 TriKEs to treat patients with CD33⁺ high-risk MDS and relapsed or refractory AML (NCT03214666, clinical trials.gov). The fully humanized 1615EpCAM TriKE specific against EpCAM⁺ ovarian cancer cells is another example of TriKEs [126]. Recently, this TriKE was improved to TetraKE, with additional specificity against cancer stem cell marker CD133 [127].

Considering all the examples above, modified antibody molecules demonstrate enormous potential in treating cancer patients.

6.3.6 Fusion proteins

Recently, fusion proteins (also known as bispecific immunoligands) have been undergoing development. These proteins are a combination of one NK cell–activating receptor (e.g. NKG2D, NKp30 or NKp80)–associated ligand and one TAA-associated ligand or antibody fragment. As an example, rG7S-MICA fusion protein cross-links NK cells with CD24⁺ HCC cells [128], while bispecific immunoligand ULBP2-aCEA cross-links NK cells with colon carcinoma cells [129], resulting in both proteins binding to cancer cells via NKG2D. AICL:HER2-scFv and B7-H6:HER2-scFv redirect NKp30- and NKp80-expressing NK cells to cancer cells [130]. NKG2D-IL-15 protein is another fusion protein that links with gastric cancer cells *via* a fragment of NKG2D domain and induces NK cell expansion by trans-presentation of IL-15 [131, 132]. In summary, fusion proteins trigger NK cell recruitment and NK cell–mediated lysis of tumor cells and enhance ADCC.

6.3.7 Immunomodulatory drugs

Some pharmacological agents, including lenalidomide, histone deacetylase, demethylating agents, DNA-damaging agents, bortezomib, histamine, imatinib and sorafenib, are applicable for restoration of NK cell–mediated anti-tumor activity [22]. Some of them can induce Fas and TRAIL receptor (bortezomib) or NK cell–activating receptor (lenalidomide). The other pharmacological agents may induce the activation of NK cell receptor (histone deacetylase, demethylating agents, DNA-damaging agents). Imatinimab activates NK cells *via* improving the NK/DC cross talk. Histamine prevents the downregulation of NK-activating receptors such as NKp46 and NKG2D. Monomethyl fumarate can enhance the cytotoxicity of CD56⁺ NK cells against tumor cells through degranulation and the upregulation of NKp46 and CD107a [133]. IL-15 superagonist augments NK cell function against ovarian cancer [134].

6.3.8 Cytokines

Certain cytokines such as IL-2, IL-12, IL-15, IL-18 and IL-21 are capable of augmenting NK cell anti-tumor activity. These cytokines are critical for NK cell differentiation, proliferation and survival. Recombinant IL-2 appeared to be a promising therapeutic biologic agent for metastatic cancer treatment as it is capable of triggering lymphokine-activated killers (LAKs) [135]. However, it did not work out after all due to its high potential to induce the Treg generation [136]. The other cytokines are still in preclinical studies as they carry multiple other functions in the immune system and hence could exhibit some unwanted adverse effects upon treatment. These cytokines in therapy should therefore be investigated in clinical trials with the greatest caution.

6.3.9 Adoptive transfer of NK cells

The major objective of NK cell–based adoptive cellular immunotherapy is to revive the patient's innate immune surveillance and control of tumor by infusion of activated NK cells. This type of therapy has shown the efficacy in treating hematopoietic malignancies.

There are a many sources for obtaining NK cells *ex vivo*. The most popular is peripheral blood mononuclear cells (PBMCs). Umbilical cord blood and induced pluripotent stem cells are other sources for obtaining NK cells [137]. The NK cell expansion rate and purity of NK cells depend on enriching the NK cells. Enriching the NK cells *via* isolation from PBMCs increases purity but limits the expansion rate. However, expansion without prior NK cell isolation increases the expansion rate but reduces purity. The NK cell expansion also depends also on used NK cell stimulators (cytokines, feeder cells, monoclonal antibodies). It is known that feeder cells engineered with membrane-bound IL-15 and 4-1BB strongly expand highly cytotoxic NK cells [138]. Anti-CD16 and anti-CD52 monoclonal antibodies also increase the expansion of NK cells [139].

NK cell–based adoptive cellular immunotherapy is based on the adoptive transfer of autologous LAK cells with high-dose IL-2 therapy, cytokine-induced killers (CIKs) that arise from PBMC cultures with stimulation of anti-CD3 monoclonal antibodies, IFN-γ and IL-2, autologous NK cells, allogenic NK cells, NK cell lines and genetically modified NK cells.

The use of LAKs is the most primitive strategy for NK cell adoptive transfer. Due to limited anti-tumor efficacy and the toxicity induced by the systemic administration of IL-2, this strategy has been abandoned.

Another source of NK cells for adoptive transfer is *ex vivo* expanded autologous NK cells. An example of this strategy is adoptive transfer of CIKs. These cells arise from PBMC cultures with stimulation of anti-CD3 monoclonal antibodies, IFN-γ and

IL-2. CIKs are characterized by $CD3^+$ $CD56^+$ phenotype and have enhanced cytotoxic activity compared to LAKs cells against cancers.

Allogenic NK cells are another option as a source of NK cells. Among the strategies of allogenic NK cell adoptive transfer, human haploidentical NK cell transplantation with KIR mismatch is most efficient in treating hematopoietic malignancy. This approach is very effective because KIR mismatch increases the NK cell–mediated killing of tumor cells [140].

NK cell lines are another source for NK cell adoptive transfer. The advantages of these cells is their stability and sufficient quantity of therapeutic cells required for clinical demands. Among the established NK cell lines, NK-92 and NKG are the most promising ones and have been used in clinical trials. In 2017, activated NK-92 therapy was qualified as an orphan drug treatment for Merkel cell carcinoma.

Genetically modified NK cells are the newest strategy of NK cell adoptive transfer therapy. Genetic modification of NK cells creates more efficient anti-tumor NK cells such as NK cells with autocrine cytokines, NK cells with overexpression of activating receptors, NK cells with reduced expression of inhibitory receptors and NK cells with chimeric antigen receptor (CAR) (CAR-NK cells). CAR arming retargets TAA expressed on tumor cells, CD16a arming improves ADCC effects and cytokine gene transfer (IL-2, IL-12, IL-15 and stem cell factor) promotes NK cell survival.

CAR-NK cells redirect NK cells to kill tumor cells. The major sources of these cells are primary NK cells, NK-92 and YT cell lines. Many studies investigate CAR-NK cells redirecting for various TAAs such as CD19 [141], CD20 [142], CD138 [143], GD2 [144], EGFR [145], human epidermal growth factor receptor 2 (HER2) [146], epithelial cell adhesion molecule (EpCAM), CD33 and carcino-embrionic antigen (CEA) [147, 148]. CAR-NK cells have shown promising preclinical efficacy in the control of tumors. Using NK-92 cell lines involving Epstein-Barr virus (EBV) infection for constructing NK-CAR cells is also necessary to irradiate these cells before adoptive transfer [149]. CAR-NK cells based on NK-92 cells lack CD16a, so this technique cannot be used in combination with therapeutic antibodies.

Arming the NK cells with a high-affinity variant of CD16a is another strategy to enhance NK cells with anti-tumor function. NK-92Fc with high-affinity CD16 has shown promising efficacy in ADCC.

Another approach for genetic modification is to reduce the expression of inhibitory NK cell receptors such as KIRs. NK-92 can be used in this strategy. However, deletion of inhibitory response is always associated with hyporesponsiveness of these cells.

6.3.10 NK cells and combination immunotherapy

Adoptive transfer with *ex vivo* expanded NK cells combined with therapeutic antibodies improves the ADCC effect. In some clinical trials, TAA-targeting antibodies were administered in combination with NK cell infusion. However, chronic stimulation of NK by therapeutic antibodies impairs NK-cell–mediated cytotoxicity, and hence, it became one of the reasons for clinical resistance development.

6.3.11 Oncolytic virus–based therapy

Another approach is use of oncolytic viruses (OVs) that specifically infect and lyse cancer cells. Additionally, OVs induce systemic anti-tumor immune response. With their potential to deliver immune-modulating factors (cytokines, modified antibody molecules) to TME [150, 151], these are an interesting candidate to consider in terms of novel therapeutic approaches, for example, an attenuated herpes simplex virus (HSV) expressing GM-CSF was US Food and Drug Administration (FDA) approved for metastatic melanoma [151]. A variety of OVs are at various stages of preclinical studies, and some of them have entered clinical trials [152]. Looking into combining OVs and checkpoint inhibitors has indicated remarkable results [153].

OV-based therapy induces innate immune response by the expression of death-associated molecule pattern (DAMP), pathogen-associated molecule pattern (PAMP), toll-like receptor (TLR) engagement and induction of the immune cell death in infected tumor cells [150]. It also modulates NK cell ligand expression on cancer cells and drives NK cell–mediated clearance [154, 155]. Additionally, OV-based therapy increases DC-mediated NK cell activation [156, 157].

To enhance anti-tumor immunity, OVs have been engineered to express NK-stimulating cytokines (IL-12, IL-15, IL-18, Regulated on Activation, Normal T-cell Expressed and Secreted (RANTES) and GM-CSF) [158–160] and cytokine superagonist such as IL-15 superagonist (IL-15 complexed with its receptor IL-15Ra). Recently, infected cell vaccine (IVC) consisting of irradiated autologous tumor cells previously infected *ex vivo* with cytokine-expressing OV has been developed. This IVC led to complete regression of tumors [161] and allowed for controlled and safe release of cytokines at the site of tumors.

OV-based therapy can be used as a monotherapy or in combination with CAR-NK cells to treat cancer. Recently, reports show promising results using combination therapy of EGFR-CAR-NK cells and HSV 1 for breast cancer brain metastasis [162].

This approach has promising therapeutic advantages, but premature clearance of an OV by the host immune system is one of the substantial limitations.

6.4 Macrophages in cancer treatment

The development of novel anti-tumor therapeutic strategies using of TAMs focus on depleting TAMs and/or on reprogramming their tumor-promoting effects to restore a favorable anti-tumor immune response within the tumor tissue [163].

6.4.1 Depleting TAMs

Strong proofs indicate that the accumulation of macrophages in tumors is due to the continuous recruitment of monocytes from the circulation in response to tumor-derived factors (TDFs). The key TDFs include colony-stimulating factor-1 (CSF-1) and chemokine ligands, such as CCL2 (monocyte chemoattractant protein-1: MCP-1) and VEGF. CCL2 has been described as the major TDF involved in monocyte recruitment (CCL2-CCR2 axis). Blocking CCR2 ligands can suppress the accumulation of TAMs in tumors [164]. Currently, CCR2 inhibitors and anti-CCL2 antibodies are used in experimental models in mice as well as in human clinical trials, both as a monotherapy- as well as chemotherapy-associated therapy. In addition, in murine breast cancer models, a rebound effect has also been described using anti-CCL2 treatment, which has been associated with increased infiltration of monocytes in the tumor and consequent acceleration of lung metastasis. Currently, clinical trials are also underway with the use of a dual CCR2/CCR5 antagonist and chemotherapy or nivolumab in patients with metastatic colorectal and pancreatic cancer. Another axis involved in monocyte recruitment and differentiation into TAMs is the CXCL12/CXCR4 axis. Studies on a breast cancer model have shown that CXCL12 expression by tumor cells correlates with its increased invasiveness. Thus, the use of the CXCR4 antagonist (AMD3100) reduces tumor spread and metastasis. AMD3100 is also used in combination with pembrolizumab in patients with refractory head and neck squamous cell carcinoma. In addition, in patients with solid tumors and patients with AML, the CXCR4 antagonist (plerixafor) is also used as a monotherapy or in combination with G-CSF treatment or with chemotherapy. The inactivation of CXCL12 by pegylated L-oligoribonucleotide (Pegol Olaptesed-NOX-A12) is used in MM patients and in combination with pembrolizumab in patients with colorectal and pancreatic cancer [163].

An important axis for reducing the amount of TAMs within a tumor or peritumor tissue is the CSF-1/CSF-1R axis. A humanized antibody (emactuzumab) targeted at CSF-1R is used, both in animal research models as well as in human clinical trials as a monotherapy or in combination with immune checkpoint blockade to treat solid tumors and in combination with cyclophosphamide in ovarian cancer patients. Inhibition of the CSF-1/CSF-1R axis is also used in human clinical studies, including the treatment of patients with metastatic breast cancer and soft-tissue sarcoma, as well as for the treatment of metastatic melanoma. Clinical trials including the combination of radiotherapy with pexidartinib (PLX3397) and temozolomide in the treatment of glioblastoma and androgen deprivation therapy for prostate cancer are also underway [163].

6.4.2 Reprograming TAMs

Currently, several methods are used to switch M2 pro-tumor macrophages into M1 macrophages with anti-tumor properties. The most commonly used are TLR agonists, nucleic acids and monoclonal antibodies [163].

6.4.2.1 Using TLRs in immunotherapy

TLRs are innate immunity pattern recognition receptors that preferentially stimulate the immune response associated with M1-macrophage activity [35]. The agonists of intracellulary located TLRs such as TLR3, TLR7, TRL8 and TLR9 have been used. Currently, imiquimod (TLR7 agonist) is used to treat the squamous and basal cell carcinoma. Additionally, experimental studies demonstrate that using poly I:C (TLR3 agonist) is highly effective in the treatment of murine melanoma models [165, 166]. Similarly, results to poly I:C are observed using the TLR9 agonist CpG in clinical trials for lymphomas in combination with ibrutinib or radiation therapy [166]. Additionally, studies have been conducted with an agonist to TLR7/8. Unfortunately, it has resulted with a high toxicity that included body inflammation, lymphopenia, anemia and flu-like symptoms [167].

6.4.2.2 RNA delivery

Currently, very promising therapies are those that use advanced technologies engaging delivery of messenger RNA (mRNA), small interfering RNA (siRNA) or microRNA (miRNA) as a new strategy, aiming to switch off macrophage activity. As an example to deliver mRNA into TAMs, charge-altering releasable transporters (CARTs) have been developed, using oligo (carbonate-b-α-amino esters), as dynamic carriers [168]. Alternative studies demonstrate the biodegradable polymeric nanoparticles (NPs) encapsulating two mRNAs, where the first mRNA encodes an IFN regulatory factor 5 (IRF5) and the second one encodes the IkB kinase (IKK) kinase that phosphorylates and activates IRF5 (a murine ovarian tumor model, a murine model of glioma and lung metastasis) [169]. siRNA has been used to silence the expression of genes in murine breast cancer cells and lung metastasis [170], but miRNA was used in a mouse sarcoma model which was associated with decreased tumor growth and prolonged survival [171]. It should be highlighted that currently no clinical trials have been initiated using RNA delivery technology to reprogram macrophages on human model.

6.4.2.3 Antibodies to reprogramming TAMs

Pharmacological applications of antibodies have been found to support immune processes, for example, phagocytosis. CD47 inhibitors as a monotherapy have been used in various preclinical advanced solid and hematological cancer models and in combination with anti-CD20 or anti-PD-1 antibodies in the treatment of solid tumors, an aggressive type of lymphoma or myeloma [172, 173]. A very promising result was obtained with the use of agonistic anti-CD40 antibodies both in a monotherapy and in combination with checkpoint immunotherapy or chemotherapy (various murine tumor models) [174, 175].

To conclude, it is very important to note that drastic reduction of TAMs, M1 and M2 macrophages, can be very harmful since all those immune cells play an important role in innate immunity. Therefore, the use of treatment that selectively eliminates or reduces the amount of M2 macrophages appears to be a better and more effective therapy. An example is using CD163 mAbs conjugated with lipid NPs loaded with doxorubicin (DOX) in a mouse melanoma model. Recent studies indicate that reprogramming of the macrophages (M2 to M1), instead of using therapy that aims to completely deplete TAMs, may be more effective and safer for the patient.

6.5 DCs in cancer treatment

There are two main concepts of the DC-based immunotherapy: DCs obtained from an autologous donor (generated *ex vivo*), pulsed with tumor Ag and used as vaccine or activated *in vivo* by exogenous factors.

6.5.1 *Ex vivo* generated DC vaccines

Peripheral blood monocytes as well as $CD34^+$ cells that originate from hematopoietic progenitor cells can both give rise to DC populations [176]. However, isolation of $CD14^+$ monocytes from the blood of a patient is the most common technique to obtain DCs. Such monocytes are stimulated by GM-CSF and IL-4 within 5–7 days. This enables multiplication and transformation of monocytes into iDCs. In addition, cytokines (TNF-α, IL-1β, IL-6), as well as other factors including PGE_2, TLR agonists or poly(I:C), are essential to differentiate iDCs into mDCs [177–182]. One of the most important aspects of DC-based vaccines is their pulsation with an appropriate Ag. Generally, to obtain an effective anti-tumor response, the DCs have been pulsed with isolated or recombinant TAAs (full-length proteins or peptides), transfected with tumor messenger RNA (mRNAs) and transduced with a TAA-coding gene.

Various TAA sources are used, and the selection of a tumor Ag includes unique (mutated) antigens as well as non-mutated self-antigens [176, 183–185]. Precisely defined peptides can elicit a strong anti-tumor response. Among the human tumor Ags recognized by T cells are prostatic acid phosphatase (PAP) – prostate cancer, CEA – many carcinomas, Wilms' tumor antigen – pancreatic ductal adenocarcinomas, HER2/neu – breast cancer, MUC-1 – breast, ovarian, pancreatic carcinomas, gp100 – melanomas, MART – melanomas, MAGE1 and MAGE3 – melanomas, carcinomas, sarcomas, other [177–182, 186].

Whole-tumor cell lysates are also used as potent tumor antigens. Alternatively, dendritomes are created, utilizing the effect of the fusion between activated tumor cells with DCs [176, 183–185]. The advantage of pulsing DCs with peptides obtained from TAA is the ability to combine Ags directly with MHC molecules to create complexes. All proteins or lysates of whole tumor cells must be absorbed by DCs by endocytosis and have to be processed for effective presentation on the DC surface. However, the use of cell lysates has a significant advantage – cells can be induced by many tumor stimulators, simultaneously [176].

Viral vectors are also an alternative method of delivering the genetic information about specific TAA. The introduction of the specific TAA code together with removing the genes responsible for virulation makes this solution very attractive. The use of vectors allows delivering additional genes into the DC genome. The genes can encode cytokines or co-stimulatory molecules that support DC maturation [176], [184]. One of the most commonly used TAA-carrying vectors is lentiviruses owing to their reduced immunogenicity.

Currently, the only FDA-approved immunotherapy is Sipuleucel-T (also known as APC 8015, Provenge) for the patient suffering from metastatic prostate cancer. This vaccine is a carefully designed combination of DCs, B lymphocytes, macrophages and NK cells cultured *ex vivo* together with a recombinant protein consisting of PAP and GM-CSF. The aim of this immunotherapy is the use of DCs to activate specific T-cells against tumors. The outcome of this technique allows for prolonged survival time of those affected with prostatic cancer (4-month – prolonged median survival in phase III trials) as compared to the placebo group [187].

The combination of few therapies can also be used with a very positive outcome. Compared to monotherapy with Sipuleucel-T, the use of combination therapy Sipuleucel-T with anti-CTLA-4 antibodies (ipilimumab) increased the level of cancer-specific antibodies and was well tolerated by patients with metastatic refractory prostate cancer [188].

Extensive clinical research is underway on cell vaccines called dendritic-cell vaccine (CVAC), which in clinical trials are used in treating patients suffering from prostate (DCVAC/PCa), lung (DCVAC/LuCa) or ovarian cancers (DCVAC/OvCa). The essence of the DCVAC vaccine is autologous DCs generated *ex vivo*, which are pulsed with cancer cells killed by high hydrostatic pressure [189], [190]. In phase I/II clinical trials in which DCVAC/PCa vaccines were used in combination with docetaxel chemotherapy, prolonged survival was observed in patients with metastatic refractory prostate cancer [191]. Currently, the third phase of clinical trials on the use of DCVAC/PCa vaccines in patients with prostate cancer is in progress. Moreover, DCVAC/LuCa decreases the percentage of Tregs and induced tumor antigen–specific $CD4^+$ and $CD8^+$ T-cell responses *in vitro* that produce IFN-γ. The above represents preclinical trials for ongoing phase I/II NSCLC clinical trial DC vaccination with chemotherapy (NCT02470468) [190].

It is therefore evident that there is great potential for these vaccines in targeting cancerous cells in diseased patients. However, they also have many limitations. Obtaining DC vaccines requires very tedious procedures, especially when autologous vaccines would have to be prepared in clinical conditions, many times or on a large scale. In addition, *ex vivo* DC culture conditions do not guarantee that their full potential is achieved.

The cause of failure of this vaccination may be also late stage of cancer and the presence of some cells (Treg, regDC) in the TME which deactivates the immune response.

6.5.2 Targeting antigen to DCs *in vivo*

A major breakthrough in the field of immunotherapy was research on the delivery of Ags for DCs activation *in vivo* [192]. This attempt to stimulate and direct the action of DCs occurring in the patient's body seemed to be an attractive opposite to the costly and laborious method of generating DCs in *ex vivo* conditions [193]. The pioneering studies of Ralph Steinman and Michel Nussenzweig have demonstrated the effectiveness of coupling Ags and antibodies specific for DCs surface receptors such as DEC-205 or DCIR on DCs *in vivo* [192], [194, 195]. The first Ag combined with the anti–DEC-205 antibody was peptide from lysozyme from hen egg, which predominantly induces a $CD4^+$ T-cell response [192–194], [196].

Combining Ags with appropriate carriers makes them easier to present to target cells. Such carriers may be monoclonal antibodies directed against precisely defined surface markers located on DCs [184]. Many antibodies against surface receptors on DC have been used to deliver Ag and induce anti-tumor responses, including anti–DEC-205, anti-CD40, anti-Clec9A, anti–DC-SIGN, anti-DCIR anti-XCR1. Besides, the antibodies conjugated with Ag co-administration of activating agents (adjuvants) are required. Some of the adjuvants used to aid in generating an immune response are anti-CD40 monoclonal antibody, poly I:C, inhibitor IDO and recombinant Flt3L [196–198]. Currently in clinical trials anti-DEC205 antibodies, NY-ESO-1tumor Ag, and poly-ICLC adjuvant (NCT00948961; phase I/II), decitabine (NCT01834248; phase I), inhibitor IDO (NCT02166905, phase I/II) or recombinant Flt3L (NCT02129075; phase II) are applied [184]. Clinical trials of anti–DEC205-NY-ESO-1 are currently ongoing in many types of tumor including ovarian cancer (NCT02166905), AML (NCT01834248) and various solid cancers (NCT01522820, NCT02661100) [199–201].

Other vaccinations that are in clinical trials (phase III, NCT00089856; NCT00133224) are GVAX. The essence of the allogeneic GM-CSF Secreting Tumor Immunotherapy (GVAX) vaccine is irradiated cancer cells modified for GM-CSF production. These cancer cells are autologous, or their source could be allogenic cell lines. GM-CSF can attract macrophages, granulocytes, T lymphocytes and DCs to tumor tissue and stimulate these cells *in situ*. The use of this vaccine in patients suffering from pancreatic, prostate and skin (melanoma) cancers induced both humoral and cellular anti-tumor response [176], [196].

Various strategies are used to obtain DCs in an *ex vivo* model for the preparation of cell vaccines. In addition, *in vivo* supporting of DCs by antibodies and other factors can induce anti-tumor immune response. Vaccines containing DCs are combined with other immunotherapy (e.g. blocking PD-1 or CTLA-4 antibodies) or with conventional treatments (e.g. chemotherapy). Despite the tremendous progress that has been made in these therapies in the last few decades, there are still many unsolved problems associated with the use of DCs.

6.6 Conclusions

The innate cells including NK cells, macrophages and DCs are important tools in combating cancers. They can rapidly recognize and eliminate transformed and malignant cells, making them an ideal target for cancer immunotherapy. Therapeutic modalities such as controlled reactivation of NK cells and DCs as well as reprogramming of M2 macrophages or selectively eliminating M2 macrophages appear maybe effective and safe for the patient. Reactivation of NK cells can be effective especially in the scenario where T cells fail. However, more research is necessary to effectively exploit the full power of these cells against cancer.

Author contribution: All the authors have accepted responsibility for the entire content of this submitted manuscript and approved submission.
Research funding: None declared.
Conflict of interest statement: The authors declare no conflicts of interest regarding this article.

References

1. Richards DM, Hettinger J, Feuerer M. Monocytes and macrophages in cancer: development and functions. *Cancer Microenviron* 2013;6:179–91. https://doi.org/10.1007/s12307-012-0123-x.
2. Lee H-W, Choi H-J, Ha S-J, Lee K-T, Kwon Y-G. Recruitment of monocytes/macrophages in different tumor microenvironments. *Biochim Biophys Acta Rev Canc* 2013;1835:170–9. https://doi.org/10.1016/j.bbcan.2012.12.007.
3. Anisiewicz A, Okła K, Wawruszak A. Tumor associated macrophages-origin, characteristic and importance in breast cancer. *Rev Res Cancer Treat* 2015;1.
4. Lin A, Schildknecht A, Nguyen LT, Ohashi PS. Dendritic cells integrate signals from the tumor microenvironment to modulate immunity and tumor growth. *Immunol Lett* 2010;127:77–84. https://doi.org/10.1016/j.imlet.2009.09.003.

5. Lutz MB, Schuler G. Immature, semi-mature and fully mature dendritic cells: which signals induce tolerance or immunity?. *Trends Immunol* 2002;23:445–9. https://doi.org/10.1016/s1471-4906(02)02281-0.
6. Lutz MB. Therapeutic potential of semi-mature dendritic cells for tolerance induction. *Front Immunol* 2012;3:123. https://doi.org/10.3389/fimmu.2012.00123.
7. Dudek AM, Martin S, Garg AD, Agostinis P. Immature, semi-mature, and fully mature dendritic cells: toward a DC-cancer cells interface that augments anticancer immunity. *Front Immunol* 2013;4:438. https://doi.org/10.3389/fimmu.2013.00438.
8. Lanzavecchia A, Sallusto F. The instructive role of dendritic cells on T cell responses: lineages, plasticity and kinetics. *Curr Opin Immunol* 2001;13:291–8. https://doi.org/10.1016/s0952-7915(00)00218-1.
9. Lijun Z, Xin Z, Danhua S, Xiaoping L, Jianliu W, Huilan W, .et al. Tumor-infiltrating dendritic cells may be used as clinicopathologic prognostic factors in endometrial carcinoma. *Int J Gynecol Canc* 2012;22:836–41. https://doi.org/10.1097/igc.0b013e31825401c6.
10. Ma Y, Shurin GV, Peiyuan Z, Shurin MR. Dendritic cells in the cancer microenvironment. *J Cancer* 2013;4:36–44. https://doi.org/10.7150/jca.5046.
11. Seliger B. Different regulation of MHC Class i antigen processing components in human tumors. *J Immunotoxicol* 2008;5:361–7. https://doi.org/10.1080/15476910802482870.
12. Gajewski TF, Schreiber H, Fu YX. Innate and adaptive immune cells in the tumor microenvironment. *Nat Immunol* 2013;14:1014–22. https://doi.org/10.1038/ni.2703.
13. Lardner A. The effects of extracellular pH on immune function. *J Leukoc Biol* 2001;69:522–30.
14. Lai P, Rabinowich H, Crowley-Nowick PA, Bell MC, Mantovani G, Whiteside TL. Alterations in expression and function of signal-transducing proteins in tumor-associated T and natural killer cells in patients with ovarian carcinoma. *Clin Canc Res* 1996;2:161–73. https://doi.org/10.1189/jlb.69.4.522.
15. Elkabets M, Ribeiro VSG, Dinarello CA, Ostrand-Rosenberg S, Di Santo JP, Apte RN. IL-1β regulates a novel myeloid-derived suppressor cell subset that impairs NK cell development and function. *Eur J Immunol* 2010;40:3347–57. https://doi.org/10.1002/eji.201041037.
16. Li T, Yang Y, Hua X, Wang G, Liu W, Jia C. Hepatocellular carcinoma-associated fibroblasts trigger NK cell dysfunction via PGE2 and IDO. *Cancer Lett* 2012;318:154–61. https://doi.org/10.1016/j.canlet.2011.12.020.
17. Hoskin D, Mader J, Furlong S, Conrad D, Blay J. Inhibition of T cell and natural killer cell function by adenosine and its contribution to immune evasion by tumor cells (Review). *Int J Oncol* 2008;32:527–35. https://doi.org/10.3892/ijo.32.3.527.
18. Ban Y, Zhao Y, Liu F, Dong B, Kong B, Qu X. Effect of indoleamine 2,3-dioxygenase expressed in HTR-8/SVneo cells on decidual NK cell cytotoxicity. *Am J Reprod Immunol* 2016;75:519–28. https://doi.org/10.1111/aji.12481.
19. Lee J-C, Lee K-M, Kim D-W, Heo DS. Elevated TGF-beta1 secretion and down-modulation of NKG2D underlies impaired NK cytotoxicity in cancer patients. *J Immunol* 2004;172:7335–40. https://doi.org/10.4049/jimmunol.172.12.7335.
20. Slattery K, Gardiner CM. NK cell metabolism and TGFβ – implications for immunotherapy. *Front Immunol* 2019;10:2915. https://doi.org/10.3389/fimmu.2019.02915.
21. Viel S, Marçais A, Guimaraes FS-F, Loftus R, Rabilloud J, Grau M. TGF-β inhibits the activation and functions of NK cells by repressing the mTOR pathway. *Sci Signal* 2016;9:ra19. https://doi.org/10.1126/scisignal.aad1884.
22. Chretien AS, Le Roy A, Vey N, Prebet T, Blaise D, Fauriat C. Cancer-induced alterations of NK-mediated target recognition: current and investigational pharmacological strategies aiming at restoring NK-mediated anti-tumor activity. *Front Immunol* 2014;5. https://doi.org/10.3389/fimmu.2014.00122.

23. Trotta R, Dal Col J, Yu, J, Ciarlariello D, Thomas B, Zhang X, et al. TGF-beta utilizes SMAD3 to inhibit CD16-mediated IFN-gamma production and antibody-dependent cellular cytotoxicity in human NK cells. *J Immunol* 2008;181:3784–92. https://doi.org/10.4049/jimmunol.181.6.3784.
24. Gao Y, Souza-Fonseca-Guimaraes F, Bald T, Ng SS, Young A, Ngiow, SF, et al. Tumor immunoevasion by the conversion of effector NK cells into type 1 innate lymphoid cells. *Nat Immunol* 2017;18:1004–15. https://doi.org/10.1038/ni.3800.
25. Cortez VS, Ulland TK, Cervantes-Barragan L, Bando JK, Robinette ML, Wang Q, et al. SMAD4 impedes the conversion of NK cells into ILC1-like cells by curtailing non-canonical TGF-β signaling. *Nat Immunol* 2017;18:995–1003. https://doi.org/10.1038/ni.3809.
26. Tufa DM, Ahmad F, Chatterjee D, Ahrenstorf G, Schmidt RE, Jacobs R. IL-1β limits the extent of human 6-sulfo LacNAc dendritic cell (slanDC)-mediated NK cell activation and regulates CD95-induced apoptosis. *Cell Mol Immunol* 2017;14:976–85. https://doi.org/10.1038/cmi.2016.17.
27. Robson NC, Wei H, McAlpine T, Kirkpatrick N, Cebon J, Maraskovsky E. Activin-A attenuates several human natural killer cell functions. *Blood* 2009;113:3218–25. https://doi.org/10.1182/blood-2008-07-166926.
28. Groh V, Wu J, Yee C, Spies T. Tumour-derived soluble MIC ligands impair expression of NKG2D and T-cell activation. *Nature* 2002;419:734–8. https://doi.org/10.1038/nature01112.
29. Textor S, Dürst M, Jansen L, Accardi R, Tommasino M, Trunk MJ, et al. Activating NK cell receptor ligands are differentially expressed during progression to cervical cancer. *Int J Canc* 2008;123:2343–53. https://doi.org/10.1002/ijc.23733.
30. Pesce S, Tabellini G, Cantoni C, Patrizi O, Coltrini D, Rampinelli F, et al. B7-H6-mediated downregulation of NKp30 in NK cells contributes to ovarian carcinoma immune escape. *OncoImmunology* 2015;4. https://doi.org/10.1080/2162402x.2014.1001224.
31. Movahedi K, Laoui D, Gysemans C, Baeten M, Stangé G, Den Van Bossche J, et al. Different tumor microenvironments contain functionally distinct subsets of macrophages derived from Ly6C(high) monocytes. *Cancer Res* 2010;70:5728–39. https://doi.org/10.1158/0008-5472.can-09-4672.
32. MacDonald KPA, Palmer JS, Cronau S, Seppanen E, Olver S, Raffelt NC, et al. An antibody against the colony-stimulating factor 1 receptor depletes the resident subset of monocytes and tissue- and tumor-associated macrophages but does not inhibit inflammation. *Blood* 2010;116:3955–63. https://doi.org/10.1182/blood-2010-02-266296.
33. Sica A, Bronte V. Altered macrophage differentiation and immune dysfunction in tumor development. *J Clin Invest Am Soc Clin Invest* 2007;117:1155–66. https://doi.org/10.1172/jci31422.
34. Hagemann T, Biswas SK, Lawrence T, Sica A, Lewis CE. Regulation of macrophage function in tumors: the multifaceted role of NF-κB. *Blood* 2009;113:3139–46. https://doi.org/10.1182/blood-2008-12-172825.
35. Mantovani A, Marchesi F, Malesci A, Laghi L, Allavena P. Tumour-associated macrophages as treatment targets in oncology. *Nat Rev Clin Oncol* 2017;14:399–416. https://doi.org/10.1038/nrclinonc.2016.217.
36. Allavena P, Mantovani A. Immunology in the clinic review series; focus on cancer: tumour-associated macrophages: undisputed stars of the inflammatory tumour microenvironment. *Clin Exp Immunol* 2012;167:195–205. https://doi.org/10.1111/j.1365-2249.2011.04515.x.
37. De Palma M, Lewis CE. Macrophage regulation of tumor responses to anticancer therapies. *Cancer Cell* 2013;23:277–86. https://doi.org/10.1016/j.ccr.2013.02.013.
38. Menetrier-Caux C, Montmain G, Dieu MC, Bain C, Favrot MC, Caux C, et al. Inhibition of the differentiation of dendritic cells from CD34+ progenitors by tumor cells: role of interleukin-6

and macrophage colony- stimulating factor. *Blood* 1998;92:4778–91. https://doi.org/10.1182/blood.v92.12.4778.
39. Bharadwaj U, Li M, Zhang R, Chen C, Yao Q. Elevated interleukin-6 and G-CSF in human pancreatic cancer cell conditioned medium suppress dendritic cell differentiation and activation. *Cancer Res* 2007;67:5479–88. https://doi.org/10.1158/0008-5472.can-06-3963.
40. Gabrilovich DI, Chen HL, Girgis KR, Cunningham HT, Meny GM, Nadaf S, et al. Production of vascular endothelial growth factor by human tumors inhibits the functional maturation of dendritic cells. *Nat Med* 1996;2:1096–103. https://doi.org/10.1038/nm1096-1096.
41. Sombroek CC, Stam AGM, Masterson AJ, Lougheed SM, Schakel MJAG, Meijer CJLM, et al. Prostanoids play a major role in the primary tumor-induced inhibition of dendritic cell differentiation. *J Immunol* 2002;168:4333–43. https://doi.org/10.4049/jimmunol.168.9.4333.
42. Stock A, Booth S, Cerundolo V. Prostaglandin E2 suppresses the differentiation of retinoic acid-producing dendritic cells in mice and humans. *J Exp Med* 2011;208:761–73. https://doi.org/10.1084/jem.20101967.
43. Liu Y, Xiao L, Joo, K Il, Hu B, Fang J, Wang P. In situ modulation of dendritic cells by injectable thermosensitive hydrogels for cancer vaccines in mice. *Biomacromolecules* 2014;15:3836–45. https://doi.org/10.1021/bm501166j.
44. López MN, Pereda C, Segal G, Muñoz L, Aguilera R, González FE, et al. Prolonged survival of dendritic cell-vaccinated melanoma patients correlates with tumor-specific delayed type IV hypersensitivity response and reduction of tumor growth factor β-expressing T cells. *J Clin Oncol* 2009;27:945–52. https://doi.org/10.1200/jco.2008.18.0794.
45. Hargadon KM. Tumor-altered dendritic cell function: implications for anti-tumor immunity. *Front Immunol* 2013;4:192. https://doi.org/10.3389/fimmu.2013.00192.
46. Galon J, Costes A, Sanchez-Cabo F, Kirilovsky A, Mlecnik B, Lagorce-Pagès C, et al. Type, density, and location of immune cells within human colorectal tumors predict clinical outcome. *Science* 2006;313:1960–4. https://doi.org/10.1126/science.1129139.
47. Steidl C, Lee T, Shah SP, Farinha P, Han G, Nayar T, et al. Tumor-associated macrophages and survival in classic Hodgkin's lymphoma. *N Engl J Med* 2010;362:875–85. https://doi.org/10.1056/nejmoa0905680.
48. DeNardo DG, Brennan DJ, Rexhepaj E, Ruffell B, Shiao SL, Madden SF, et al. Leukocyte complexity predicts breast cancer survival and functionally regulates response to chemotherapy. *Cancer Discov* 2011;1:54–67. https://doi.org/10.1158/2159-8274.cd-10-0028.
49. Kurahara H, Shinchi H, Mataki Y, Maemura K, Noma H, Kubo F, et al. Significance of M2-polarized tumor-associated macrophage in pancreatic cancer. *J Surg Res* 2011;167:e211–9. https://doi.org/10.1016/j.jss.2009.05.026.
50. Zhang BC, Gao J, Wang J, Rao ZG, Wang BC, Gao JF. Tumor-associated macrophages infiltration is associated with peritumoral lymphangiogenesis and poor prognosis in lung adenocarcinoma. *Med Oncol* 2011;28:1447–52. https://doi.org/10.1007/s12032-010-9638-5.
51. Curiel TJ, Wei S, Dong H, Alvarez X, Cheng P, Mottram P, et al. Blockade of B7-H1 improves myeloid dendritic cell-mediated antitumor immunity. *Nat Med* 2003;9:562–7. https://doi.org/10.1038/nm863.
52. Krempski J, Karyampudi L, Behrens MD, Erskine CL, Hartmann L, Dong H, et al. Tumor-infiltrating programmed death receptor-1 + dendritic cells mediate immune suppression in ovarian cancer. *J Immunol* 2011;186:6905–13. https://doi.org/10.4049/jimmunol.1100274.
53. Ma Y, Shurin G V., Gutkin DW, Shurin MR. Tumor associated regulatory dendritic cells. *Semin Cancer Biol* 2012;22:298–306. https://doi.org/10.1016/j.semcancer.2012.02.010.

54. Freund-Brown J, Chirino L, Kambayashi T. Strategies to enhance NK cell function for the treatment of tumors and infections. *Crit Rev Immunol* 2018;38:105–30. https://doi.org/10.1615/critrevimmunol.2018025248.
55. Li Y, Sun R, Li Y, Sun R. Tumor immunotherapy: new aspects of natural killer cells. *Chinese J Cancer Res* 2018;30:173–96. https://doi.org/10.21147/j.issn.1000-9604.2018.02.02.
56. Ruggeri L, Capanni M, Casucci M, Volpi I, Tosti A, Perruccio K, et al. Role of natural killer cell alloreactivity in HLA-mismatched hematopoietic stem cell transplantation. *Blood* 1999;94:333–9. https://doi.org/10.1182/blood.v94.1.333.413a31_333_339.
57. Guillerey C, Huntington ND, Smyth MJ. Targeting natural killer cells in cancer immunotherapy. *Nat Immunol* 2016;17:1025–36. https://doi.org/10.1038/ni.3518.
58. Romagné F, André P, Spee P, Zahn S, Anfossi N, Gauthier L, et al. Preclinical characterization of 1-7F9, a novel human anti-KIR receptor therapeutic antibody that augments natural killer-mediated killing of tumor cells. *Blood* 2009;114:2667–77. https://doi.org/10.1182/blood-2009-02-206532.
59. Benson DM, Hofmeister CC, Padmanabhan S, Suvannasankha A, Jagannath S, Abonour R, et al. A phase 1 trial of the anti-KIR antibody IPH2101 in patients with relapsed/refractory multiple myeloma. *Blood* 2012;120:4324–33. https://doi.org/10.1182/blood-2012-06-438028.
60. Korde N, Carlsten M, Lee MJ, Minter A, Tan E, Kwok M, et al. A phase II trial of pan-KIR2D blockade with IPH2101 in smoldering multiple myeloma. *Haematologica* 2014;99. https://doi.org/10.3324/haematol.2013.103085.
61. Benson DM, Cohen AD, Jagannath S, Munshi NC, Spitzer G, Hofmeister CC, et al. A phase I trial of the anti-KIR antibody IPH2101 and lenalidomide in patients with relapsed/refractory multiple myeloma. *Clin Canc Res* 2015;21:4055–61. https://doi.org/10.1158/1078-0432.ccr-15-0304.
62. Vey N, Bourhis JH, Boissel N, Bordessoule D, Prebet T, Charbonnier A, et al. A phase 1 trial of the anti-inhibitory KIR mAb IPH2101 for AML in complete remission. *Blood* 2012;120:4317–23. https://doi.org/10.1182/blood-2012-06-437558.
63. Kohrt HE, Thielens A, Marabelle A, Sagiv-Barfil, Sola C, Chanuc F, et al. Anti-KIR antibody enhancement of anti-lymphoma activity of natural killer cells as monotherapy and in combination with anti-CD20 antibodies. *Blood* 2014;123:678–86. https://doi.org/10.1182/blood-2013-08-519199.
64. Nijhof IS, Lammerts van Bueren JJ, van Kessel B, Andre P, Morel Y, Lokhorst HM, et al. Daratumumab-mediated lysis of primary multiple myeloma cells is enhanced in combination with the human anti-KIR antibody IPH2102 and lenalidomide. *Haematologica* 2015;100:263–8. https://doi.org/10.3324/haematol.2014.117531.
65. Wieten L, Mahaweni NM, Voorter CEM, Bos GMJ, Tilanus MGJ. Clinical and immunological significance of HLA-E in stem cell transplantation and cancer. *Tissue Antigens* 2014;84:523–35. https://doi.org/10.1111/tan.12478.
66. Kochan G, Escors D, Breckpot K, Guerrero-Setas D. Role of non-classical MHC class I molecules in cancer immunosuppression. *Oncoimmunology* 2013;2. https://doi.org/10.4161/onci.26491.
67. lo Monaco E, Tremante E, Cerboni C, Melucci E, Sibilio L, Zingoni A, et al. Human leukocyte antigen E contributes to protect tumor cells from lysis by natural killer cells. *Neoplasia* 2011;13:822–30. https://doi.org/10.1593/neo.101684.
68. Mamessier E, Sylvain A, Thibult ML, Houvenaeghel G, Jacquemier J, Castellano R, et al. Human breast cancer cells enhance self tolerance by promoting evasion from NK cell antitumor immunity. *J Clin Invest* 2011;121:3609–22. https://doi.org/10.1172/jci45816.

69. André P, Denis C, Soulas C, Bourbon-Caillet C, Lopez J, Arnoux T, et al. Anti-NKG2A mAb is a checkpoint inhibitor that promotes anti-tumor immunity by unleashing both T and NK cells. *Cell* 2018;175:1731–43.e13. https://doi.org/10.1016/j.cell.2018.10.014.
70. van Montfoort N, Borst L, Korrer MJ, Sluijter M, Marijt KA, Santegoets SJ, et al. NKG2A blockade potentiates CD8 T cell immunity induced by cancer vaccines. *Cell* 2018;175:1744–55e15. https://doi.org/10.1016/j.cell.2018.10.028.
71. Segal NH, Naidoo J, Curigliano G, Patel S, Sahebjam S, Papadopoulos KP, et al. First-in-human dose escalation of monalizumab plus durvalumab, with expansion in patients with metastatic microsatellite-stable colorectal cancer. *J Clin Oncol* 2018;36:3540. https://doi.org/10.1200/jco.2018.36.15_suppl.3540.
72. Kamiya T, Seow SV, Wong D, Robinson M, Campana D. Blocking expression of inhibitory receptor NKG2A overcomes tumor resistance to NK cells. *J Clin Invest* 2019;129:2094–106. https://doi.org/10.1172/jci123955.
73. González Á, Rebmann V, Lemaoult J, Horn PA, Carosella ED, Alegre E. The immunosuppressive molecule HLA-G and its clinical implications. *Crit Rev Clin Lab Sci* 2012;49:63–84. https://doi.org/10.3109/10408363.2012.677947.
74. Heidenreich S, Zu Eulenburg C, Hildebrandt Y, Stübig T, Sierich H, Badbaran A, et al. Impact of the NK cell receptor LIR-1 (ILT-2/CD85j/LILRB1) on cytotoxicity against multiple myeloma. *Clin Dev Immunol* 2012. 2012. https://doi.org/10.1155/2012/652130.
75. Ndhlovu LC, Lopez-Vergès S, Barbour JD, Brad Jones R, Jha AR, Long BR, et al. Tim-3 marks human natural killer cell maturation and suppresses cell-mediated cytotoxicity. *Blood* 2012;119:3734–43. https://doi.org/10.1182/blood-2011-11-392951.
76. Gleason MK, Lenvik TR, McCullar V, Felices M, O'Brien MS, Cooley SA, et al. Tim-3 is an inducible human natural killer cell receptor that enhances interferon gamma production in response to galectin-9. *Blood* 2012;119:3064–72. https://doi.org/10.1182/blood-2011-06-360321.
77. Da Silva IP, Gallois A, Jimenez-Baranda S, Khan S, Anderson AC, Kuchroo VK, et al. Reversal of NK-cell exhaustion in advanced melanoma by Tim-3 blockade. *Cancer Immunol Res* 2014;2:410–22. https://doi.org/10.1158/2326-6066.cir-13-0171.
78. Wang Z, Zhu J, Gu H, Yuan Y, Zhang B, Zhu D, et al. The clinical significance of abnormal tim-3 expression on NK cells from patients with gastric cancer. *Immunol Invest* 2015;44:578–89. https://doi.org/10.3109/08820139.2015.1052145.
79. Komita H, Koido S, Hayashi K, Kan S, Ito M, Kamata Y, et al. Expression of immune checkpoint molecules of T cell immunoglobulin and mucin protein 3/galectin-9 for NK cell suppression in human gastrointestinal stromal tumors. *Oncol Rep* 2015;34:2099–105. https://doi.org/10.3892/or.2015.4149.
80. Xu L, Huang Y, Tan L, Yu W, Chen D, Lu C, et al. Increased Tim-3 expression in peripheral NK cells predicts a poorer prognosis and Tim-3 blockade improves NK cell-mediated cytotoxicity in human lung adenocarcinoma. *Int Immunopharmacol* 2015;29:635–41. https://doi.org/10.1016/j.intimp.2015.09.017.
81. Zhu C, Anderson AC, Schubart A, Xiong H, Imitola J, Khoury SJ, et al. The Tim-3 ligand galectin-9 negatively regulates T helper type 1 immunity. *Nat Immunol* 2005;6:1245–52. https://doi.org/10.1038/ni1271.
82. Chiba S, Baghdadi M, Akiba H, Yoshiyama H, Kinoshita I, Dosaka-Akita H, et al. Tumor-infiltrating DCs suppress nucleic acid-mediated innate immune responses through interactions between the receptor TIM-3 and the alarmin HMGB1. *Nat Immunol* 2012;13:832–42. https://doi.org/10.1038/ni.2376.
83. DeKruyff RH, Bu X, Ballesteros A, Santiago C, Chim Y-LE, Lee H-H, et al. T cell/transmembrane, ig, and mucin-3 allelic variants differentially recognize phosphatidylserine

and mediate phagocytosis of apoptotic cells. *J Immunol* 2010;184:1918–30. https://doi.org/10.4049/jimmunol.0903059.
84. So EC, Khaladj-Ghom A, Ji Y, Amin J, Song Y, Burch E, et al. NK cell expression of Tim-3: first impressions matter. *Immunobiology* 2019;224:362–70. https://doi.org/10.1016/j.imbio.2019.03.001.
85. Xu L, Huang Y, Tan L, Yu W, Chen D, Lu C, et al. Increased Tim-3 expression in peripheral NK cells predicts a poorer prognosis and Tim-3 blockade improves NK cell-mediated cytotoxicity in human lung adenocarcinoma. *Int Immunopharmacol* 2015;29:635–41. https://doi.org/10.1016/j.intimp.2015.09.017.
86. Zhang B, Zhao W, Li H, Chen Y, Tian H, Li L, et al. Immunoreceptor TIGIT inhibits the cytotoxicity of human cytokine-induced killer cells by interacting with CD155. *Cancer Immunol Immunother* 2016;65:305–14. https://doi.org/10.1007/s00262-016-1799-4.
87. Stanietsky N, Simic H, Arapovic J, Toporik A, Levy O, Novik A, et al. The interaction of TIGIT with PVR and PVRL2 inhibits human NK cell cytotoxicity. *Proc Natl Acad Sci USA* 2009;106:17858–63. https://doi.org/10.1073/pnas.0903474106.
88. Stein N, Tsukerman P, Mandelboim O. The paired receptors TIGIT and DNAM-1 as targets for therapeutic antibodies. *Hum Antibodies* 2017;25:111–9. https://doi.org/10.3233/hab-160307.
89. Xu F, Sunderland A, Zhou Y, Schulick RD, Edil BH, Zhu Y. Blockade of CD112R and TIGIT signaling sensitizes human natural killer cell functions. *Cancer Immunol Immunother* 2017;66:1367–75. https://doi.org/10.1007/s00262-017-2031-x.
90. Yu X, Harden K, Gonzalez LC, Francesco M, Chiang E, Irving B, et al. The surface protein TIGIT suppresses T cell activation by promoting the generation of mature immunoregulatory dendritic cells. *Nat Immunol* 2009;10:48–57. https://doi.org/10.1038/ni.1674.
91. Roman Aguilera A, Lutzky VP, Mittal D, Li XY, Stannard K, Takeda K, et al. CD96 targeted antibodies need not block CD96-CD155 interactions to promote NK cell anti-metastatic activity. *OncoImmunology* 2018;7. https://doi.org/10.1080/2162402x.2018.1424677.
92. Stojanovic A, Fiegler N, Brunner-Weinzierl M, Cerwenka A. CTLA-4 is expressed by activated mouse NK cells and inhibits NK cell IFN-γ production in response to mature dendritic cells. *J Immunol* 2014;192:4184–91. https://doi.org/10.4049/jimmunol.1302091.
93. Laurent S, Queirolo P, Boero S, Salvi S, Piccioli P, Boccardo S, et al. The engagement of CTLA-4 on primary melanoma cell lines induces antibody-dependent cellular cytotoxicity and TNF-α production. *J Transl Med* 2013;11:108. https://doi.org/10.1186/1479-5876-11-108.
94. Benson DM, Bakan CE, Mishra A, Hofmeister CC, Efebera Y, Becknell B, et al. The PD-1/PD-L1 axis modulates the natural killer cell versus multiple myeloma effect: a therapeutic target for CT-011, a novel monoclonal anti-PD-1 antibody. *Blood* 2010;116:2286–94. https://doi.org/10.1182/blood-2010-02-271874.
95. Beldi-Ferchiou A, Lambert M, Dogniaux S, Vély F, Vivier E, Olive D, et al. PD-1 mediates functional exhaustion of activated NK cells in patients with Kaposi sarcoma. *Oncotarget* 2016;7:72961–77. https://doi.org/10.18632/oncotarget.12150.
96. Wiesmayr S, Webber SA, Macedo C, Popescu I, Smith L, Luce J, et al. Decreased NKp46 and NKG2D and elevated PD-1 are associated with altered NK-cell function in pediatric transplant patients with PTLD. *Eur J Immunol* 2012;42:541–50. https://doi.org/10.1002/eji.201141832.
97. Pesce S, Greppi M, Tabellini G, Rampinelli F, Parolini S, Olive D, et al. Identification of a subset of human natural killer cells expressing high levels of programmed death 1: a phenotypic and functional characterization. *J Allergy Clin Immunol* 2017;139:335–46.e3. https://doi.org/10.1016/j.jaci.2016.04.025.
98. Berger R, Rotem-Yehudar R, Slama G, Landes S, Kneller A, Leiba M, et al. Phase i safety and pharmacokinetic study of CT-011, a humanized antibody interacting with PD-1, in patients

with advanced hematologic malignancies. *Clin Canc Res* 2008;14:3044–51. https://doi.org/10.1158/1078-0432.ccr-07-4079.

99. Chen R, Zinzani PL, Fanale MA, Armand P, Johnson NA, Brice P, et al. Phase II study of the efficacy and safety of pembrolizumab for relapsed/refractory classic Hodgkin Lymphoma. *J Clin Oncol* 2017;35:2125–32. https://doi.org/10.1200/jco.2016.72.1316.
100. WestinJR, Chu F, Zhang M, Fayad LE, Kwak LW, Fowler N, et al. Safety and activity of PD1 blockade by pidilizumab in combination with rituximab in patients with relapsed follicular lymphoma: a single group, open-label, phase 2 trial. *Lancet Oncol* 2014;15:69–77. https://doi.org/10.1016/s1470-2045(13)70551-5.
101. Xu F, Liu J, Liu D, Liu B, Wang M, Hu Z, et al. LSECtin expressed on melanoma cells promotes tumor progression by inhibiting antitumor T-cell responses. *Cancer Res* 2014;74:3418–28. https://doi.org/10.1158/0008-5472.can-13-2690.
102. Wang J, Sanmamed MF, Datar I, Su TT, Ji L, Sun J, et al. Fibrinogen-like protein 1 is a major immune inhibitory ligand of LAG-3. *Cell* 2019;176:334–47.e12. https://doi.org/10.1016/j.cell.2018.11.010.
103. Miyazaki T, Dierich A, Benoist C, Mathis D. Independent modes of natural killing distinguished in mice lacking Lag3. *Science* 1996;272:405–8. https://doi.org/10.1126/science.272.5260.405.
104. Huard B, Tournier M, Triebel F. LAG-3 does not define a specific mode of natural killing in human. *Immunol Lett* 1998;61:109–12. https://doi.org/10.1016/s0165-2478(97)00170-3.
105. Andrews LP, Marciscano AE, Drake CG, Vignali DAA. LAG3 (CD223) as a cancer immunotherapy target. *Immunol Rev* 2017;276:80–96. https://doi.org/10.1111/imr.12519.
106. Zhang J, Basher F, Wu JD. NKG2D ligands in tumor immunity: two sides of a coin. *Front Immunol* 2015;6. https://doi.org/10.3389/fimmu.2015.00097.
107. Bastid J, Regairaz A, Bonnefoy N, Dejou C, Giustiniani J, Laheurte C, et al. Inhibition of CD39 enzymatic function at the surface of tumor cells alleviates their immunosuppressive activity. *Cancer Immunol Res* 2015;3:254–65. https://doi.org/10.1158/2326-6066.cir-14-0018.
108. Raskovalova T, Huang X, Sitkovsky M, Zacharia LC, Jackson EK, Gorelik E. Gs protein-coupled adenosine receptor signaling and lytic function of activated NK cells. *J Immunol* 2005;175:4383–91. https://doi.org/10.4049/jimmunol.175.7.4383.
109. Allard B, Longhi MS, Robson SC, Stagg J. The ectonucleotidases CD39 and CD73: novel checkpoint inhibitor targets. *Immunol Rev* 2017;276:121–44. https://doi.org/10.1111/imr.12528.
110. Häusler SF, del Barrio IM, Diessner J, Stein RG, Strohschein J, Hönig A, et al. Anti-CD39 and anti-CD73 antibodies A1 and 7G2 improve targeted therapy in ovarian cancer by blocking adenosine-dependent immune evasion. *Am J Transl Res* 2014;6:129.
111. Seidel UJE, Schlegel P, Lang P. Natural killer cell mediated antibody-dependent cellular cytotoxicity in tumor immunotherapy with therapeutic antibodies. *Front Immunol* 2013;4. https://doi.org/10.3389/fimmu.2013.00076.
112. Bartkowiak T, Curran MA. 4-1BB agonists: multi-potent potentiators of tumor immunity. *Front Oncol* 2015;5. https://doi.org/10.3389/fonc.2015.00117.
113. Baessler T, Charton JE, Schmiedel BJ, Grünebach F, Krusch M, Wacker A, et al. CD137 ligand mediates opposite effects in human and mouse NK cells and impairs NK-cell reactivity against human acute myeloid leukemia cells. *Blood* 2010;115:3058–69. https://doi.org/10.1182/blood-2009-06-227934.
114. Diefenbach A, Jensen ER, Jamieson AM, Raulet DH. Rae1 and H60 ligands of the NKG2D receptor stimulate tumour immunity. *Nature* 2001;413:165–71. https://doi.org/10.1038/35093109.

115. Holliger P, Hudson PJ. Engineered antibody fragments and the rise of single domains. *Nat Biotechnol* 2005;23:1126–36. https://doi.org/10.1038/nbt1142.
116. Vallera DA, Zhang B, Gleason MK, Oh S, Weiner LM, Kaufman DS, et al. Heterodimeric bispecific single-chain variable-fragment antibodies against EpCAM and CD16 induce effective antibody-dependent cellular cytotoxicity against human carcinoma cells. *Cancer Biother Radiopharm* 2013;28:274–82. https://doi.org/10.1089/cbr.2012.1329.
117. Reiners KS, Kessler J, Sauer M, Rothe A, Hansen HP, Reusch U, et al. Rescue of impaired NK cell activity in hodgkin lymphoma with bispecific antibodies in vitro and in patients. *Mol Ther* 2013;21:895–903. https://doi.org/10.1038/mt.2013.14.
118. JU Schmohl, Gleason MK, Dougherty PR, Miller JS, Vallera DA. Heterodimeric bispecific single chain variable fragments (scFv) killer engagers (BiKEs) enhance NK-cell activity against CD133+ colorectal cancer cells. *Target Oncol* 2016;11:353–61. https://doi.org/10.1007/s11523-015-0391-8.
119. Reusch U, Burkhardt C, Fucek I, Le Gall, F Le Gall M, Hoffmann K, et al. A novel tetravalent bispecific TandAb (CD30/CD16A) efficiently recruits NK cells for the lysis of CD30+ tumor cells. *MAbs* 2014;6:728–39. https://doi.org/10.4161/mabs.28591.
120. Gleason MK, Ross JA, Warlick ED, Lund TC, Verneris MR, Wiernik A, et al. CD16xCD33 bispecific killer cell engager (BiKE) activates NK cells against primary MDS and MDSC CD33+ targets. *Blood* 2014;123:3016–26. https://doi.org/10.1182/blood-2013-10-533398.
121. Turini M, Chames P, Bruhns P, Baty D, Kerfelec B. A FcγRIII-engaging bispecific antibody expands the range of HER2-expressing breast tumors eligible to antibody therapy. *Oncotarget* 2014;5:5304–19. https://doi.org/10.18632/oncotarget.2093.
122. Dong B, Zhou C, He P, Li J, Chen S, Miao J, et al. A novel bispecific antibody, BiSS, with potent anti-cancer activities. *Cancer Biol Ther* 2016;17:364–70. https://doi.org/10.1080/15384047.2016.1139266.
123. Osaki T, Fujisawa S, Kitaguchi M, Kitamura M, Nakanishi T. Development of a bispecific antibody tetramerized through hetero-associating peptides. *FEBS J* 2015;282:4389–401. https://doi.org/10.1111/febs.13505.
124. Wiernik A, Foley B, Zhang B, Verneris MR, Warlick E, Gleason MK, et al. Targeting natural killer cells to acute myeloid leukemia in vitro with a CD16x33 bispecific killer cell engager and ADAM17 inhibition. *Clin Canc Res* 2013;19:3844–55. https://doi.org/10.1158/1078-0432.ccr-13-0505.
125. Rothe A, Sasse S, Topp MS, Eichenauer DA, Hummel H, Reiners KS, et al. A phase 1 study of the bispecific anti-CD30/CD16A antibody construct AFM13 in patients with relapsed or refractory Hodgkin lymphoma. *Blood* 2015;125:4024–31. https://doi.org/10.1182/blood-2014-12-614636.
126. JU Schmohl, Felices M, Taras E, Miller JS, Vallera DA. Enhanced ADCC and NK cell activation of an anticarcinoma bispecific antibody by genetic insertion of a modified IL-15 cross-linker. *Mol Ther* 2016;24:1312–22. https://doi.org/10.1038/mt.2016.88.
127. JU Schmohl, Felices M, Todhunter D, Taras E, Miller JS, Vallera DA. Tetraspecific scFv construct provides NK cell mediated ADCC and self-sustaining stimuli via insertion of IL-15 as a cross-linker. *Oncotarget* 2016;7:73830–44. https://doi.org/10.18632/oncotarget.12073.
128. Wang T, Sun F, Xie W, Tang M, He H, Jia X, et al. A bispecific protein rG7S-MICA recruits natural killer cells and enhances NKG2D-mediated immunosurveillance against hepatocellular carcinoma. *Canc Lett* 2016;372:166–78. https://doi.org/10.1016/j.canlet.2016.01.001.
129. Rothe A, Jachimowicz RD, Borchmann S, Madlener M, Keßler J, Reiners KS, et al. The bispecific immunoligand ULBP2-aCEA redirects natural killer cells to tumor cells and reveals

potent anti-tumor activity against colon carcinoma. *Int J Canc* 2014;134:2829–40. https://doi.org/10.1002/ijc.28609.
130. Peipp M, Derer S, Lohse S, Staudinger M, Klausz K, Valerius T, et al. HER2-specific immunoligands engaging NKp30 or NKp80 trigger NK-cell-mediated lysis of tumor cells and enhance antibodydependent cell-mediated cytotoxicity. *Oncotarget* 2015;6:32075–88. https://doi.org/10.18632/oncotarget.5135.
131. Xia Y, Chen B, Shao X, Xiao W, Qian L, Ding Y, et al. Treatment with a fusion protein of the extracellular domains of NKG2D to IL-15 retards colon cancer growth in mice. *J Immunother* 2014;37:257–66. https://doi.org/10.1097/cji.0000000000000033.
132. Chen Y, Chen B, Yang T, Xiao W, Qian L, Ding Y, et al. Human fused NKG2D-IL-15 protein controls xenografted human gastric cancer through the recruitment and activation of NK cells. *Cell Mol Immunol* 2017;14:293–307. https://doi.org/10.1038/cmi.2015.81.
133. Vego H, Sand KL, Høglund RA, Fallang LE, Gundersen G, Holmøy T, et al. Monomethyl fumarate augments NK cell lysis of tumor cells through degranulation and the upregulation of NKp46 and CD107a. *Cell Mol Immunol* 2016;13:57–64. https://doi.org/10.1038/cmi.2014.114.
134. Felices M, Chu S, Kodal B, Bendzick L, Ryan C, Lenvik AJ, et al. IL-15 super-agonist (ALT-803) enhances natural killer (NK) cell function against ovarian cancer. *Gynecol Oncol* 2017;145:453–61. https://doi.org/10.1016/j.ygyno.2017.02.028.
135. Rosenberg SA, Lotze MT, Muul LM, Leitman S, Chang AE, Ettinghausen SE, et al. Observations on the systemic administration of autologous lymphokine-activated killer cells and recombinant interleukin-2 to patients with metastatic cancer. *N Engl J Med* 1985;313:1485–92. https://doi.org/10.1056/nejm198512053132327.
136. Malek TR. The main function of IL-2 is to promote the development of T regulatory cells. *J Leukoc Biol* 2003;74:961–5. https://doi.org/10.1189/jlb.0603272.
137. Cheng M, Chen Y, Xiao W, Sun R, Tian Z. NK cell-based immunotherapy for malignant diseases. *Cell Mol Immunol* 2013;10:230–52. https://doi.org/10.1038/cmi.2013.10.
138. Fujisaki H, Kakuda H, Shimasaki N, Imai C, Ma J, Lockey T, et al. Expansion of highly cytotoxic human natural killer cells for cancer cell therapy. *Cancer Res* 2009;69:4010–7. https://doi.org/10.1158/0008-5472.can-08-3712.
139. Masuyama J ichi, Murakami T, Iwamoto S, Fujita S. Ex vivo expansion of natural killer cells from human peripheral blood mononuclear cells co-stimulated with anti-CD3 and anti-CD52 monoclonal antibodies. *Cytotherapy* 2016;18:80–90. https://doi.org/10.1016/j.jcyt.2015.09.011.
140. Miller JS, Soignier Y, Panoskaltsis-Mortari A, McNearney SA, Yun GH, Fautsch SK, et al. Successful adoptive transfer and in vivo expansion of human haploidentical NK cells in patients with cancer. *Blood* 2005;105:3051–7. https://doi.org/10.1182/blood-2004-07-2974.
141. Romanski A, Uherek C, Bug G, Seifried E, Klingemann H, Wels WS, et al. CD19-CAR engineered NK-92 cells are sufficient to overcome NK cell resistance in B-cell malignancies. *J Cell Mol Med* 2016;20:1287–94. https://doi.org/10.1111/jcmm.12810.
142. Müller T, Uherek C, Maki G, Chow KU, Schimpf A, Klingemann HG, et al. Expression of a CD20-specific chimeric antigen receptor enhances cytotoxic activity of NK cells and overcomes NK-resistance of lymphoma and leukemia cells. *Cancer Immunol Immunother* 2008;57:411–23. https://doi.org/10.1007/s00262-007-0383-3.
143. Jiang H, Zhang W, Shang P, Zhang H, Fu W, Ye F, et al. Transfection of chimeric anti-CD138 gene enhances natural killer cell activation and killing of multiple myeloma cells. *Mol Oncol* 2014;8:297–310. https://doi.org/10.1016/j.molonc.2013.12.001.
144. Esser R, Müller T, Stefes D, Kloess S, Seidel D, Gillies SD, et al. NK cells engineered to express a GD 2-specific antigen receptor display built-in ADCC-like activity against tumour

cells of neuroectodermal origin. *J Cell Mol Med* 2012;16:569–81. https://doi.org/10.1111/j.1582-4934.2011.01343.x.
145. Han J, Chu J, Keung Chan W, Zhang J, Wang Y, Cohen JB, et al. CAR-engineered NK cells targeting wild-type EGFR and EGFRvIII enhance killing of glioblastoma and patient-derived glioblastoma stem cells. *Sci Rep* 2015;5. https://doi.org/10.1038/srep11483.
146. Schönfeld K, Sahm C, Zhang C, Naundorf S, Brendel C, Odendahl M, et al. Selective inhibition of tumor growth by clonal NK cells expressing an ErbB2/HER2-specific chimeric antigen receptor. *Mol Ther* 2015;23:330–8. https://doi.org/10.1038/mt.2014.219.
147. Schirrmann T, Pecher G. Human natural killer cell line modified with a chimeric immunoglobulin T-cell receptor gene leads to tumor growth inhibition in vivo. *Cancer Gene Ther* 2002;9:390–8. https://doi.org/10.1038/sj.cgt.7700453.
148. Schirrmann T, Pecher G. Specific targeting of CD33+ leukemia cells by a natural killer cell line modified with a chimeric receptor. *Leuk Res* 2005;29:301–6. https://doi.org/10.1016/j.leukres.2004.07.005.
149. Morvan MG, Lanier LL. NK cells and cancer: you can teach innate cells new tricks. *Nature Reviews Cancer. Nat Rev Cancer* 2016;16:7–19. https://doi.org/10.1038/nrc.2015.5.
150. Aitken AS, Roy DG, Bourgeois-Daigneault MC. Taking a stab at cancer; Oncolytic virus-mediated anti-cancer vaccination strategies. *Biomed MDPI AG* 2017;5. https://doi.org/10.3390/biomedicines5010003.
151. Andtbacka RHI, Kaufman HL, Collichio F, Amatruda T, Senzer N, Chesney J, et al. Talimogene laherparepvec improves durable response rate in patients with advanced melanoma. *J Clin Oncol* 2015;33:2780–8. https://doi.org/10.1200/jco.2014.58.3377.
152. Hodgins JJ, Khan ST, Park MM, Auer RC, Ardolino M. Killers 2.0: NK cell therapies at the forefront of cancer control. *J Clin Invest Am Soc Clin Invest* 2019;129:3499–510. https://doi.org/10.1172/jci129338.
153. Chesney J, Puzanov I, Collichio F, Singh P, Milhem MM, Glaspy J, et al. Randomized, open-label phase II study evaluating the efficacy and safety of talimogene laherparepvec in combination with ipilimumab versus ipilimumab alone in patients with advanced, unresectable melanoma. *J Clin Oncol* 2018;36:1658–67. https://doi.org/10.1200/jco.2017.73.7379.
154. Bhat R, Dempe S, Dinsart C, Rommelaere J. Enhancement of NK cell antitumor responses using an oncolytic parvovirus. *Int J Cancer* 2011;128:908–19. https://doi.org/10.1002/ijc.25415.
155. Bhat R, Rommelaere J. NK-cell-dependent killing of colon carcinoma cells is mediated by natural cytotoxicity receptors (NCRs) and stimulated by parvovirus infection of target cells. *BMC Cancer* 2013:13. https://doi.org/10.1186/1471-2407-13-367.
156. Errington F, Steele L, Prestwich R, Harrington KJ, Pandha HS, Vidal L, et al. Reovirus activates human dendritic cells to promote innate antitumor immunity. *J Immunol* 2008;180:6018–26. https://doi.org/10.4049/jimmunol.180.9.6018.
157. Boudreau JE, Bridle BW, Stephenson KB, Jenkins KM, Brunellière J, Bramson JL, et al. Recombinant vesicular stomatitis virus transduction of dendritic cells enhances their ability to prime innate and adaptive antitumor immunity. *Mol Ther* 2009;17:1465–72. https://doi.org/10.1038/mt.2009.95.
158. Lapteva N, Aldrich M, Weksberg D, Rollins L, Goltsova T, Chen SY, et al. Targeting the intratumoral dendritic cells by the oncolytic adenoviral vaccine expressing RANTES elicits potent antitumor immunity. *J Immunother* 2009;32:145–56. https://doi.org/10.1097/cji.0b013e318193d31e.
159. Stephenson KB, Barra NG, Davies E, Ashkar AA, Lichty BD. Expressing human interleukin-15 from oncolytic vesicular stomatitis virus improves survival in a murine metastatic colon

adenocarcinoma model through the enhancement of anti-tumor immunity. *Cancer Gene Ther* 2012;19:238–46. https://doi.org/10.1038/cgt.2011.81.
160. Choi KJ, Zhang SN, Choi IK, Kim JS, Yun CO. Strengthening of antitumor immune memory and prevention of thymic atrophy mediated by adenovirus expressing IL-12 and GM-CSF. *Gene Ther* 2012;19:711–23. https://doi.org/10.1038/gt.2011.125.
161. Alkayyal AA, Tai L-H, Kennedy MA, de Souza CT, Zhang J, Lefebvre C, et al. NK-cell recruitment is necessary for eradication of peritoneal carcinomatosis with an IL12-expressing Maraba virus cellular vaccine. *Cancer Immunol Res* 2017;5:211–21. https://doi.org/10.1158/2326-6066.cir-16-0162.
162. Chen X, Han J, Chu J, Zhang L, Zhang J, Chen C, et al. A combinational therapy of EGFR-CAR NK cells and oncolytic herpes simplex virus 1 for breast cancer brain metastases. *Oncotarget* 2016;7:27764–77. https://doi.org/10.18632/oncotarget.8526.
163. Anfray C, Aldo U, Andón FT, Allavena P. Current strategies to target tumor-associated-macrophages to improve anti-tumor immune responses. *Cells* 2019;9:46. https://doi.org/10.3390/cells9010046.
164. Canè S, Ugel S, Trovato R, Marigo I, De Sanctis F, Sartoris S, et al. The endless saga of monocyte diversity. *Front Immunol* 2019;10:1786. https://doi.org/10.3389/fimmu.2019.01786.
165. Liu L, He H, Liang R, Yi H, Meng X, Chen Z, et al. ROS-inducing Micelles sensitize tumor-associated macrophages to TLR3 stimulation for potent immunotherapy. *Biomacromolecules* 2018;19:2146–55. https://doi.org/10.1021/acs.biomac.8b00239.
166. Zhao J, Zhang Z, Xue Y, Wang G, Cheng Y, Pan Y, et al. Anti-tumor macrophages activated by ferumoxytol combined or surface-functionalized with the TLR3 agonist poly (I : C) promote melanoma regression. *Theranostics* 2018;8:6307–21. https://doi.org/10.7150/thno.29746.
167. Perkins H, Khodai T, Mechiche H, Colman P, Burden F, Laxton C, et al. Therapy with TLR7 agonists induces lymphopenia: correlating pharmacology to mechanism in a mouse model. *J Clin Immunol* 2012;32:1082–92. https://doi.org/10.1007/s10875-012-9687-y.
168. Haabeth OAW, Blake TR, McKinlay CJ, Tveita AA, Sallets A, Waymouth RM, et al. Local delivery of Ox40l, Cd80, and Cd86 mRNA kindles global anticancer immunity. *Cancer Res* 2019;79:1624–34. https://doi.org/10.1158/0008-5472.can-18-2867.
169. Zhang F, Parayath NN, Ene CI, Stephan SB, Koehne AL, Coon ME, et al. Genetic programming of macrophages to perform anti-tumor functions using targeted mRNA nanocarriers. *Nat Commun* 2019;10:3974. https://doi.org/10.1038/s41467-019-11911-5.
170. Song Y, Tang C, Yin C. Combination antitumor immunotherapy with VEGF and PlGF siRNA via systemic delivery of multi-functionalized nanoparticles to tumor-associated macrophages and breast cancer cells. *Biomaterials* 2018;185:117–32. https://doi.org/10.1016/j.biomaterials.2018.09.017.
171. Zang X, Zhang X, Hu H, Qiao M, Zhao X, Deng Y, et al. Targeted delivery of zoledronate to tumor-associated macrophages for cancer immunotherapy. *Mol Pharm* 2019;16:2249–58. https://doi.org/10.1021/acs.molpharmaceut.9b00261.
172. Yang H, Shao R, Huang H, Wang X, Rong Z, Lin Y. Engineering macrophages to phagocytose cancer cells by blocking the CD47/SIRPα axis. *Cancer Med* 2019;8:4245–53. https://doi.org/10.1002/cam4.2461.
173. Advani R, Flinn I, Popplewell L, Forero A, Bartlett NL, Ghosh N, et al. CD47 blockade by Hu5F9-G4 and rituximab in non-Hodgkin's lymphoma. *N Engl J Med* 2018;379:1711–21. https://doi.org/10.1056/nejmoa1807315.
174. Beatty GL, Chiorean EG, Fishman MP, Saboury B, Teitelbaum UR, Sun W, et al. CD40 agonists alter tumor stroma and show efficacy against pancreatic carcinoma in mice and humans. *Science* 2011;331:1612–6. https://doi.org/10.1126/science.1198443.

175. Vonderheide RH. CD40 agonist antibodies in cancer immunotherapy. *Annu Rev Med* 2020;71:47–58. https://doi.org/10.1146/annurev-med-062518-045435.
176. Sabado RL, Balan S, Bhardwaj N. Dendritic cell-based immunotherapy. *Cell Res* 2017;27:74–95. https://doi.org/10.1038/cr.2016.157.
177. Rini B. Future approaches in immunotherapy. *Semin Oncol* 2014;41:S30–40. https://doi.org/10.1053/j.seminoncol.2014.09.005.
178. Wolchok JD, Hoos A, O'Day S, Weber JS, Hamid O, Lebbé C, et al. Guidelines for the evaluation of immune therapy activity in solid tumors: immune-related response criteria. *Clin Canc Res* 2009;15:7412–20. https://doi.org/10.1158/1078-0432.ccr-09-1624.
179. Anguille S, Smits EL, Bryant C, Van Acker HH, Goossens H, Lion E, et al. Dendritic cells as pharmacological tools for cancer immunotherapys. *Pharmacol Rev* 2015;67:731–53. https://doi.org/10.1124/pr.114.009456.
180. Galluzzi L, Senovilla L, Vacchelli E, Eggermont A, Fridman WH, Galon J, et al. Trial watch: dendritic cell-based interventions for cancer therapy. *Oncoimmunology* 2012;1:1111–34. https://doi.org/10.4161/onci.21494.
181. Palucka K, Banchereau J. Cancer immunotherapy via dendritic cells. *Nat Rev Cancer* 2012;12:265–77. https://doi.org/10.1038/nrc3258.
182. Pizzurro GA, Barrio MM. Dendritic cell-based vaccine efficacy: aiming for hot spots. *Front Immunol* 2015;6:91. https://doi.org/10.3389/fimmu.2015.00091.
183. Bol KF, Schreibelt G, Gerritsen WR, de Vries IJM, Figdor CG. Dendritic cell-based immunotherapy: state of the art and beyond. *Clin Canc Res* 2016;22:1897–906. https://doi.org/10.1158/1078-0432.ccr-15-1399.
184. Constantino J, Gomes C, Falcão A, Cruz MT, Neves BM. Antitumor dendritic cell-based vaccines: lessons from 20 years of clinical trials and future perspectives. *Transl Res* 2016;168:74–95. https://doi.org/10.1016/j.trsl.2015.07.008.
185. Galluzzi L, Vacchelli E, Bravo-San Pedro JM, Buqué A, Senovilla L, Baracco EE, et al. Classification of current anticancer immunotherapies. *Oncotarget* 2014;5:12472–508. https://doi.org/10.18632/oncotarget.2998.
186. Timmerman JM, Levy R. Dendritic cell vaccines for cancer immunotherapy. *Annu Rev Med* 1999;50:507–29. https://doi.org/10.1146/annurev.med.50.1.507.
187. Kantoff, PW, Higano, CS, Shore, ND, Berger, ER, Small, EJ, Penson, DF, et al. Sipuleucel-T immunotherapy for castration-resistant prostate cancer. *N Engl J Med* 2010;363:411–22. https://doi.org/10.1056/nejmoa1001294.
188. Scholz M, Yep S, Chancey M, Kelly C, Chau K, Turner J, et al. Phase I clinical trial of sipuleucel-T combined with escalating doses of ipilimumab in progressive metastatic castrate-resistant prostate cancer. *ImmunoTargets Ther* 2017;6:11–6. https://doi.org/10.2147/itt.s122497.
189. Urbanova L, Hradilova N, Moserova I, Vosahlikova S, Sadilkova L, Hensler M, et al. High hydrostatic pressure affects antigenic pool in tumor cells: implication for dendritic cell-based cancer immunotherapy. *Immunol Lett* 2017;187:27–34. https://doi.org/10.1016/j.imlet.2017.05.005.
190. Hradilova N, Sadilkova L, Palata O, Mysikova D, Mrazkova H, Lischke R, et al. Generation of dendritic cell-based vaccine using high hydrostatic pressure for non-small cell lung cancer immunotherapy, Shiku H, editor. *PloS ONE* 2017;12:e0171539. https://doi.org/10.1016/j.jtho.2017.11.079.
191. Podrazil M, Horvath R, Becht E, Rozkova D, Bilkova P, Sochorova K, et al. Phase I/II clinical trial of dendritic-cell based immunotherapy (DCVAC/PCa) combined with chemotherapy in patients with metastatic, castration-resistant prostate cancer. *Oncotarget* 2015;6:18192–205. https://doi.org/10.18632/oncotarget.4145.

192. Steinman RM. Decisions about dendritic cells: past, present, and future. *Annu Rev Immunol* 2012;30:1–22. https://doi.org/10.1146/annurev-immunol-100311-102839.
193. Tacken PJ, Figdor CG. Targeted antigen delivery and activation of dendritic cells in vivo: steps towards cost effective vaccines. *Semin Immunol* 2011;23:12–20. https://doi.org/10.1016/j.smim.2011.01.001.
194. Hawiger D, Inaba K, Dorsett Y, Guo M, Mahnke K, Rivera M, et al. Dendritic cells induce peripheral T cell unresponsiveness under steady state conditions in vivo. *J Exp Med* 2001;194:769–79. https://doi.org/10.1084/jem.194.6.769.
195. Bonifaz L, Bonnyay D, Mahnke K, Rivera M, Nussenzweig MC, Steinman RM. Efficient targeting of protein antigen to the dendritic cell receptor DEC-205 in the steady state leads to antigen presentation on major histocompatibility complex class I products and peripheral CD8+ T cell tolerance. *J Exp Med* 2002;196:1627–38. https://doi.org/10.1084/jem.20021598.
196. Palucka K, Banchereau J. Dendritic-cell-based therapeutic cancer vaccines. *Immunity* 2013;39:38–48. https://doi.org/10.1016/j.immuni.2013.07.004.
197. Bonifaz LC, Bonnyay DP, Charalambous A, Darguste DI, Fujii SI, Soares H, et al. In vivo targeting of antigens to maturing dendritic cells via the DEC-205 receptor improves T cell vaccination. *J Exp Med* 2004;199:815–24. https://doi.org/10.1084/jem.20032220.
198. Caminschi I, Maraskovsky E, Heath WR. Targeting dendritic cells in vivo for cancer therapy. *Front Immunol* 2012;3:13. https://doi.org/10.3389/fimmu.2012.00013.
199. Hartung E, Becker M, Bachem A, Reeg N, Jäkel A, Hutloff A, et al. Induction of potent CD8 T cell cytotoxicity by specific targeting of antigen to cross-presenting dendritic cells in vivo via murine or human XCR1. *J Immunol* 2015;194:1069–79. https://doi.org/10.4049/jimmunol.1401903.
200. Terhorst D, Fossum E, Baranska A, Tamoutounour S, Malosse C, Garbani M, et al. Laser-assisted intradermal delivery of adjuvant-free vaccines targeting XCR1 + dendritic cells induces potent antitumoral responses. *J Immunol* 2015;194:5895–902. https://doi.org/10.4049/jimmunol.1500564.
201. Fossum E, Grødeland G, Terhorst D, Tveita AA, Vikse E, Mjaaland S, et al. Vaccine molecules targeting Xcr1 on cross-presenting DCs induce protective CD8+ T-cell responses against influenza virus. *Eur J Immunol* 2015;45:624–35. https://doi.org/10.1002/eji.201445080.

Łukasz Kiraga, Paulina Kucharzewska, Damian Strzemecki,
Tomasz P. Rygiel and Magdalena Król

7 Non-radioactive imaging strategies for *in vivo* immune cell tracking

Abstract: *In vivo* tracking of administered cells chosen for specific disease treatment may be conducted by diagnostic imaging techniques preceded by cell labeling with special contrast agents. The most commonly used agents are those with radioactive properties, however their use in research is often impossible. This review paper focuses on the essential aspect of cell tracking with the exclusion of radioisotope tracers, therefore we compare application of different types of non-radioactive contrast agents (cell tracers), methods of cell labeling and application of various techniques for cell tracking, which are commonly used in preclinical or clinical studies. We discuss diagnostic imaging methods belonging to three groups: (1) Contrast-enhanced X-ray imaging, (2) Magnetic resonance imaging, and (3) Optical imaging. In addition, we present some interesting data from our own research on tracking immune cell with the use of discussed methods. Finally, we introduce an algorithm which may be useful for researchers planning leukocyte targeting studies, which may help to choose the appropriate cell type, contrast agent and diagnostic technique for particular disease study.

Keywords: BLI, cell labeling, cell tracking, CT, diagnostic imaging, FLI, leukocytes, MRI

7.1 Introduction

The effectiveness of cell-based therapy depends on the successful targeting of pathological lesions by adoptively transferred leukocytes. To determine the relevance of the cell-based therapies in pre-clinical and clinical studies, the efficiency of targeting the organ of interest is the key issue [1]. *In vivo* tracking of administered cells chosen for specific disease treatment may be conducted by diagnostic imaging techniques preceded by cell labeling with special contrast agents. The most commonly used agents are those with radioactive properties and they are often described in scientific articles [2–4]. Therefore, this review paper focuses on the essential aspect of cell tracking with the exclusion of radioisotope tracers. Here, we compare methods of cell labeling, application of different types of contrast agents (cell tracers), and application of various techniques for cell tracking, which are commonly used in preclinical or clinical studies. Diagnostic imaging methods belonging to three groups are discussed: (1) Contrast-enhanced X-ray imaging, (2) Magnetic resonance imaging, and (3) Optical imaging.

7.2 X-ray computed tomography (CT)

The first use of X-rays for medical purposes was performed by Wilhelm Roentgen in 1895. Since then, X-ray imaging has grown into one of the most important diagnostic techniques. Many years later, in 1979, the Nobel Prize in Physiology and Medicine was jointly awarded to Allan M. Cormack and Godfrey N. Hounsfield "for the development of computer-assisted tomography" [5]. In contrast to classical X-ray radiography, CT imaging generates many X-ray images obtained from different angles around the body and makes use of computer-based reconstruction methods to create cross-sectional images of the examined object inside the body. Over the last decades, immense research has been performed to improve speed, resolution, and patient comfort, providing excellent imaging quality applying the lowest possible X-ray dose. Due to its broad clinical accessibility, low costs and high temporal resolution, CT has developed into one of the most frequently used imaging methods in medicine and a potential imaging technique for cell tracking [6, 7].

The CT image contrast results from differential X-ray attenuation by tissue. Bone absorbs X-rays strongly, soft tissues to some extent, whereas air absorbs X-rays very weakly. This strong contrast between bones, soft tissues and air makes CT imaging excellent for imaging certain body structures, such as the lungs, the skeleton, and calcified tissue (e.g., kidney stones). Low sensitivity and limited contrast of CT images of soft tissues result from small differences in X-ray attenuation of these tissues, and therefore without an appropriate contrast agent, soft tissues cannot be visualized with this imaging technique.

7.2.1 CT contrast agents

CT contrast agents are divided into two main groups: 1) small molecule and 2) nanoparticle CT contrast agents. For visualization of blood vessels and other organs, administrations of exogenous small molecule CT contrast agents are used. The most commonly used contrast agents in the clinic are barium sulfate suspensions, which are used for digestive tract imaging, and iodinated contrast agents, employed for vascular and digestive track imaging [8, 9]. Iodinated contrast agents, using since 1950s [10], generate low contrast because of their low payloads. Moreover, they are characterized by short blood half-life, which requires rapid CT imaging after injection. Finally, they are quickly cleared via kidneys what can cause renal damage in patients with impaired kidney function. To overcome these shortcomings of small molecule CT contrast agents, recent research focus has been on the development of other CT contrast agents, especially nanoparticle CT contrast agents.

Nanoparticles are very small chemical structures that have between 1 and 1000 nm in one dimension. They are often circular, but can be of different shapes, such as cages, stars, and rods [11]. Nanoparticles are now widely used in the field of nanomedicine as therapeutics (e.g., Doxil is the FDA-approved liposomal formulation of doxorubicin for the treatment of certain cancers [12]) or siRNA delivery systems. Nanoparticles can also be applied as tracking or contrasting compounds in various medical imaging techniques, such as MRI, PET and fluorescence imaging [13]. Iron oxide nanoparticles used for MRI imaging have been clinically approved since the mid-1990s [14]. The knowledge obtained from the use of nanoparticles in these biomedical imaging modalities is now being applied to the development of contrast agents for CT. The material for generating contrast for CT is usually located in the particle's core. The most frequently used contrast generating elements are iodine, gold, or bismuth, however the use of other elements has also been reported [11]. The core is coated with various compounds, such as a lipid, polymer, silica, protein, that provide the desired circulation times, biocompatibility, biodistribution and solubility in biological media. Additionally, the coating can be changed with various targeting moieties, such as proteins, antibodies, aptamers and others. CT nanoparticles may also be loaded with tracers for other imaging modalities, such as fluorophores for optical imaging or gadolinium chelates for MRI [13]. Eventually, in the coating or core of nanoparticles there can be encapsulated drugs or nucleic acid for the development of 'theranostic' nanoparticles, showing both therapeutic and diagnostic effects. Different types of nanoparticles have been used as contrast agents for CT, such as solid core nanoparticles (metal, metal salt or metal alloy), lipid-based structures (emulsions, liposomes, lipoproteins or micelles), or combinations of two of these.

Nanoparticle contrast agents for CT imaging have several, potential, interesting applications, including *in vivo* cell tracking in pre-clinical and clinical settings. There are some criteria to consider to design and select appropriate nanoparticle CT contrast agents for *in vivo* cell tracking. Firstly, nanoparticles should be biocompatible, which means that their uptake should not undermine cell viability and any other cellular activity, such as migration, expression of cell surface markers, and response to the stimuli. Secondly, the selected nanoparticles have to preserve their chemical properties inside the labeled cells to avoid adverse effects on cells. Thirdly, the nanoparticles should deliver large numbers of contrast agent to allow visualization of the injected cells. Among various nanoparticle contrast agents, gold nanoparticles (AuNPs) have so far been the most studied as CT factors for cell tracking because of their high density (d = 19.3 g/cm^3), biocompatibility, stability, sustained contrast and no adverse effects on cells. Given these properties, AuNPs are the next generation of contrast agents for CT imaging, with a wide range of clinical and pre-clinical applications [15].

7.2.2 CT-based immune cell tracking

Until now, most cell tracking studies have been focused on imaging methods, such radionuclide imaging, MRI and optical imaging [4]. However, CT is a promising technique for cell tracking that has several strengths, including wide clinical availability, low cost, high spatial and temporal resolution, and exceptional quantitative capabilities. Despite the low sensitivity of CT, the role of CT in non-invasive cell tracking is expected to grow, owing to the recent developments in novel contrast agents, reconstruction algorithms, and CT scanners.

The present studies of immune cell tracking with CT technique use only direct labeling of cells. This approach involves incubation of target cells with contrast agents *in vitro* before transplantation. The contrast agents either enter the cells in a process of endocytosis or phagocytosis or are bound to the cell surface. Before being administered into the subject, the labeled cells are purified from excess contrast agents. This labeling approach has several limitations, including loss of signal due to cell division and exocytosis, and inability to distinguish between living cells and dead cells, which can result in wrong interpretation of imaging results. These shortcomings may be potentially overcome with the development of indirect cell labeling methods based on reporter genes encoding receptors, enzymes or transporters that mediate accumulation of naturally occurring CT attenuating factors in the body (for example iodine) or nanoparticles.

Immune cell tracking may lead to understanding the mechanisms underlying the development of disorders, such as atherosclerosis and arthritis, in which immune cells play a crucial role. Furthermore, it can improve immune cell-based therapy approaches, which have gained attention as innovative antitumor therapy. Recent study by Meier et al. demonstrated the successful use of CT imaging for immune cell tracking [16]. T cells transduced to express melanoma-specific T cell receptors were loaded with glucose-coated AuNPs to noninvasively track these cells using conventional X-ray CT. Importantly, AuNPs labeling of T cells did not affect their function as shown by IFNγ secretion by these cells upon co-culture with target human melanoma cells. For *in vivo* studies, AuNP loaded T cells that also expressed GFP, were injected intravenously into melanoma-bearing mice. After 24 h of cell administration, micro CT revealed substantial contrast in the tumor site, indicating on the recruitment of labeled T cells to the tumor tissue. Quantitative analysis of CT images showed that about 460 000 cells targeted the tumor site after 48 h. Moreover, the results obtained from the CT analysis were corroborated by the fluorescence imaging of GFP, indicating that the CT contrast reflected immune cell trafficking to the tumor tissue. Trafficking of engineered T cells to the tumor site was further confirmed by increased tumor regression compared to control animals, which were injected with non-targeting T cells.

Monocyte recruitment plays an important role in atherosclerotic plaque progression, therefore monocyte tracking with CT imaging seems to be of high importance.

Migration of AuNP-labeled monocytes into atherosclerotic plaques was recently noninvasively tracked using CT by Chhour et al. [17]. In this study, AuNPs of 15 nm size were coated with various ligands (e.g., 11-mercaptoundecanoic acid (11-MUA), 16-mecaptohexadecanoic acid (16-MHA), poly(ethylenimine) (PEI), 4-mercapto-1-butanol (4-MB), 11-mercaptoundecyl-tetra(ethylene gycol) (MTEG) and others) to analyze which coating provides stability of nanoparticle in biological media without adverse effect on monocyte cell viability and function upon internalization. Monocyte incubation with various AuNPs for 24 h showed that AuNP preparations did not affect viability, except for PEI-coated AuNP. Furthermore, measurement of TNF-α and IL-6 cytokine secretion from labeled monocytes showed that apart from 4-MB coated AuNPs, these AuNPs did not affect cytokine release. Transmission electron microscopy (TEM) analysis of sections of monocytes after AuNP incubation revealed that AuNPs localized in vesicles within the cells. Finally, CT analysis of labeled cell pellets showed that the most efficient cell uptake was observed for 11-MUA and 4-MB coated AuNPs. Based on this *in vitro* data, 11-MUA coated AuNPs were chosen for *in vivo* evaluation. Primary mouse spleen-derived monocytes were incubated with 11-MUA AuNPs without any negative effect of AuNPs on monocyte activity. Upon labeling with 11-MUA AuNPs, primary monocytes were intravenously administered into a mouse model of atherosclerosis. Five days later, the mice were imaged with CT, and image analysis revealed attenuation in the aorta, reflecting monocyte trafficking to atherosclerotic plaques. These *in vivo* CT imaging results were validated by TEM analysis of the aortic sections showing AuNPs accumulation within monocytes in the plaques of mice, confirming that the *in vivo* CT imaging results showed monocyte recruitment.

There was also an interesting approach which concerned small molecule contrast agent application in CT cell tracking study. The alveolar macrophages were incubated with Micropaque CT (Guerbet, France) – a clinically used CT contrast agent containing barium sulfate, which was simply phagocytosed. No signs of cellular toxicity were shown in 24 h after intratracheal instillation, macrophages loaded with barium sulfate were clearly visible in CT images as cluster structures in the whole asthmatic lung. This approach presented a potential technique of asthma diagnosing [18].

In summary, despite the low sensitivity of CT, immune cell tracking with this modality is feasible. The majority of studies evaluating immune cell tracking with the CT technique have used AuNPs as contrast agents, but given the multiple reports on other CT tracers, it is anticipated that immune cell tracing will be performed with other labels. The issue of CT low sensitivity can also be addressed with research focused on the development of novel cell labeling methods, including indirect cell labeling techniques with reporter genes. Given recent developments in imaging systems and contrast agents, immune cell tracking with CT modality will likely be reported in the near future.

7.3 Magnetic resonance imaging (MRI)

MRI technique is based on the relaxation properties of hydrogen atoms in water. When a tissue is exposed to a strong magnetic field generated by the MRI scanner, the magnetic moments of the protons in water molecules are synchronized with the direction of the field. Then, a radio frequency of electromagnetic waves causes protons to obtain a higher energy level. The magnetic moments of protons with increased energy differs from the moment of the applied magnetic field. When the radio waves are turned off, the nuclei return to their equilibrium state through Brownian motion (relaxation). The protons produce a radio-frequency signal detected by the scanner. This signal is used to generate a detailed image of the tissues. The resolution of MRI imaging mainly depends on the type of tissue and the optimization of the scanner settings. However, the enhancement of the magnetic field through technical improvements allows a better signal-to-noise ratio and greater spatial specificity in 7T or more powerful MRI scanners [19].

Currently, clinical applications do not use MRI scanners with a stronger field than 3T. Therefore, it is necessary to use contrast agents. There are several ways of grouping contrasting agents for MRI use. Depending on the imaging method, we divide the contrast media into those operating in T1 and T2 sequences. Most of the compounds used in both groups are extracellular agents. They are imaged immediately after intravenous administration and are quickly eliminated [20].

7.3.1 Paramagnetic contrast agents (T1 – shortening)

In clinical application, the most popular contrast agents are those accelerating the rate of return of water molecule protons to equilibrium. This rate is usually represented by the T1 relaxation time constant, so such contrasting factors are referred to as T1 agents [21]. This group includes numerous paramagnetic contrast agents based on gadolinium ions (Gd^{3+}). Gd^{3+} is the most effective paramagnetic ion in terms of T1 relaxivity but has a high toxicity in the free state. Gd^{3+} ions have been widely used as a tracking beacon for many types of cells. Labeled monocytic cells, endothelial progenitor cells, or mesenchymal stem cells were tested in cell transplantation animal studies. In a rat model, intramuscularly injected hematopoietic progenitor cells, efficiently labeled with dextran mono-N-succinimidyl 1,4,7,10-tetraazacyclododecane-1,4,7,10-tetraacetate-gadolinium3+ (Dex-DOTA-Gd^{3+}) were successfully tracked to hind limb ischemic regions [22]. In another approach, mesenchymal stem cells were labeled with Gd-diethylene triaminepenta acetic acid (Gd-DTPA) incorporated in cationic liposomes. The cells were injected s.c. or i.m. to naïve rats and imaged through the 14-day period [23]. These contrast agents in the chelate form are taken up through pinocytosis [24]. The major drawback concerning this contrast agent, however, is low uptake by the cells. For this reason, it is necessary to implement new methods of more effective internalization. There is an interesting study published, showing Gd_2O_3 nanoparticles uptake by the hematopoietic progenitor cell line Ba/F3 and the monocytic cell line THP-1 using protamine sulfate as a transfecting factor [25].

7.3.2 Superparamagnetic contrast agents (T2 – shortening)

The second significant groups of MRI contrasts are superparamagnetic compounds. These contrast agents create 20-µm local magnetic fields which violate the main magnetic field of the MRI device. Local magnetic fields can disturb the MR frequencies of individual water protons, causing incoherence of the MR frequencies of the protons. This rate of signal reduction is often represented by the T2 relaxation time constant; thus such contrast agents are known as T2 agents [26].

Superparamagnetic iron oxide nanoparticles (SPIONs) and ultra-small superparamagnetic iron oxide (USPIONs) are nanoparticles used as contrast agents that make labeled cells dark in the T2 imaging sequences. Iron nanoparticle size varies from <50 nm (USPIONs) to 3.5 µm (SPIONs) [27].

Unfortunately, SPIONs are not viable in an aqueous medium and non-phagocytic cells cannot internalize them. However, phagocytic cells such as macrophages uptake the SPIOs efficiently in *in vitro* conditions [28]. Even if these particles form aggregates, they can be sequestrated by macrophages [29]. There is also a possibility of direct labeling of dendritic cells (DCs) with SPIOs. Tavare et al. labeled bone marrow-derived DCs by incubating them with SPIONs and their phenotype, viability and functions were comparable to unlabeled DCs. After a subcutaneous injection to mice, their tracking by MRI was efficient in the 96 h imaging time period [30].

In our own studies we successfully detected RAW264.7 macrophages loaded with ferritin encapsulated with ferrihydrite (Fn-Fh) in mammary EMT6 tumors after their administration into tail vein of BALB/c mice. The mice were imaged before administration of macrophages and then 22 h later. As a control, we used macrophages loaded with FITC-conjugated ferritin (Fn-FITC) which cannot be detected in MRI. Reduction of T2 relaxation time in the tumors was observed, due to the accumulation of ferritin encapsulated with Fh, which proved efficient targeting of macrophages to the cancer tissue (Figure 7.1).

The most commonly used approach to non-phagocytic cell labeling is the use of transfection agents. The surface of SPIONs can be modified to enhance biocompatibility. Two common methods are: creating magneto liposomes in the process of encapsulation [31] and *in situ* coating with various factors during the process of synthesis and post-synthesis [32].

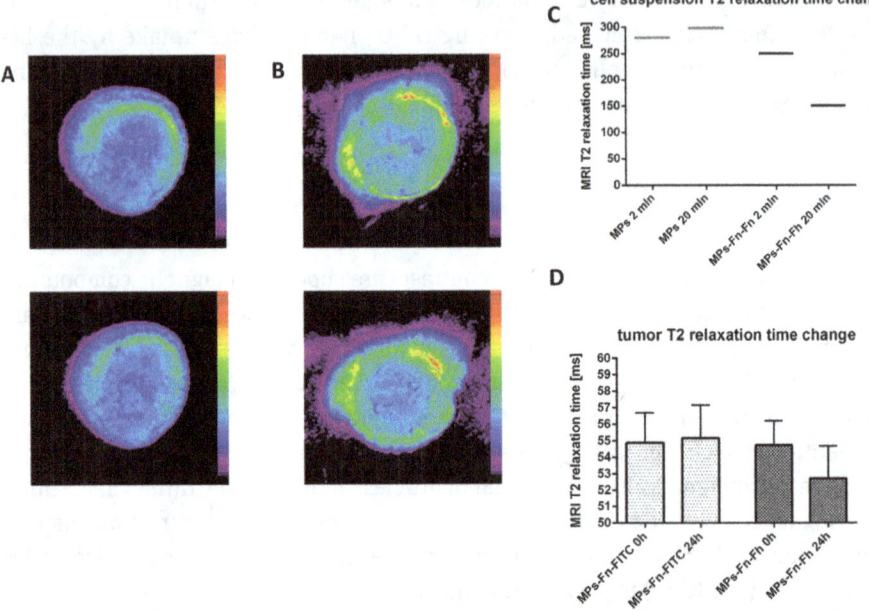

Figure 7.1: Representative MRI images of mammary tumor cross-sections. **A**–of a mouse treated with macrophages loaded with Fn-FITC as a control; upper – before *i.v.* injection and lower – 24 h after *i.v.* injection. **B**–of a mouse treated with macrophages loaded with Fn-Fh; upper – before *i.v.* injection and lower – 24 h after *i.v.* injection. In the image lower-B, obtained after the injection of Fn-Fh macrophages, slight decrease of color intensity of the center of the tumor can be noted, which represents the shortening of T2 relaxation time, specific to Fh contrasting agent. In the case of the control approach (lower-A), after the administration of Fn-FITC macrophages, the tumor's color intensity is unchanged or slightly higher. The phenomenon shown in the images corresponds with quantitative analysis results: *in vitro* macrophage (MPs) suspensions' relaxation time changes (**C**) and mouse tumors' relaxation time changes after different samples of macrophages administration (**D**).

7.3.3 "Hot-spot" MRI

Two techniques described below are termed "hot-spot magnetic resonance imaging". Their physical mode of action completely differs from paramagnetics – they do not affect the relaxation time of endogenous water nuclei, but induce signal independently – directly from the nucleus of fluorine-19 (^{19}F) or by acting on the nuclei in the local structure of the molecule that chelates dysprosium or thulium ions (in highly-shifted proton MRI). Their greatest value is in the property of rendering even small, focal shaped regions of contrast without the interference of background signal from endogenous ^{1}H. This feature makes them ideal contrasting agents for cell tracking approaches, although mainly in preclinical studies.

Soon after the invention of proton MRI, the utility of ^{19}F MRI was demonstrated in 1977 by Holland et al. [33]. The ^{19}F is a natural halogen, a non-radioactive isotope of fluorine. The physical principles behind the detection and image formation of ^{19}F-based contrasts are analogous to endogenous water. ^{19}F has a relative sensitivity of 83% compared to ^{1}H and is absent in biological tissues providing background-free imaging of ^{19}F-based probes [34]. Simply stated, ^{19}F is a versatile nuclear magnetic resonance active nucleus. Therefore, ^{19}F acts completely differently compared to metal-ion-based contrast agents, which are detectable through their indirect effects on the surrounding water protons. There are various compounds containing ^{19}F suitable for cell labeling, chemically belonging to perfluorocarbons (PFC): perfluoropolyether, perfluoro-15-crown-5-ether, hexafluoroethane, sulfur hexafluoride, perfluoropropane [35]. PFC nanoemulsions are specifically engineered to be endocytosed, even by non-phagocytic cells [36]. Easy application of PFC emulsion leads to their wide usage. Therefore, many types of cells labeled this way were tracked with MRI. *In vivo* cell tracking using PFCs was first used to track dendritic cells' migration in mice [37]. Another approach involved tracking of transgenic DO11.10 mouse-derived Tc lymphocytes. They were efficiently labeled with perfluoropolyether and then intraperitoneally injected into mice with induced inflammation. MRI revealed a signal only from the inguinal draining lymph node on the side of the antigen transfer, which was increasing up to day 7 and gradually decreased until the 21st day after transplantation [38]. Bouchlaka et al. stated that human NK cells can be detected with the use of longitudinal MRI for up to 8 days post intatumoral administration in NSG mice bearing human xenograft tumors [39].

7.3.4 Highly-shifted proton (HSP) MRI

The physical mode of action of another group of MRI contrast agents is far different. Certain lanthanides, in particular thulium (Tm) and dysprosium (Dy), cause a chemical shift of the proton resonance frequency far away from the water peak, from which the MR imaging signal is normally collected. These two lanthanates acting as contrasts are chelated by molecules of 1,4,7,10-tetraazacyclododecane-$\alpha,\alpha',\alpha'',\alpha'''$-tetramethyl-1,4,7,10-tetraacetic acid (DOTMA). If cells are labeled with Tm-DOTMA, an MRI image of the labeled cells can be acquired without a signal from tissue water [40]. In the context of the cell labeling method, MRI signal comes only from labeled cells. Schmidt et al. evaluated the functionality and viability of bone marrow-derived macrophages (BMDMs) marked with Tm-DOTMA for the purpose of targeting inflammation induced by subcutaneous injection of polyacrylamide gel pellets. The labeled cells were injected intravenously, and MRI analysis showed strong Tm-DOTMA signal intensity observable for 7 days in the sites of inflammation. *In vitro* phantom imaging determined a detection threshold of approximately 600 cells.

The authors claim that the labeling of macrophages most likely occurs by pinocytosis and this process does not significantly affect their viability [41].

7.3.5 CEST MRI

Contrast Enhancement by Saturation Transfer (CEST) MRI is a different MRI technique where a narrow pre-saturation pulse is used to target a pool of chemically exchangeable protons (i.e., O–H, N–H_2, N–H) within a narrow chemical shift range. If the saturation pulse is applied for long enough this pool of protons will chemically exchange with the bulk water protons in the surrounding tissue transferring their saturation and causing a slight dimming of MRI signal in the region of the CEST agent. A subtraction of the CEST MRI image from the corresponding non CEST image will give bright spots only in the location of the CEST agent, which can be overlaid on the non-CEST image [42]. CEST is a multicolor imaging technique where several individual CEST agents can be detected within the same sample provided their chemically exchangeable protons have a different chemical shift range. A new appropriately targeted CEST scan is required for each CEST agent to be detected. Early implementations of CEST imaging involved using lysine-rich (N–H_2), serine-rich (O–H), and arginine-rich (N–H) proteins as the proton source [43] and was successfully used to visualize a graft of glioma cells genetically engineered to produce lysine-rich reporter protein (LRP) in rats [44]. However, CEST for cell tracking with diamagnetic proton sources is hampered by low sensitivity and it can be challenging to engineer mammalian cells to produce the nM-mM concentrations of protein required for detection. Progress in the use of CEST for cellular tracking has been reviewed by McMahan and Gilad [45].

7.3.6 PARACEST MRI

The CEST saturation pulses can be targeted to the highly-shifted protons induced by the paramagnetism of thelanthanide complexes used for HSP MRI to generate PARACEST images [46, 47]. The advantage of this approach for cell tracking is that there is no need to genetically engineer cells to express the CEST reporter, instead the cells need to be encouraged to take up the lanthanide paramagnetic shift complexes. Another significant value of this technique is the possibility to track two different types of cells at the same time, provided that two different contrasting agents are used for the cell labeling [48]. However, as with the protein-based CEST imaging of cells, sensitivity is an issue with a similar1 to 10 mM concentration range of PARACEST agent required to generate an image. For the reader interested in further detail the advantages and disadvantages of the key MRI-based cell tracking methodologies are expertly reviewed and compared by Srivasta et al. [49].

An evidence of the effectiveness of this method in cell tracing is provided by Ferrauto et al. Mouse macrophages (J774.A1) and melanoma cells (B16-F10) were labeled with Yb- and Eu-HPDO3A complexes, respectively. After injection of the cellular mixture into mice flanks, two differently labeled populations of cells were detectable *in vivo* using appropriate radiofrequency and that distinguishing each of the two types of cells in the same area was possible [47].

7.4 Optical imaging

There are two subtypes of this technique: bioluminescence imaging (BLI) and fluorescence imaging (FLI). Both techniques are only applicable in pre-clinical studies. They are both based on the detection of photons emitted by an examined object. The significant difference between these two techniques is that in the case of the BLI, the imaged object must contain genetically modified cells and, comparing with the FLI, the cells do not need to contain an additional gene, as they can be labeled with certain fluorophores. Another difference is that the BLI requires a substrate for the enzyme to be administrated and the FLI needs exposure to light of a specific wavelength.

7.4.1 Bioluminescence Imaging (BLI)

BLI is a technique of optical molecular imaging based on the detection of a biological process called bioluminescence. This phenomenon requires the presence of a luciferase enzyme, a substrate and oxygen. Some luciferases require other cofactors, such as ATP or Mg^{2+}. Since mammalian cells do not have luciferase genes, they need to be modified genetically in order to express this enzyme. In a summary, bioluminescence imaging requires three pivotal elements: a luciferase gene-based reporter constructs, which is expressed in selected animal cells; administering the substrate to the transgenic animal, and lastly, obtaining and processing light signals using dedicated low-light imaging systems [50]. Some of the most frequently used bioluminescent reporters with their substrates and emission peaks are: North American firefly *(Photinus pyralis)* luciferase (d-Luciferin, 600 nM), Sea pansy *(Renilla reniformis)* luciferase (coelenterazine, 480 nM), click beetle *(Pyrophorus plagiophthalamus)* luciferase (d-Luciferin 613 nM or 537 nM), *Gaussia princeps* luciferase (coelenterazine, 480–600 nM) [50].

Bioluminescence imaging is cheap, easy to use and the costs of instrumentation are relatively low. BLI may provide the highest sensitivity and highest quality data, especially in small animals. This is because of the absence of natural, endogenous luciferase expression in mammalian cells. BLI can also contribute to reducing the number of animals required for experiments, since multiple measurements can be taken from the same animal over time, minimizing the effects of biological variation [51], [, 52]. On the other hand, bioluminescence imaging has a few drawbacks. First of all, it has very poor spatial resolution compared to other methods described in this article. Moreover, BLI has a limited photon penetration depth, in contrast to MRI and SPECT/PET, which achieve unlimited tissue penetration [53]. Limited photon penetration is followed by low quantification accuracy related to loss and diffraction of light in the body. Importantly, BLI does not require an excitatory light source, in contrast to fluorescence imaging. However, the light signal provided by BLI is usually dim, which makes the subsequent microscopy and micro endoscopy more challenging.

Many important studies have already been conducted with the help of bioluminescence imaging. Nguyen et al. for instance, conducted a longitudinal, noninvasive study, in which luciferase-expressing regulatory T cells (Tregs) were tracked in an allogeneic bone marrow transplant model [54]. The aim was to demonstrate their co-localization with effector T cells, and also to show their spread in secondary lymphoid organs before migration into inflamed tissues. Shin et al. in turn, indicated the distribution of macrophages to the atherosclerotic lesion in an *in vivo* mouse model, using bioluminescence imaging. Their observations indicated that macrophage chemotaxis is far stronger in mechanical atherosclerosis than in lipogenic atherosclerosis [55]. Lee et al. adopted BLI to observe the recruitment of macrophages to chemically induced inflammatory changes in a mouse model. Bioluminescence turned out to be the most efficient method amongst the ones used in the experiment [56].

7.4.2 Fluorescence imaging (FLI)

In vivo FLI is a method of detecting fluorescence emission from fluorophores in the whole bodies of small, living animals, which involves the use of a sensitive, low-light camera with appropriate filters [57]. The FLI method relies on light emission from either a fluorophore or a fluorescent protein (FP) after excitation by a light source at the appropriate wavelength, since every fluorophore and FP has their unique excitation and emission profile.

FLI, along with BLI are the two most commonly used optical imaging methods, but since they both exploit different energy sources to generate the signal, they have different biological and technical applications. FLI is less sensitive than BLI with a common high background signal. FLI does not require any preparatory procedures, which makes it straightforward and uniquely suited for visualization in live tissue. FLI is also believed to be a better alternative for visualizing tumors than BLI, because it is still not clear whether luciferase reporter genes used in BLI can function in a stable manner over significant time periods in cancer cell lines, solid tumors and their metastases [58, 59].

Cells can exhibit fluorescence based on their expression of fluorescent proteins (FP). The most frequent practice is to introduce a transgene (reporter gene) into the cell. The transgene encodes for fluorescent protein, which acts as a reporter probe.

Transcription of the gene results in the production of fluorescent protein, which can subsequently be detected with optical imaging techniques [60]. Another method used in FLI is direct labeling of cells. Cells can be simply labeled with an affinity ligand specific to a certain target (e.g., an antibody), conjugated to a fluorophore [61]. A characteristic feature of fluorophores is that they fluoresce even when they are not bound to their target, which may lead to nonspecific background signals [62].

Some of the most frequently used FPs and fluorophores are listed below Table 7.1.

Table 7.1: The most frequently used fluorescent proteins and fluorophores with their characteristics.

Protein (Acronym)	Fluorescence Color	Excitation Maximum (nm)	Emission Maximum (nm)
EBFP	Blue	383	445
Azurite	Blue	384	450
GFP	Green	395/475	509
mTurquoise	Cyan	434	474
ECFP	Cyan	439	476
EGFP	Green	484	507
EYFP	Yellow	514	527
mOrange	Orange	548	562
mRuby	Red	558	605
Fluorophore	**Fluorescence color**	**Excitation Maximum (nm)**	**Emission Maximum (nm)**
Alexa Fluor 405	Blue	401	421
FITC	Green	490	525
Alexa Fluor 488	Green	495	519
Rhodamine	Yellow	530	580
Alexa Fluor 546	Yellow	556	573
Alexa Fluor 647	Far red	650	665
DiR	Near infrared	748	780

FLI is used to visualize diverse leukocyte behaviors in many contexts, including their homing abilities [63], cell proliferation [64], cell death [65], and cell–cell interactions [66]. What is worth mentioning is that FLI is nowadays commonly used to track monocytes, macrophages, DCs and T cells in tumors [53, 67].

Nakajima et al. studied CD4 T cells transfected with GFP to track them in mice with induced arthritis. One of the outcomes was that T cells inhibited collagen-induced arthritis by suppressing autoimmune response at the site of inflammation [68]. Lim et al. labeled human NK cells with anti-human CD56 antibody-coated quantum dots and injected them directly into tumors to observe their cytotoxicity [69]. The aim of the study was to demonstrate that cells used in immunotherapy can be labeled with fluorescent crystals and maintain their therapeutic effects and viability, which is essential in monitoring cell-based cancer therapies. Chtanova et al. in turn, used FLI to investigate a host-pathogen interaction by using GFP-expressing neutrophils and observing their colocalization with an intracellular parasite, *Toxoplasma gondii* [70]. They observed that neutrophils can form either small and transient or large and long-lasting clusters via impressively coordinated migration patterns. Fluorescence imaging allowed to investigate the role of macrophages in preventing the neurotropic vesicular stomatitis virus (VSV) from accessing cells. In this study, GFP was expressed by macrophages and cells were infected with VSV. Thanks of that, it was possible to observe that macrophages prevent lymph-borne neurotropic viruses from infecting the central nervous system (CNS) [71].

Fluorescent proteins and dyes produce free radicals and other highly reactive breakdown products in their excited state and while photobleaching, FLI is potentially dangerous to observed cells and other *in vitro* samples. Free radicals can easily impair the biological functions and chemical structure of biomolecules [72]. Naturally, all cells have numerous ways to cope with free radicals, and as long as the mechanisms are not overwhelmed, cells can tolerate fluorescence excitation. Interestingly, the intensity of phototoxicity depends on a large extent on the fluorophore. Fluorescent proteins, for instance, have a tendency to be the least phototoxic, because the photobleaching chemistry is enclosed within the β-barrel structure. The best way to reduce photobleaching and its consequences is simply reducing light exposure and its intensity [73]. The accuracy of the FLI result, however, can be affected by a dye influence on the antibodies binding affinity. It was noted that conjugation of fluorescent dyes and antibodies decreases antibody affinity in most cases [74].

Our own studies of tumor homing of fluorescently labeled macrophages showed that they can be easily detected using IVIS® Spectrum *In Vivo* Imaging System. We labeled BMDMs with DIR and administered *i.v.* into the tail vein of BALB/c mice with EMT6 mammary tumors. Analyzing the images 24 h post BMDM administration, we were able to detect clear signals from the tumor (right thigh) and some signal from the liver and the spleen (Figure 7.2).

7.5 Summary

As presented in the manuscript, there are various techniques for efficient cell tracking without the use of radio-isotopes. If SPECT and PET are excluded, *in vivo* preclinical studies concerning cell tracking may be successfully conducted using MRI, BLI, FLI, and even CT techniques, however, only MRI and CT can be translated into clinical studies. Therefore, if the use of diagnostic techniques based on

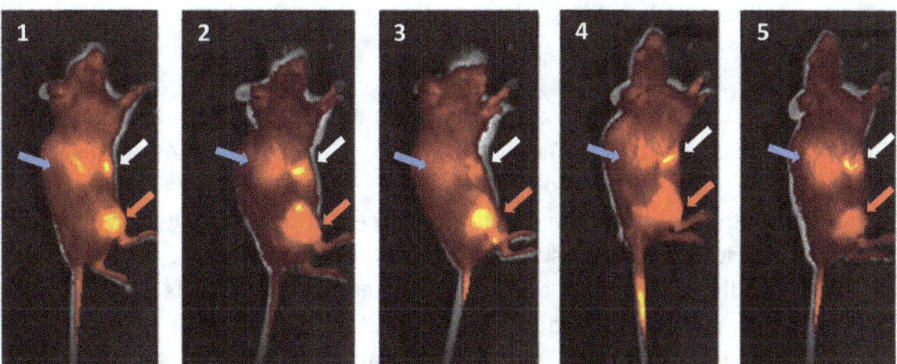

Figure 7.2: FLI images of five BALB/c mice 24 h after BMDMs *i.v.* administration; the arrows indicate the regions with high signal: red – from tumors, white – from spleens, and blue – from livers.

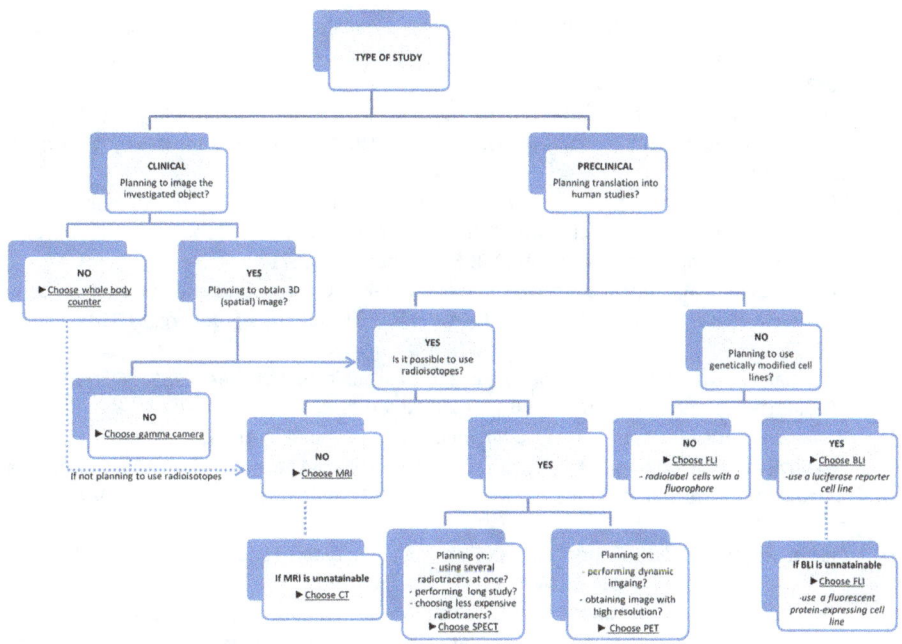

Figure 7.3: Diagram presenting an algorithm of choosing diagnostic technique and cell labeling method, relevant to the study plan.

the application of radioisotopes is not considered, there is still the possibility to effectively evaluate cell tracking using different, but not less effective methods. Finally, we present a flow diagram (Figure 7.3), which is a guideline to find a proper diagnostic tool among all potential, applicable imaging techniques, depending on the specifics of the research.

Acknowledgments: We would like to kindly acknowledge Mr. Kijan Crowley for his linguistic verification and grammar corrections. We thank MSc Jarosław Olszewski for researching articles related to MRI.

Author contributions: All the authors have accepted responsibility for the entire content of this submitted manuscript and approved submission.

Research funding: This paper and research cited in this article is funded by the European Research Council Starting Grant McHAP: 715048 (MKról).

Conflict of interest statement: The authors declare no conflicts of interest regarding this article.

References

1. Bulte JWM, Daldrup-Link HE. Clinical tracking of cell transfer and cell transplantation: trials and tribulations. *Radiology* 2018;289:604–15. doi:https://doi.org/10.1148/radiol.2018180449.
2. Farahi N, Loutsios C, Tregay N, Summers C, Lok LSC, Ruparelia P, et al., Radiolabelled leucocytes in human pulmonary disease. *Br Med Bull* 2018;127:69–82. doi:https://doi.org/10.1093/bmb/ldy022.
3. Weist MR, Starr R, Aguilar B, Chea J, Miles JK, Poku E, et al., PET of adoptively transferred chimeric antigen receptor T cells with 89Zr-oxine. *J Nucl Med: Off Publ Soc Nucl Med* 2018;59:1531–7. doi:https://doi.org/10.2967/jnumed.117.206714.
4. Perrin J, Capitao M, Mougin-Degraef M, Guérard F, Faivre-Chauvet A, Rbah-Vidal L, et al., Cell tracking in cancer immunotherapy. *Front Med* 2020;7:34. doi:https://doi.org/10.3389/fmed.2020.00034.
5. Goodman LR. The beatles, the Nobel prize, and CT scanning of the chest. *Radiol Clin* 2010;48:1–7. https://doi.org/10.1016/j.rcl.2009.09.008.
6. Bernstein AL, Dhanantwari A, Jurcova M, Cheheltani R, Naha PC, Ivanc T, et al., Improved sensitivity of computed tomography towards iodine and gold nanoparticle contrast agents via iterative reconstruction methods. *Sci Rep* 2016;6:26177. doi:https://doi.org/10.1038/srep26177.
7. Pelc NJ. Recent and future directions in CT imaging. *Ann Biomed Eng* 2014;42:260–8. https://doi.org/10.1007/s10439-014-0974-z.
8. Oliva MR, Erturk SM, Ichikawa T, Rocha T, Ros PR, Silverman SG, et al., *Gastrointestinal tract wall visualization and distention during abdominal and pelvic multidetector CT with a neutral barium sulphate suspension: comparison with positive barium sulphate suspension and with water*. JBR-BTR 2012;95:237–42. https://doi.org/10.5334/jbr-btr.628.
9. Kurihara O, Takano M, Uchiyama S, Fukuizumi I, Shimura T, Matsushita M, et al., Microvascular resistance in response to iodinated contrast media in normal and functionally impaired kidneys. *Clin Exp Pharmacol Physiol* 2015;42:1245–50. doi:https://doi.org/10.1111/1440-1681.12479.

10. Pasternak JJ, Williamson EE. Clinical pharmacology, uses, and adverse reactions of iodinated contrast agents: a primer for the non-radiologist. *Mayo Clin Proc* 2012;87:390–402. https://doi.org/10.1016/j.mayocp.2012.01.012.
11. Kim J, Chhour P, Hsu J, Litt HI, Ferrari VA, Popovtzer R, et al., Use of nanoparticle contrast agents for cell tracking with computed tomography. *Bioconjugate Chem* 2017;28:1581–97. doi:https://doi.org/10.1021/acs.bioconjchem.7b00194.
12. Gabizon AA. Pegylated liposomal doxorubicin: metamorphosis of an old drug into a new form of chemotherapy. *Canc Invest* 2001;19:424–36. https://doi.org/10.1081/CNV-100103136.
13. Bernsen MR, Guenoun J, van Tiel ST, Krestin GP. Nanoparticles and clinically applicable cell tracking. *BJR* 2015;88:20150375. https://doi.org/10.1259/bjr.20150375.
14. Callera F, de Melo CMTP. Magnetic resonance tracking of magnetically labeled autologous bone marrow CD34+ cells transplanted into the spinal cord via lumbar puncture technique in patients with chronic spinal cord injury: CD34+ cells' migration into the injured site. *Stem Cell Dev* 2007;16:461–6. https://doi.org/10.1089/scd.2007.0083.
15. Mieszawska AJ, Mulder WJM, Fayad ZA, Cormode DP. Multifunctional gold nanoparticles for diagnosis and therapy of disease. *Mol Pharm* 2013;10:831–47. https://doi.org/10.1021/mp3005885.
16. Meir R, Shamalov K, Betzer O, Motiei M, Horovitz-Fried M, Yehuda R, et al., Nanomedicine for cancer immunotherapy: tracking cancer-specific T-cells in vivo with gold nanoparticles and CT imaging. *ACS Nano* 2015;9:6363–72. doi:https://doi.org/10.1021/acsnano.5b01939.
17. Serban MA. Translational biomaterials—the journey from the bench to the market—think 'product'. *Curr Opin Biotechnol* 2016;40:31–4. https://doi.org/10.1016/j.copbio.2016.02.009.
18. Dullin C, dal Monego S, Larsson E, Mohammadi S, Krenkel M, Garrovo C, et al., Functionalized synchrotron in-line phase-contrast computed tomography: a novel approach for simultaneous quantification of structural alterations and localization of barium-labelled alveolar macrophages within mouse lung samples. *J Synchrotron Radiat* 2015;22:143–55. doi: https://doi.org/10.1107/S1600577514021730.
19. Usselman RJ, Qazi S, Aggarwal P, Eaton S, Eaton G, Russek S, et al., Gadolinium-loaded viral capsids as magnetic resonance imaging contrast agents. *Appl Magn Reson* 2015;46:349–55. doi:https://doi.org/10.1007/s00723-014-0639-y.
20. Pierre VC, Allen MJ, Caravan P. Contrast agents for MRI: 30+ years and where are we going? *J Biol Inorg Chem* 2014;19:127–31. https://doi.org/10.1007/s00775-013-1074-5.
21. Sinharay S, Pagel MD. Advances in magnetic resonance imaging contrast agents for biomarker detection. *Annu Rev Anal Chem* 2016;9:95–115. https://doi.org/10.1146/annurev-anchem-071015-041514.
22. Agudelo CA, Tachibana Y, Hurtado AF, Ose T, Iida H, Yamaoka T. The use of magnetic resonance cell tracking to monitor endothelial progenitor cells in a rat hindlimb ischemic model. *Biomaterials* 2012;33:2439–48. https://doi.org/10.1016/j.biomaterials.2011.11.075.
23. Guenoun J, Koning GA, Doeswijk G, Bosman L, Wielopolski PA, Krestin GP, et al., Cationic Gd-DTPA liposomes for highly efficient labeling of mesenchymal stem cells and cell tracking with MRI. *Cell Transplant* 2012;21:191–205. doi:https://doi.org/10.3727/096368911X593118.
24. Navya PN, Kaphle A, Srinivas SP, Bhargava SK, Rotello VM, Daima HK. Current trends and challenges in cancer management and therapy using designer nanomaterials. *Nano Convergence* 2019;6:23. https://doi.org/10.1186/s40580-019-0193-2.
25. Bhorade R, Weissleder R, Nakakoshi T, Moore A, Tung C-H. Macrocyclic chelators with paramagnetic cations are internalized into mammalian cells via a HIV-tat derived membrane

translocation peptide. *Bioconjugate Chem* 2000;11:301–5. https://doi.org/10.1021/bc990168d.
26. Laurent S, Elst LV, Muller RN. Superparamagnetic iron oxide nanoparticles for MRI. In: Merbach A, Helm L, Tóth É, editors. *The chemistry of contrast agents in medical magnetic resonance imaging*. Chichester, UK: John Wiley & Sons, Ltd; 2013.
27. Elias A, Tsourkas A. Imaging circulating cells and lymphoid tissues with iron oxide nanoparticles. *Hematology* 2009;2009:720–6. https://doi.org/10.1182/asheducation-2009.1.720.
28. Walter GA, Cahill KS, Huard J, Feng H, Douglas T, Sweeney HL, et al., Noninvasive monitoring of stem cell transfer for muscle disorders. *Magn Reson Med* 2004;51:273–7. doi:https://doi.org/10.1002/mrm.10684.
29. Seifalian A, Bull E, Madani S, Green M, Seifalian A. Stem cell tracking using iron oxide nanoparticles. *IJN* 2014;9:1641. https://doi.org/10.2147/IJN.S48979.
30. Tavaré R, Sagoo P, Varama G, Tanriver Y, Warely A, Diebold SS, et al., Monitoring of in vivo function of superparamagnetic iron oxide labelled murine dendritic cells during anti-tumour vaccination. *PloS One* 2011;6:e19662. doi:https://doi.org/10.1371/journal.pone.0019662.
31. De Cuyper M, Joniau M. Magnetoliposomes: formation and structural characterization. *Eur Biophys J* 1988;15:311–9. https://doi.org/10.1007/BF00256482.
32. Berry CC, Wells S, Charles S, Aitchison G, Curtis ASG. Cell response to dextran-derivatised iron oxide nanoparticles post internalisation. *Biomaterials* 2004;25:5405–13. https://doi.org/10.1016/j.biomaterials.2003.12.046.
33. Holland GN, Bottomley PA, Hinshaw WS. 19F magnetic resonance imaging. *J Magn Reson* 1969;28:133–6.
34. Bachert P. Pharmacokinetics using fluorine NMR in vivo. *Prog Nucl Magn Reson Spectrosc* 1998;33:1–56. https://doi.org/10.1016/S0079-6565(98)00016-8.
35. Fox MS, Gaudet JM, Foster PJ. Fluorine-19 MRI contrast agents for cell tracking and lung imaging. *Magn Reson Insights* 2015;8:53–67. doi:https://doi.org/10.4137/MRI.S23559.
36. Janjic JM, Ahrens ET. Fluorine-containing nanoemulsions for MRI cell tracking: fluorine-containing nanoemulsions for MRI cell tracking. *WIREs Nanomed Nanobiotechnol* 2009;1:492–501. https://doi.org/10.1002/wnan.35.
37. Ahrens ET, Flores R, Xu H, Morel PA. In vivo imaging platform for tracking immunotherapeutic cells. *Nat Biotechnol* 2005;23:983–7. https://doi.org/10.1038/nbt1121.
38. Srinivas M, Turner MS, Janjic JM, Morel PA, Laidlaw DH, Ahrens ET. In vivo cytometry of antigen-specific t cells using 19 F MRI. *Magn Reson Med* 2009;62:747–53. https://doi.org/10.1002/mrm.22063.
39. Bouchlaka MN, Ludwig KD, Gordon JW, Kutz MP, Bednarz BP, Fain SB, et al., 19 F-MRI for monitoring human NK cells in vivo. *OncoImmunology* 2016;5:e1143996. doi:https://doi.org/10.1080/2162402X.2016.1143996.
40. Senanayake PK, Rogers NJ, Finney KN, Harvey P, Funk AM, Wilson JI, et al., A new paramagnetically shifted imaging probe for MRI: PARASHIFT imaging for molecular MRI. *Magn Reson Med* 2017;77:1307–17. doi:https://doi.org/10.1002/mrm.26185.
41. Schmidt R, Nippe N, Strobel K, Masthoff M, Reifschneider O, Castelli DD, et al., Highly shifted proton MR imaging: cell tracking by using direct detection of paramagnetic compounds. *Radiology* 2014;272:785–95. doi:https://doi.org/10.1148/radiol.14132056.
42. Liu G, Song X, Chan KWY, McMahon MT. Nuts and bolts of chemical exchange saturation transfer MRI. *NMR Biomed* 2013. https://doi.org/10.1002/nbm.2899.
43. McMahon MT, Gilad AA, DeLiso MA, Cromer Berman SM, Bulte JWM, van Zijl PCM. "New 'multicolor' polypeptide diamagnetic chemical exchange saturation transfer (DIACEST)

contrast agents for MRI. *Magn Reson Med* 2008;60:803–12. https://doi.org/10.1002/mrm.21683.
44. Gilad AA, McMahon MT, Walczak P, Winnard PT Jr, Raman V, van Laarhoven HW, et al., Artificial reporter gene providing MRI contrast based on proton exchange. *Nat Biotechnol* 2007;25:217–9. doi:https://doi.org/10.1038/nbt1277.
45. McMahon MT, Gilad AA. Cellular and molecular imaging using chemical exchange saturation transfer (CEST). *Top Magn Reson Imag* 2016;25:197–204. https://doi.org/10.1097/RMR.0000000000000105.
46. Zhang S, Merritt M, Woessner DE, Lenkinski RE, Sherry AD. "PARACEST Agents: modulating MRI contrast via water proton exchange. *Acc Chem Res* 2003;36:783–90. https://doi.org/10.1021/ar020228m.
47. Ferrauto G, Castelli DD, Terreno E, Aime S. In vivo MRI visualization of different cell populations labeled with PARACEST agents. *Magn Reson Med* 2013;69:1703–11. https://doi.org/10.1002/mrm.24411.
48. Aime S, Carrera C, Delli Castelli D, Geninatti Crich S, Terreno E. Tunable imaging of cells labeled with MRI-PARACEST agents. *Angew Chem Int Ed* 2005;44:1813–5. https://doi.org/10.1002/anie.200462566.
49. Srivastava AK, Kadayakkara DK, Bar-Shir A, Gilad AA, McMahon MT, Bulte JWM. Advances in using MRI probes and sensors for in vivo cell tracking as applied to regenerative medicine. *Dis Model Mech* 2015;8:323–36. https://doi.org/10.1242/dmm.018499.
50. Mezzanotte L, van 't Root M, Karatas H, Goun EA, Löwik CWGM. In vivo molecular bioluminescence imaging: new tools and applications. *Trends Biotechnol* 2017;35:640–52. https://doi.org/10.1016/j.tibtech.2017.03.012.
51. Kim JE, Kalimuthu S, Ahn B-C. In vivo cell tracking with bioluminescence imaging. *Nucl Med Mol Imaging* 2015;49:3–10. https://doi.org/10.1007/s13139-014-0309-x.
52. Sadikot RT. Bioluminescence imaging. *Proc Am Thorac Soc* 2005;2:537–40. https://doi.org/10.1513/pats.200507-067DS.
53. Liu Z, Li Z. Molecular imaging in tracking tumor-specific cytotoxic T lymphocytes (CTLs). *Theranostics* 2014;4:990–1001. https://doi.org/10.7150/thno.9268.
54. Nguyen VH, Zeiser R, Dasilva DL, Chang DS, Beilhack A, Contag CH, et al., In vivo dynamics of regulatory T-cell trafficking and survival predict effective strategies to control graft-versus-host disease following allogeneic transplantation. *Blood* 2007;109:2649–56. doi:https://doi.org/10.1182/blood-2006-08-044529.
55. Shin IJ, Shon SM, Schellingerhout D, Park JY, Kim JY, Lee SK, et al., Characterization of partial ligation-induced carotid atherosclerosis model using dual-modality molecular imaging in ApoE knock-out mice. *PloS One* 2013;8:e73451. doi:https://doi.org/10.1371/journal.pone.0073451.
56. Lee HW, Jeon YH, Hwang MH, Kim JE, Park TI, Ha JH, et al., Dual reporter gene imaging for tracking macrophage migration using the human sodium iodide symporter and an enhanced firefly luciferase in a murine inflammation model. *Mol Imag Biol* 2013;15:703–12. doi:https://doi.org/10.1007/s11307-013-0645-8.
57. Rao J, Dragulescu-Andrasi A, Yao H. Fluorescence imaging in vivo: recent advances. *Curr Opin Biotechnol* 2007;18:17–25. https://doi.org/10.1016/j.copbio.2007.01.003.
58. Hoffman RM. Whole-body fluorescence imaging with green fluorescence protein. *Green fluorescent protein*, Hicks BW New Jersey: Humana Press; 2002, 183.
59. Mosaad EO, Futrega K, Seim I, Gloss B, Chambers KF, Clements JA, et al., Constraints to counting bioluminescence producing cells by a commonly used transgene promoter and its implications for experimental design. *Sci Rep* 2019;9:11334. doi:https://doi.org/10.1038/s41598-019-46916-z.

60. Ntziachristos V. Fluorescence molecular imaging. *Annu Rev Biomed Eng* 2006;8:1–33. https://doi.org/10.1146/annurev.bioeng.8.061505.095831.
61. Kremers G-J, Gilbert SG, Cranfill PJ, Davidson MW, Piston DW. Fluorescent proteins at a glance. *J Cell Sci* 2011;124:2676. https://doi.org/10.1242/jcs.095059.
62. Funovics M, Weissleder R, Tung C-H. Protease sensors for bioimaging. *Anal Bioanal Chem* 2003;377:956–63. https://doi.org/10.1007/s00216-003-2199-0.
63. Germain RN, Robey EA, Cahalan MD. A decade of imaging cellular motility and interaction dynamics in the immune system. *Science* 2012;336:1676–81. https://doi.org/10.1126/science.1221063.
64. Stoll S. Dynamic imaging of T cell-dendritic cell interactions in lymph nodes. *Science* 2002;296:1873–6. https://doi.org/10.1126/science.1071065.
65. Mempel TR, Pittet MJ, Khazaie K, Weninger W, Weissleder R, von Boehmer H, et al., Regulatory T cells reversibly suppress cytotoxic T cell function independent of effector differentiation. *Immunity* 2006;25:129–41. doi:https://doi.org/10.1016/j.immuni.2006.04.015.
66. Cahalan MD, Parker I. Choreography of cell motility and interaction dynamics imaged by two-photon microscopy in lymphoid organs. *Annu Rev Immunol* 2008;26:585–626. https://doi.org/10.1146/annurev.immunol.24.021605.090620.
67. Laviron M, Combadière C, Boissonnas A. Tracking monocytes and macrophages in tumors with live imaging. *Front Immunol* 2019;10:1201. https://doi.org/10.3389/fimmu.2019.01201.
68. Nakajima A, Seroogy CM, Sandora MR, Tarner IH, Costa GL, Taylor-Edwards C, et al., Antigen-specific T cell–mediated gene therapy in collagen-induced arthritis. *J Clin Invest* 2001;107:1293–301. doi:https://doi.org/10.1172/JCI12037.
69. Lim YT, Cho MY, Noh Y-W, Chung JW, Chung BH. Near-infrared emitting fluorescent nanocrystals-labeled natural killer cells as a platform technology for the optical imaging of immunotherapeutic cells-based cancer therapy. *Nanotechnology* 2009;20:475102. https://doi.org/10.1088/0957-4484/20/47/475102.
70. Chtanova T, Schaeffer M, Han SJ, van Dooren GG, Nollmann M, Herzmark P, et al., Dynamics of neutrophil migration in lymph nodes during infection. *Immunity* 2008;29:487–96. doi: https://doi.org/10.1016/j.immuni.2008.07.012.
71. Iannacone M, Moseman EA, Tonti E, Bosurgi L, Junt T, Henrickson SE, et al., Subcapsular sinus macrophages prevent CNS invasion on peripheral infection with a neurotropic virus. *Nature* 2010;465:1079–83. doi:https://doi.org/10.1038/nature09118.
72. Laissue PP, Alghamdi RA, Tomancak P, Reynaud EG, Shroff H. Assessing phototoxicity in live fluorescence imaging. *Nat Methods* 2017;14:657–61. https://doi.org/10.1038/nmeth.4344.
73. Ettinger A, Wittmann T. Fluorescence live cell imaging. *Methods in cell biology*, Waters, JC, Wittmann, T Cambridge: Academic Press; 2014, 123.
74. Szabó Á, Szendi-Szatmári T, Ujlaky-Nagy L, Rádi I, Vereb G, Szöllősi J, et al., The effect of fluorophore conjugation on antibody affinity and the photophysical properties of dyes. *Biophys J* 2018;114:688–700. doi:https://doi.org/10.1016/j.bpj.2017.12.011.

Xavier Montané, Karolina Matulewicz, Karolina Balik,
Paulina Modrakowska, Marcin Łuczak, Yaride Pérez Pacheco,
Belen Reig Vano, Josep M. Montornés, Anna Bajek and
Bartosz Tylkowski

8 Present trends in the encapsulation of anticancer drugs

Abstract: Different nanomedicine devices that were developed during the recent years can be suitable candidates for their application in the treatment of various deadly diseases such as cancer. From all the explored devices, the nanoencapsulation of several anticancer medicines is a very promising approach to overcome some drawbacks of traditional medicines: administered dose of the drugs, drug toxicity, low solubility of drugs, uncontrolled drug delivery, resistance offered by the physiological barriers in the body to drugs, among others. In this chapter, the most important and recent progress in the encapsulation of anticancer medicines is examined: methods of preparation of distinct nanoparticles (inorganic nanoparticles, dendrimers, biopolymeric nanoparticles, polymeric micelles, liposomes, polymersomes, carbon nanotubes, quantum dots, and hybrid nanoparticles), drug loading and drug release mechanisms. Furthermore, the possible applications in cancer prevention, diagnosis, and cancer therapy of some of these nanoparticles have been highlighted.

Keywords: anticancer drugs, cancer therapy, drug delivery systems, encapsulation, nanomedicine, nanoparticles (NPs)

8.1 Introduction

The current pandemic situation caused by SARS-CoV-2, which has rapidly spread throughout the world, has highlighted the importance of medical research. At present, numerous investigations are mainly focused on obtaining a vaccine for the COVID-19 disease, among which some very encouraging results have been obtained [1].

Despite this, humanity does not have to forget that there are other diseases that still do not have a defined therapy. One of these diseases is cancer. Cancer includes a group of diseases that are characterized by an anomalous growth of malignant cells that can form a tumor with the potential to invade or propagate to other parts

---

This article has previously been published in the journal Physical Sciences Reviews. Please cite as: Montané, X., Matulewicz, K., Balik, K., Modrakowska, P., Łuczak, M., Pérez Pacheco, Y., Reig-Vano, B., Montornés, J. M., Bajek, A.,Tylkowski, B. Present trends in the encapsulation of anticancer drugs *Physical Sciences Reviews* [Online] 2021, 6. DOI: 10.1515/psr-2020-0080

https://doi.org/10.1515/9783110662306-008

of the body [2, 3]. There are various causes associated to the development of cancer, among which the most important are strongly influenced by lifestyle and habits. Some of these factors are smoking, obesity, processed meat consumption, radiation, family history, stress environmental factors, etc. [4]. More than 100 types of cancer are known to affect humans [5]. An estimation recently published by the American Cancer Society predicted that during the year 2020 there were 1,806,590 new cancer cases diagnosed and 606,520 cancer deaths only in the United States, making cancer the second leading cause of death in the country after hearth disease [6]. As for the entire planet, the predictions of the World Health Organization projected that the deaths from cancer worldwide will exceed 13 million by the year 2030 [7]. These data proves that in the present cancer is one of the most serious health issues in the world.

In fact, research has also played a very important role that has allowed an enormous progress in the prevention, detection, and treatment of cancer. However, it is still very difficult to assign an appropriate therapy for the different types of cancer that exist due to the late-stage diagnosis, inadequate strategies for addressing aggressive metastasis, and a lack of clinical procedures for overcoming multidrug resistant cancer [8, 9].

One of the research tools that is responsible of some advances in the development of more effective therapies against cancer has been nanomedicine. Nanomedicine is the medical application of nanotechnology and includes devices in the nanometer size, which are comparable to many biological macromolecule, such as enzymes, antibodies, hormones, DNA plasmids, or antibodies [10, 11]. The devices used in nanomedicine comprise nanoparticles (NPs), biological devices, and nanoelectronic biosensors. It is a crucial factor that NPs used in medical applications have a size very similar to most biological molecules and structures so that NPs can be useful for both *in vitro* and *in vivo* biomedical purposes.

As shown in Figure 8.1, there has been a growing interest in the research and publications associated to the encapsulation in cancer therapy over the last decade [12, 13]. To highlight the importance of research associated to cancer, *Nature* editorial group started publishing last year the journal *Nature Cancer* to cover one of the most active fields in biomedical research [9].

8.2 Encapsulated anticancer drug delivery systems

Distinct nanostructures with various forms and sizes have been prepared and applied to encapsulate different kinds of drugs: inorganic NPs, dendrimers, biopolymeric NPs, polymeric micelles, liposomes, polymersomes, carbon nanotubes (CNTs), and quantum dots (QDs) (Figure 8.2). All these nanostructures, their specific properties, and some examples of their current applications related to cancer diseases are discussed further [14, 15].

8.2 Encapsulated anticancer drug delivery systems — 195

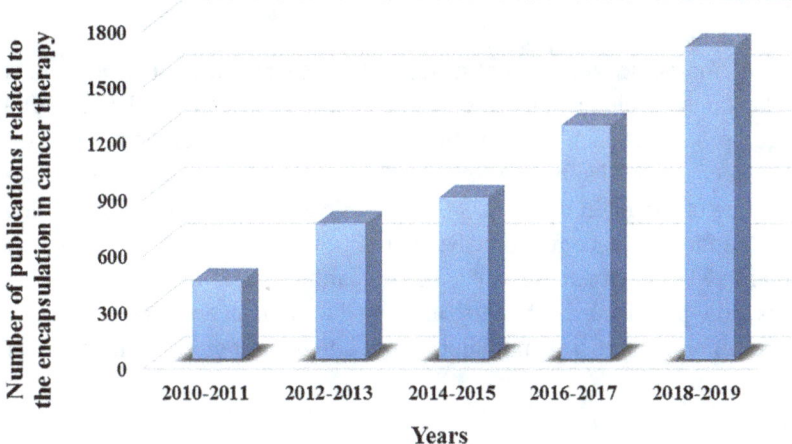

Figure 8.1: Number of peer-reviewed articles published during the last decade in the field of "encapsulation in cancer therapy".

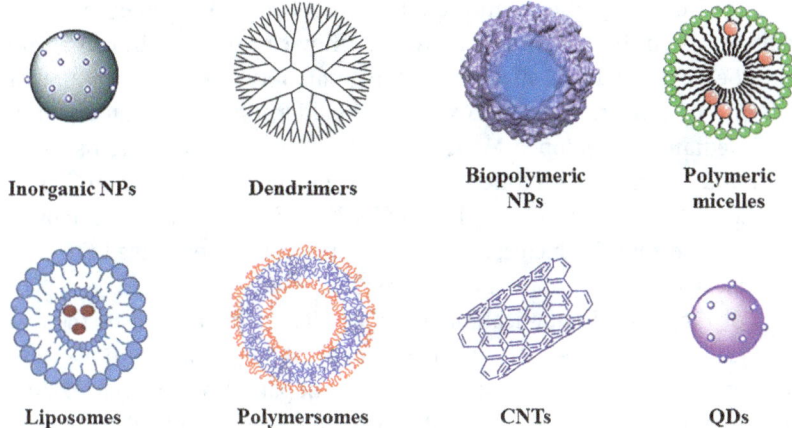

Figure 8.2: Examples of different types of nanoparticles used as drug delivery systems: inorganic NPs, dendrimers, biopolymeric NPs, polymeric micelles, liposomes, polymersomes, CNTs, and QDs.

To choose which type of NPs would be best for application in cancer diagnosis and drug delivery applications, numerous factors are taken into account:
- The biocompatibility and biodegradability of the NPs.
- The size and the shape of the NPs.
- The toxicity of both the nanomaterial and the encapsulated drug.
- The desired drug release profile.
- The required surface charge of the NPs.
- The properties of the encapsulated drug into the NPs such as their stability and aqueous solubility.

8.2.1 Inorganic NPs

In recent years, inorganic NPs have attracted great research interest in numerous medical applications (in bioimaging, as biosensors, or in targeted drug delivery) due to their unique properties such as their strong optical absorption and scattering (in the case of gold or silver NPs) or their excellent magnetic properties (for iron oxide NPs) [16, 17]. One of the main drawbacks of inorganic NPs is their intrinsic toxicity. Nonetheless, the facile surface functionalization of inorganic NPs limits the toxicity of the inorganic NPs and increases their biocompatibility and versatility, making these systems more promising for their use in biomedical applications [18].

There are different kinds of inorganic nanoparticles, among which the most common are:
- Gold NPs
- Silver NPs
- Iron oxide NPs
- Silica NPs.

Among all inorganic nanoparticles, iron oxide nanoparticles are widely used for the encapsulation of anticancer drugs. Moreover, they can be used in cancer radiotherapy, a therapy based on ionized radiation commonly used in the treatment of different types of cancer: breast, neck, head, bladder, and lung cancers. In a recent study, Medianchi developed $Mg_{(1-x)}Cu_xFe_2O_4$ superparamagnetic particles with a size around 30 nm that exhibited an efficient cell death of Michigan Cancer Foundation-7 (MCF-7) breast cancer cell lines under X-ray irradiation [19]. Moreover, the author mentioned that the increase of cupper concentration enhanced the apoptotic effect of the synthesized nano-radiosensitizers.

Examples of the possible applications of silica NPs in breast cancer therapy were reported just in recent times. This year, Sheena et al. [20] proposed the encapsulation of cupper complex in mesoporous silica nanoparticles that were coated with folic acid to build up inorganic NPs with low toxicity and good biodistribution. Once tested *in vitro*, the (FA@MSNPs-[Cu(L)(dppz)]$^+$) NPs induced in a more intense way the death of MCF-7 and MDA-MB-231 breast cancer cell lines compared with the nonencapsulated [Cu(L)(dppz)]$^+$. Moreover, the folate ligand conferred to these NPs the tumor targeting ability, because their delivery mechanism is mediated by the folate receptor (FR), a biomarker for tumor cells due to its overexpression on a great number of tumors.

8.2.2 Dendrimers

Dendrimers are repetitively hyperbranched polymers that consist of a central core with repeated branches of interior layers (or also called generations) and the outer surface moieties, which contain the terminal functionalities [21]. The term dendrimer comes from the Greek words "Dendron," which means a tree, and "Meros," which means a part. The 3D spherical shape of dendrimers, their nanometer size, their good chemical and biological stability, and their ability to penetrate cell walls make dendrimers exceptional candidates to build up nontoxic and biodegradable NPs [22, 23]. The most common dendrimers, the use of which had been reported in distinct biomedical applications, are polyamidoamine (PAMAM) dendrimers, poly-propylene imine dendrimers, and poly-L-Lysine dendrimers [24, 25].

A highly exploited advantage of dendrimers is the functionalization of their surface moieties, which can be achieved by different methods in order to favor the drug loading in dendrimers. Besides, medicines can be incorporated into the dendrimers by three distinct methodologies:
- Through a covalent interaction between the drug and the dendrimer.
- Via ionic coordination of the drug to the outer functional groups located in the surface moieties of the dendrimer.
- Simply encapsulation of the drugs into dendrimer cavities.

Two of the last possibilities were evaluated recently by Alfei et al. [26]. The authors studied during the last decade the incorporation in dendrimers of gallic acid, a natural phenolic compound found in gallnuts, tea leaves among other plants, which have shown potential properties as anticancer medicine due to its excellent antioxidant properties. Precisely, the comparison between the two dendrimer systems (the first one includes the covalent interaction of gallic acid with polyester-based dendrimers while in the second one gallic acid is incorporated physically into the dendrimer cavities) did not show significant differences in *in vitro* studies. However, the addition of gallic acid to biocompatible and biodegradable dendrimers results in more effective systems that activated the reactive oxygen species production inducing the apoptosis of HTLA-230 and HTLA-ER neuroblastoma cell lines. Moreover, the coordination of gallic acid with these bio-based dendrimers increased the solubility of gallic acid.

One of the most recent studies involved the encapsulation of bortezomib (BTZ) and anticancer drug used in the therapy of multiple myeloma and mantle cell lymphoma, to overcome its poor solubility in aqueous media. The authors considered two different approaches for the encapsulation of BTZ: a physical entrapment of the drug and the chemical conjugation to the hydroxyl groups of the dendrimer. Particularly, Sahoo et al. [27] functionalized polyamidoamine (PAMAM) dendrimers with polyethylene glycol (PEG), which increased the biocompatibility of BTZ and improved their encapsulation efficiency. When the release of the encapsulated BZT in the dendrimers was evaluated *in vitro*, a gradual and more sustained release of BZT was achieved. Furthermore, the encapsulated BZT in PEG-PAMAM dendrimer showed the higher efficacy in killing A549 lung cancer cells and MCF-7 breast cancer cells.

8.2.3 Biopolymeric NPs

Taking benefits from the materials found in nature, the design and construction of biopolymeric NPs are a powerful tool for the development of nanomedical devices [28]. The first biopolymeric NPs were designed in the early 1970s by Scheffel et al. [29] using albumin protein, basically as substitutes for liposomes to overcome the liposomal poor stability in biological fluids.

Natural biopolymers could be obtained from various resources:
- Higher plants: starch, cellulose, gum Arabic, guar gum.
- Algae: alginate, galactans, carrageenan.
- Animals: chitin, chitosan, hyaluronic acid (HA), glycosaminoglycans.
- Microorganisms: dextran, gellan gum, xanthan gum, bacterial cellulose.

Biopolymeric NPs offer several advantages [28]:
- Easy preparation from biocompatible and biodegradable polymers.
- Lack of toxicity.
- High stability in biological fluids and during storage.
- High-value products with low environmental impact.
- High specific surface area.
- Good mechanical and barrier properties.
- Abundant and usually cheap raw materials.

All these properties have increased the interest of biopolymeric NPs in medicine during the last decades. For instance, Pal et al. [30] reported the encapsulation of curcumin (CUR), a natural anticancer drug obtained from the turmeric plant that is extensively employed in the treatment of many cancers such as those of breast, lung, prostate, liver, and cervices, in gum acacia microspheres. Moreover, the conjugation of folic acid to these microspheres enabled the obtention of targeted drugs due to the overexpression of FRs in cancer cells. Once tested in *in vitro* and *in vivo* experiments with triple negative breast cancer cells (TNBC), these biopolymeric NPs triggered the apoptosis of cancerous cells by damaging their DNA and their mitochondrial membrane.

Additionally, the design of complex biopolymers from natural resources could be an interesting opportunity to design NPs with enhanced therapeutic effects. Thus, the chemical modification of poly(y-glutamic acid) (y-PGA) with phenylalanine and the subsequent loading of y-PGA NPs with paclitaxel (PTX), an anticancer drug used in the treatment of lung, cervical, pancreatic, ovarian, and breast cancers among others, was presented by Kim et al. [31]. Additionally, the y-PGA NPs were coated with polydopamine (PDA), a biopolymer inspired by mussels that can be used in the preparation of drug delivery systems [32]. Definitely, the synthesized NPs could be used in cancer phototherapies due to the light absorption of PDA upon irradiation with near-infrared (NIR), which decreases the tumor proliferation of CT26 murine colorectal carcinoma cell lines in *in vivo* studies.

8.2.4 Polymeric micelles

Polymeric micelles are formed by the self-assembly of amphiphilic block copolymers in selective solvents above a certain concentration, the critical micelle concentration. Different shapes have been reported for the assembled amphiphilic block copolymers into micelles: spherical micelles, cylindrical micelles, and vesicles [33, 34].

The use of polymeric micelles has gained a lot of interest in biomedical applications related to cancer (biomedical imaging and drug delivery applications) because they enable the encapsulation of common hydrophobic medicines in the micelle core, constituted by hydrophobic polymers. Some of the polymers typically used as core in micellar structures are:
- Poly(D,L-lactic acid).
- Poly(α-amino acid).
- Pluronics.
- Polycaprolactones (PCL).

On the other hand, PEG is usually used as shell of the micelle due to their low toxicity [35, 36]. Other advantages of polymeric micelles are their good chemical stability and the enhanced permeability and retention effect (EPR) [37, 38]. Precisely, the EPR effect, which is the basis for passive tumor targeting, makes the accumulation of NPs of certain sizes into the solid tumor easier because the rapid overgrowth of epithelial cells in the vascularization to the proximity of cancer cells produces gaps between the junctions of these epithelial cells through which NPs can pass [39].

In an interesting example, Liu et al. [40] built up mixed micelles with naturally derived polymers composed of a *Bletilla striata* polysaccharide modified with stearic acid and folic acid as a core, which are encircled by a D-α-tocopheryl polyethylene glycol succinate (TPGS). The encapsulated anthracycline doxorubicin (DOX), an anticancer drug applied in the treatment of various human tumors (bladder, stomach, ovaries, lung, and thyroid, etc.), in these micellar structures exhibited a higher cytotoxicity against 4T1 breast cancer cell lines in *in vitro* and *in vivo* studies than free DOX. Besides, the complex micelles showed a greater release of DOX in acidic environments via the endocytosis mediated by FRs and the clathrin mechanisms.

On the other hand, Zhou et al. [41] constructed spherical polymeric micelles in the nanometer size that could be used for drug delivery of PTX. In this study, PTX is encapsulated in a micelle by a covalent bond between PTX and biocompatible and biodegradable aliphatic polycarbonates, which are surrounded by a PEG shell. Taking advantage of the pH-sensitive acetal bond between PTX and the polycarbonates, the authors achieved a targeted and controlled drug delivery system that has the ability to stop the proliferation of A549 lung carcinogenic cells.

8.2.5 Liposomes

Liposomes are one of the first "biodevices" that have been studied in medical applications. In fact, Gregory Gregoriadis suggested in the early 1970s the use of liposomes for drug delivery applications [42]. Liposomes are spherical vesicles with a phospholipid bilayer [43]. The particular properties of liposomes such as their cell-like membrane structure, their high biocompatibility, low immunogenicity, and increasing efficiency of the encapsulated drugs make them ideal candidates for drug delivery applications [44, 45].

Nevertheless, liposomes are usually functionalized with PEG or with other ligands such as peptides, antibodies, and carbohydrates to reduce their instability, insufficient drug loading, their fast drug release, and short circulation times in blood [46].

The encapsulation of more than one anticancer drug in NPs is usually employed in cancer therapy due to their higher efficacy compared with the utilization of one anticancer drug [47]. As an example, Li et al. [48] used liposomes to encapsulate DOX and simvastatin. In recent years, it has been shown that some compounds used as drugs from the statins family such as simvastatin, which is a drug generally used to decrease the lipid levels and the risk of heart problems, have beneficial effects such as the inhibition of the growth of several solid tumors (breast, ovarian, lung, colon, and prostate cancers and leukemia) [49]. In the last step, the liposomes are coated with trastuzumab, an antibody used as anticancer drug used to suppress the proliferation of carcinogenic cells, in order to build up targeted nanodrugs that clearly stop the proliferation of PC-3 prostate tumor cells by the combination therapy via multiple mechanisms.

On the other hand, chemically modified liposomes can be used in the coencapsulation of both indocyanine green, a cyanine dye approved by the U. S. Food and Drug Administration (FDA), and DOX. It has been proved that the addition of a folate and conjugated gadolinium (Gd) chelate into liposomal shells intensifies the targeting and magnetic resonance performance of the tumor tissue. Therefore, after intravenous injection of these nanoparticles in nude mice, the combination of both indocyanine green and DOX allowed a precise diagnosis by the use of imaging-guided NIR-triggered phototherapy together with chemotherapy, which leads to HeLa cervical tumor suppression [50].

8.2.6 Polymersomes

Polymersomes are synthetic vesicles of a nanometer size like liposomes that are built from amphiphilic block copolymers [51]. They are generally constituted by a hydrophobic bilayer that encircles a hydrophilic core. These types of polymeric structures have aroused much interest in the field of medicine since they have been reported for the first time by Hest et al. [52]. Polymersomes present many of the properties of natural liposomes combined with increased stability and reduced permeability. Furthermore, one of the great advantages that polymersomes have over liposomes is the design of synthetic polymers to easily tune the final properties of polymersomes, allowing, for example, the control of the release of encapsulated drugs [53, 54].

Among the most widespread applications of the use of polymersomes in the field of cancer, Nehate et al. [55] developed methoxypoly(ethylene) glycol polycaprolactone diblock (mPEG-SS-PCL-OH) polymersomes with encapsulated DOX that allows a drug release when a redox stimulus is applied by degrading the polymersomes. In this investigation, the authors reported a high encapsulation efficiency of DOX in the biocompatible polymersomes. The results demonstrated that the encapsulation of DOX in these polymersomes induced a decrease in the proliferation of cancer cells *in vitro* and the reduction of tumor weight *in vivo* in Swiss albino mice compared with administered free DOX against HeLa and MDA-MB-231 cancerous cell lines, suggesting that these polymersomes could become an interesting candidate for clinical applications that release the drug when a stimulus is applied.

In another recent study, Hossainzadeh et al. [56] reported the encapsulation of silibinin, a natural flavonoid medicine without side effects, which is studied in the treatment of some cancers such as prostate and lung cancer, in poly(ethylene glycol$_{400}$)-oleate polymersome NPs to mainly overcome the poor solubility of silibinin in water and its biological stability in physiological pH. Moreover, the improvement of the controlled release of silibinin results in the death of MDA-MB-231 breast cancer cells by the upregulation of genes that stimulate the apoptosis of carcinogenic cells such as P53 (tumor protein p53), caspase 9, and BAX (Bcl-2-associated X protein) and by reducing the expression of two microRNA correlated to various types of cancer (miR-125b and miR-182).

8.2.7 CNTs

CNTs are carbon cylinders of graphene sheets characterized by presenting a diameter in the nanometer size. CNTs are classified based on the number of graphene layers that constituted a single nanotube:
- Single-wall carbon nanotubes (SWCNTs), which have a single cylindrical layer of atoms.
- Multiwall carbon nanotubes (MWCNTs), which consist of two or more nanotubes linked by weak intermolecular forces or may be a single graphene-like sheet rolled up several times around a cylindrical hollow.

The potential applications of CNTs have been examined in different fields due to their excellent optical and electronic properties and their exceptional surface area. Moreover, CNTs exhibited a great cell membrane permeability [57]. In medicine, CNTs have been studied as promising carrier for drug delivery because they can be attached to some molecules such as radioisotopes, antibodies, or anticancer drugs [58]. Moreover, the possibility of the application of CNTs in biosensing, bioimaging, and tissue engineering has been investigated in the last decades [59, 60].

One of the properties that increases the compatibility of CNTs with drugs and with human cells is the ease with which they can be functionalized by using different approaches. Lately, Mohseni-Dargah et al. [61] chemically modified MWCNTs with pyridine to obtain biocompatible nanocarriers loaded with the inducible caspase-9 (iC9) suicide gene, which induce the expression of caspase-9 enzyme and give rise to cancer cell death. By the combination of chemotherapy and gene therapy, the authors proved the efficient delivery of iC9 suicide gene by the functionalized MWCNTs, leading to the apoptosis of MCF-7 breast cancer cells *in vitro* by means of a targeting therapy that overcomes the limitations of conventional cancer therapies such as drug resistance.

A recent example showing that SWCNTs could be employed in cancer treatment and diagnosis has been reported by Sundaran et al. [62]. In this work, the walls of HA-SWCNTs were easily coated by a strong noncovalent $\pi-\pi$ interactions with chlorin e6 (Ce6), a tetrapyrrole organic compound used as photosensitizer in cancer photodiagnosis and photodynamic therapies due to their intense absorption wavelength at 650–670 nm combined with their fast and selective accumulation in the target tissue [63]. Although the authors suggested that further studies are needed to investigate the mechanisms that induce cancer cell death, they demonstrated that these bionanocarriers caused the apoptosis of Caucasian colon adenocarcinoma human cell lines (Caco-2 cells).

8.2.8 QDs

QDs are semiconductor crystal particles in the nanometer size, which are typically constructed by a core that comprises elements of the groups II–VI or III–V that are enclosed by a polymeric shell [64]. Since their discovery in the early 1980s, the excellent optical and electronic properties of QDs made these materials exceptional candidates for their use in a wide range of applications: photocatalysis, photodectectors, solar energy conversion, light emitting diodes, and display applications [64, 65]. QDs are also widely used in biomedical applications such as in cancer therapies, especially in cancer diagnosis and cell imaging, although they are also used in drug delivery [66]. In contrast, due to the actual pandemic, some authors studied the possible use of QDs as biosensors and antiviral agents against COVID-19 disease [67].

One of the problems that QDs present is their poor water solubility, which is usually solved upon surface modification. By coating black phosphorus QDs with PDA, Li et al. [68] greatly increased the stability in water of biocompatible and biodegradable QDs. Moreover, the strong absorption of PDA modified QDs in the NIR combined with laser irradiation could be the differential factor that causes A375 melanoma cells death.

Furthermore, 2,3-dimethylmaleic anhydride (DMMA) was also used to cover ZnO QDs filled with DOX anticancer drug and phenylsulfonyl furoxan (PSF), a compound used as nitric oxide (NO) donor, a molecule that has shown a great potential in cancer therapy (NO reduces the multidrug resistance of cancerous cells among other functions) [69]. The resulting nanospheres, which have an average size of 7.9 nm, present a long blood circulation that facilitates their accumulation in tumors. Besides, the acidic pH of tumors facilitates the dissolution of these NPs in the targeting tissue, inducing a higher release of NO combined with Zn^{2+} cations and DOX that together stopped the proliferation of human gastric carcinoma SGC7901 cells and human gastric carcinoma drug-resistant SGC7901/ADR cells and finally led to their apoptosis *in vitro* and *in vivo* experiments.

8.3 Hybrid encapsulation systems

The great number of possibilities that nanomedicine offered to encapsulate drugs encourage scientists in the research to develop mechanisms for the early detection of cancer as well as the design of more effective cancer therapies [70]. The discussion of some recent trends of each type of NPs was carried out in the previous sections of this review. Nevertheless, in the strategy to increase the efficacy of cancer therapies, the utilization of all the available tools is required. Therefore, much of the recent research related to cancer diseases has merged different types of NPs in order to take advantage of their properties. These types of composite materials are often called hybrid NPs [71].

During the past years, researchers combined different encapsulation systems to build up hybrid NPs to take advantage of each component. Among the combinations that have been studied against various types of cancer, the most popular one is the encapsulation of inorganic NPs in dendrimers or polymers to increase the solubility of inorganic NPs. As described further, the scientific community has explored further in recent years by obtaining novel NPs that offer promising results in cancer therapy, in particular in the field of drug delivery applications [72, 73].

One of the hybrid nanostructures that have been explored as drug delivery systems are metal–organic frameworks (MOFs) because they present an easy functionalization to obtain stimuli-responsive vehicles [74]. Nevertheless, one of the typical drawbacks of these structures is that MOFs are not stable in aqueous solutions. Taking advantage of the fact that the pH of cancerous environments is acidic,

Javanbakht et al. [75] tailored a drug delivery system with sensitization to pH to release the drug in the proximity by coating Zn-based MOFs with carboxylatemethylcellulose, a biocompatible polysaccharide. The encapsulated anticancer drug in the porous of the MOF, 5-fluorouracil (5-FU) is a potent antimetabolite that is employed in the treatment of colon, esophageal, stomach, pancreatic, cervical, and breast cancers, among others. As expected, the polysaccharide shell permitted a better control of 5-FU release, demonstrating that the new MOFs could be used as targeted nanocarriers in the treatment of HeLa cervical cancer cells.

Another possibility has been published by Karimi et al. [76], which reported the synthesis of maltose and 3-aminopropyltrimethoxysilane (APTMS) functionalized QDs that are covalently bonded to a triazine-derived dendrimer. The holes in the dendrimer structure were used to encapsulate DOX, obtaining a loading efficiency around 63%. *In vitro* studies against A549 lung cancer cells evidenced that the new hybrid NPs exhibited a pH-dependent release and are more effective in inducing cancer cell death compared with the administration of free DOX.

8.4 Overview

Some of the NPs described in this chapter, the loaded anticancer drug, its applications in cancer therapy, and the corresponding reference are summarized in Table 8.1.

8.4 Conclusions

As summarized earlier, distinct scientific strategies to design novel NPs give solutions to progress in the detection and diagnostic methodologies of several types of cancer as well as for the delivery of drugs in cancer therapy. The different types of cancer, the state of the cancer, and the way to attack the cancer (chemotherapy or radiotherapy) make necessary the development of different NPs systems. The new NPs usually limit the drawbacks of conventional drugs: reduce the toxicity, the drug resistance, the dose requirement, and low absorption of drugs, increase the tumor site targeting, the solubility, and the controlled release of the medicines, among others. Since the 1990s, numerous NPs such as synthetic polymer particles, liposomes, micellar NPs, protein NPs, etc., have been approved in clinical tests by the FDA. Furthermore, various new nanoencapsulated systems that are currently under development show promising properties, thereby being able to provide new tools in cancer therapy. Nowadays, hybrid NPs plays an important role in recent advances in the delivery of anticancer drug.

Consequently, the use of nanomedicine revolutionizes our ability to diagnose cancer and the cancer therapy. However, the complexity of cellular interactions

Table 8.1: Examples of each type of NPs, the encapsulated anticancer drug, and their possible applications in cancer therapy.

NPs	Encapsulated anticancer drug	Applications in cancer therapy	References
$Mg_{(1-x)}Cu_xFe_2O_4$ superparamagnetic NPs	–	As radiosensitizers in cancer radiotherapy. Induce the apoptosis of Michigan cancer Foundation-7 (MCF-7) breast cancer cells by X-ray irradiation.	[19]
Folic acid–coated mesoporous silica NPs encapsulated cupper (II) complex (FA@MSNPs-[Cu(L)(dppz)]$^+$)	–	Tumor targeting NPs. Apoptosis of MCF-7 and MDA-MB-231 breast cancer cell lines via the folate receptor (FR) mechanism.	[20]
Gallic acid–loaded polyester-based dendrimers	Gallic acid	Promote apoptosis of HTLA-230 and HTLA-ER neuroblastoma cell lines by the activation of reactive oxygen species production.	[26]
PEG-PAMAM dendrimers	Bortezomib (BTZ)	Obtention of controlled drug delivery system, causing the death of A549 lung cancer cells and MCF-7 breast cancer cells.	[27]
Gum acacia microspheres coated with folic acid	Curcumin (CUR)	Tumor targeting particles. Apoptosis of triple negative breast cancer cells (TNBC) by damaging their DNA and their mitochondrial membrane.	[30]
Polydopamine (PDA) γ-PGA NPs	Paclitaxel (PTX)	Stop the tumor proliferation of CT26 murine colorectal carcinoma cell line in cancer phototherapy.	[31]
D-α-Tocopheryl polyethylene glycol succinate – Bletilla striata polysaccharide micelle	Doxorubicin (DOX)	Controlled drug delivery system in acidic environments. Induce the apoptosis of 4T1 breast cancer cells.	[40]
PEG – aliphatic polycarbonates micelles	PTX	Tumor targeting NPs with a controlled release of anticancer drug. Suppress the proliferation of A549 lung cancer cells.	[41]
Liposome	DOX, simvastatin, and trastuzumab	Tumor targeting NPs. Stop the proliferation of PC-3 prostate tumor cells.	[48]

Table 8.1 (continued)

NPs	Encapsulated anticancer drug	Applications in cancer therapy	References
Liposome	DOX and indocyanine green	Strengthen the targeting and magnetic resonance performance of the NPs. Enable a precise diagnosis in imaging-guided near-infrared (NIR)-triggered phototherapy. Suppression of HeLa cervical cancer cells when phototherapy and chemotherapy are combined.	[50]
Methoxypoly(ethylene) glycol polycaprolactone diblock (mPEG-SS-PCL-OH) polymersome	DOX	Reduce the proliferation of HeLa cervical cancer cells and MDA-MB-231 breast cancer cells.	[55]
Poly(ethylene glycol$_{400}$)-oleate polymersome	Silibinin	Controlled drug delivery system under pH stimulus. Increase the activity of genes involved in the apoptosis of MDA-MB-231 breast cancer cells: tumor protein P53 (p53), caspase 9, and Bcl-2-associated X protein (BAX). Reduces the expression of 2 microRNA correlated to some cancers (miR-125b and miR-182).	[56]
Pyridine-MWCNTs	Inducible caspase-9 (iC9)	Tumor targeting NPs. Efficient apoptosis of MCF-7 breast cancer cells when chemotherapy and gene therapy are employed together.	[61]
Hyaluronic acid (HA)-SWCNTs	Chlorin e6 (Ce6)	Tumor targeting NPs. Apoptosis of Caucasian colon adenocarcinoma human cell lines (Caco-2 cells).	[62]
Polydopamine (PDA)-black phosphorus QDs	–	Combined with laser irradiation, they cause A375 melanoma cells death.	[68]
2,3-Dimethylmaleic anhydride (DMMA)-coated ZnO QDs	DOX and phenylsulfonyl furoxan (PSF)	Tumor targeting and pH-sensitive NPs. Reduce the proliferation of human gastric carcinoma SGC7901 cells and human gastric carcinoma drug-resistant SGC7901/ADR cells and induce their apoptosis.	[69]

Table 8.1 (continued)

NPs	Encapsulated anticancer drug	Applications in cancer therapy	References
Carboxylatemethylcellulose-coated Zn-based metal–organic frameworks	5-Fluorouracil (5-FU)	Tumor targeting NPs with a controlled release of anticancer drug. Apoptosis of HeLa cervical cancer cells.	[75]
3-Aminopropyltrimethoxysilane (APTMS)-functionalized QDs covalently bonded to a triazine-derived dendrimer	DOX	Controlled drug delivery system under pH stimulus. Enhanced apoptotic effect against A549 lung cancer cells vs. free administered DOX.	[76]

with the distinct NPs and their fate still require more extensive studies to enhance the great impact of NPs on cancer treatment approaches.

Acknowledgment: This work was cofinanced by the European Union from the European Social Fund under the Knowledge Education Development 2014–2020 Operational Program. Project implemented as part of the competition of the National Center for Research and Development: for Interdisciplinary Programs of Doctoral Studies on increasing the quality and effectiveness of education at doctoral studies (enrollment number: POWR.03.02.00-IP.08-00-DOK/16).

Author contribution: All the authors have accepted responsibility for the entire content of this submitted manuscript and approved submission.

Research funding: None declared.

Conflict of interest statement: The authors declare no conflicts of interest regarding this article.

References

1. Moderna's. Moderna work on a COVID-19 vaccine candidate. Available from: https://www.modernatx.com/modernas-work-potential-vaccine-against-covid-19 [Accessed 14 Aug 2020].
2. Shewach DS, Kuchta RD. Introduction to cancer chemotherapeutics. *Chem Rev* 2009;109:2859–61, https://doi.org/10.1021/cr900208x.
3. Costa J. Cancer. Availabe from: https://www.britannica.com/science/cancer-disease [Accessed 02 Jan 2020].
4. Blackadar CB. Historical review of the causes of cancer. *World J Clin Oncol* 2016;7:54–86, https://doi.org/10.5306/wjco.v7.i1.54.
5. Afshar M, Madani S, Tarazoj AA, Papi SH, Otroshi O, Gandomani HS, et al. Physical activity and types of cancer. *WCR* 2018;5:1–11.

6. Cancer facts and figures 2020. Available from: https://www.cancer.org/research/cancer-facts-statistics/all-cancer-facts-figures/cancer-facts-figures-2019.html [Accessed 10 Aug 2020].
7. *World cancer report 2014*. Lyon: International Agency for Research on Cancer; 2014.
8. Patel SP, Patel PB, Parekh BB. Application of nanotechnology in cancers prevention, early detection and treatment. *J Canc Res Therapeut* 2014;10:479–86.
9. Bernards R, Jafee E, Joyce JA, Lowe SW, Mardis ER, Morrison SJ, et al. A roadmap for the next decade in cancer research. *Nat Cancer* 2020;1:12–17, https://doi.org/10.1038/s43018-019-0015-9.
10. Freitas RA. *Nanomedicine, volume I: basic capabilities*, 1st ed. Texas: Landes Bioscience; 1999.
11. Kawasaki ES, Player A. Nanotechnology, nanomedicine, and the development of new, effective therapies for cancer. *Nanomed Nanotechnol* 2005;1:101–9, https://doi.org/10.1016/j.nano.2005.03.002.
12. Alsehli M. Polymeric nanocarriers as stimuli-responsive systems for targeted tumor (cancer) therapy: recent advances in drug delivery. *Saudi Pharmaceut J* 2020;28:255–65, https://doi.org/10.1016/j.jsps.2020.01.004.
13. Chi XQ, Liu K, Luo XJ, Yin ZY, Lin HY, Gao JH. Recent advances of nanomedicines for liver cancer therapy. *J Mater Chem B* 2020;8:3747–71, https://doi.org/10.1039/c9tb02871d.
14. Wilczewska AZ, Niemirowicz K, Markiewicz KH, Car H. Nanoparticles as drug delivery systems. *Pharmacol Rep* 2012;64:1020–37, https://doi.org/10.1016/s1734-1140(12)70901-5.
15. Montané X, Bajek A, Roszkowski K, Montornés JM, Giamberini M, Roszkowski S, et al. Encapsulation for cancer therapy. *Molecules* 2020;25:1–26, https://doi.org/10.3390/molecules25071605.
16. Murakami T, Tsuchida K. Recent advances in inorganic nanoparticle-based drug delivery systems. *Mini Rev Med Chem* 2008;8:175–83.
17. Fernandes N, Rodrigues CF, Moreira AF, Correia IJ. Overview of the application of inorganic nanomaterials in cancer photothermal therapy. *Biomater Sci* 2020;8:2990–3020, https://doi.org/10.1039/d0bm00222d.
18. Paris JL, Baeza A, Vallet-Regi M. Overcoming the stability, toxicity, and biodegradation challenges of tumor stimuli-responsive inorganic nanoparticles for delivery of cancer therapeutics. *Expet Opin Drug Deliv* 2019;16:1095–112, https://doi.org/10.1080/17425247.2019.1662786.
19. Meidanchi A. $Mg_{(1-x)}Cu_xFe_2O_4$ superparamagnetic nanoparticles as nano-radiosensitizer agents in radiotherapy of MCF-7 human breast cancer cells. *Nanotechnology* 2020;31:1–9, https://doi.org/10.1088/1361-6528/ab8cf2.
20. Sheenaa TS, Dhivya R, Rajiu V, Jeganathan K, Palaniandavar M, Mathan G, et al. Folate-engineered mesoporous silica-encapsulated copper (II) complex [Cu(L)(dppz)]+: an active targeting cell-specific platform for breast cancer therapy. *Inorg Chim Acta* 2020;510:1–7, https://doi.org/10.1016/j.ica.2020.119783.
21. Hirao A, Yoo HS. Dendrimer-like star-branched polymers: novel structurally well-defined hyperbranched polymers. *Polym J* 2011;43:2–17, https://doi.org/10.1038/pj.2010.109.
22. Tripathy S, Das MK. Dendrimers and their applications as novel drug delivery carriers. *J Appl Pharmaceut Sci* 2013;3:142–9.
23. Sandoval-Yanez C, Rodriguez CC. Dendrimers: amazing platforms for bioactive molecule delivery systems. *Materials* 2020;13:1–20, https://doi.org/10.3390/ma13030570.
24. Mendes LP, Pan J, Torchilin VP. Dendrimers as nanocarriers for nucleic acid and drug delivery in cancer therapy. *Molecules* 2017;22:1–21.

25. Yousefi M, Narmani A, Jafari SM. Dendrimers as efficient nanocarriers for the protection and delivery of bioactive phytochemicals. *Adv Colloid Interfac* 2020;278:1–13, https://doi.org/10.1016/j.cis.2020.102125.
26. Alfei S, Marengo B, Zuccari G, Turrini F, Domenicotti C. Dendrimer nanodevices and gallic acid as novel strategies to fight chemoresistance in neuroblastoma cells. *Nanomaterials-Basel* 2020;10:1–30, https://doi.org/10.3390/nano10061243.
27. Sahoo RK, Gothwal A, Rani S, Nakhate KT, Ajazuddin Gupta, U. PEGylated dendrimer mediated delivery of bortezomib: drug conjugation versus encapsulation. *Int J Pharm* 2020;584:1–13, https://doi.org/10.1016/j.ijpharm.2020.119389.
28. Bassas-Galia M, Follonier S, Pusnik M, Zinn M. Natural polymers: a source of inspiration. In: Perale G, Hilborn J, editors. *Bioresorbable polymers for biomedical applications: from fundamentals to translational medicine*. Cambridge: Woodhead Publishing; 2017. pp. 31–64.
29. Scheffel U, Rhodes BA, Natarajan TK, Wagner HNJr. Albumin microspheres for study of the reticuloendothelial system. *J Nucl Med* 1972;13:498–503.
30. Pal K, Roy S, Parida PK, Dutta A, Bardhan S, Das S, et al. Folic acid conjugated curcumin loaded biopolymeric gum acacia microsphere for triple negative breast cancer therapy in invitro and invivo model. *Mater Sci Eng C-Mater* 2019;95:204–16, https://doi.org/10.1016/j.msec.2018.10.071.
31. Kim D, Le QV, Kim YB, Oh YK. Safety and photochemotherapeutic application of poly(γ-glutamic acid)-based biopolymeric nanoparticle. *Acta Pharm Sin B* 2019;9:565–74, https://doi.org/10.1016/j.apsb.2019.01.005.
32. Lynge ME, van der Westen R, Postma A, Stadler B. Polydopamine-a nature-inspired polymer coating for biomedical science. *Nanoscale* 2011;3:4916–28, https://doi.org/10.1039/c1nr10969c.
33. Gothwal A, Khan I, Gupta U. Polymeric micelles: recent advancements in the delivery of anticancer drugs. *Pharm Res (NY)* 2016;33:18–39, https://doi.org/10.1007/s11095-015-1784-1.
34. Keskin D, Tezcaner A. Micelles as delivery system for cancer treatment. *Curr Pharmaceut Des* 2017;23:5230–41.
35. Batrakova EV, Bronich TK, Vetro JA, Kabanov AV. Polymer micelles as drug carriers. In: Torchilin, VP, editor. *Nanoparticulates as drug carriers*, 1st ed. London: Imperial College Press London; 2006. pp. 57–9s3.
36. Yu GP, Ning Q, Mo ZC, Tang SS. Intelligent polymeric micelles for multidrug codelivery and cancer therapy. *Artif Cell Nanomed B* 2019;47:1476–87, https://doi.org/10.1080/21691401.2019.1601104.
37. Aziz ZABA, Ahmad A, Mohd-Setapar SH, Hassan H, Lokhat D, Kamal MA, et al. Recent advances in drug delivery of polymeric nano-micelles. *Curr Drug Metabol* 2017;18:16–29, https://doi.org/10.2174/1389200217666160921143616.
38. Dai Y, Chen X, Zhang XJ. Recent advances in stimuli-responsive polymeric micelles via click chemistry. *Polym Chem* 2019;10:34–44, https://doi.org/10.1039/c8py01174e.
39. Sun TM, Zhang YS, Pang B, Hyun DC, Yang MX, Xia YN. Engineered nanoparticles for drug delivery in cancer therapy. *Angew Chem Int Ed* 2014;53:12320–64.
40. Liu YR, Wu J, Huang L, Qiao J, Wang N, Yu D, et al. Synergistic effects of antitumor efficacy via mixed nano-size micelles of multifunctional Bletilla striata polysaccharide-based copolymer and D-α-tocopheryl polyethylene glycol succinate. *Int J Biol Macromol* 2020;154:499–510, https://doi.org/10.1016/j.ijbiomac.2020.03.136.
41. Zhou SY, Fu SW, Wang HL, Deng YH, Zhou X, Sun W, et al. Acetal-linked polymeric prodrug micelles based on aliphatic polycarbonates for paclitaxel delivery: preparation, characterization,in vitrorelease and anti-proliferation effects. *J Biomater Sci Polym Ed* 2020;31:1–32, https://doi.org/10.1080/09205063.2020.1792046.

42. Gregoriadis G. Liposomes as carriers of enzymes or drugs: new approach to treatment of storage diseases. *Biochem J* 1971;124:58, https://doi.org/10.1042/bj1240058p.
43. Bozzuto G, Molinari A. Liposomes as nanomedical devices. *Int J Nanomed* 2015;10:975–99, https://doi.org/10.2147/ijn.s68861.
44. Gao A, Hu XL, Saeed M, Chen BF, Li YP, Yu HJ. Overview of recent advances in liposomal nanoparticle-based cancer immunotherapy. *Acta Pharmacol Sin* 2019;40:1129–37, https://doi.org/10.1038/s41401-019-0281-1.
45. Kiaie SH, Mojarad-Jabali S, Khaleseh F, Allahyari S, Taheri E, Zakeri-Milani P, et al. Axial pharmaceutical properties of liposome in cancer therapy: recent advances and perspectives. *Int J Pharm* 2020;581:1–18, https://doi.org/10.1016/j.ijpharm.2020.119269.
46. Allen TM, Cullis PR. Liposomal drug delivery systems: from concept to clinical applications. *Adv Drug Deliv Rev* 2013;65:36–48, https://doi.org/10.1016/j.addr.2012.09.037.
47. Dorababu A. Recent advances in nanoformulated chemotherapeutic drug delivery (2015-2019). *Chemistry* 2019;4:8731–44, https://doi.org/10.1002/slct.201901064.
48. Li N, Xie X, Hu YX, He HD, Fu X, Fang TT, et al. Herceptin-conjugated liposomes co-loaded with doxorubicin and simvastatin in targeted prostate cancer therapy. *Am J Transl Res* 2019;11:1255–69.
49. Safwat S, Ishak RA, Hathout RM, Mortada ND. Statins anticancer targeted delivery systems: re-purposing an old molecule. *J Pharm Pharmacol* 2017;69:613–24, https://doi.org/10.1111/jphp.12707.
50. Dai YN, Su JZ, Wu K, Ma WK, Wang B, Li MX, et al. Multifunctional thermosensitive liposomes based on natural phase-change material: near-infrared light-triggered drug release and multimodal imaging-guided cancer combination therapy. *ACS Appl Mater Interfaces* 2019;11:10540–53, https://doi.org/10.1021/acsami.8b22748.
51. Discher DE, Ahmed F. Polymersomes. *Annu Rev Biomed Eng* 2006;8:323–41, https://doi.org/10.1146/annurev.bioeng.8.061505.095838.
52. Van Hest JC, Delnoye DA, Baars MW, van Genderen MH, Meijer EW. Polystyrene-dendrimer amphiphilic block copolymers with a generation-dependent aggregation. *Science* 1995;268:1592–5, https://doi.org/10.1126/science.268.5217.1592.
53. Lee JS, Feijen J. Polymersomes for drug delivery: design, formation and characterization. *J Control Release* 2012;161:473–83, https://doi.org/10.1016/j.jconrel.2011.10.005.
54. Matoori S, Leroux JC. Twenty-five years of polymersomes: lost in translation? *Mater Horiz* 2020;7:1297–309, https://doi.org/10.1039/c9mh01669d.
55. Nehate C, Nayal A, Koul V. Redox responsive polymersomes for enhanced doxorubicin delivery. *ACS Biomater Sci Eng* 2019;5:70–80, https://doi.org/10.1021/acsbiomaterials.8b00238.
56. Hossainzadeh S, Ranji N, Sohi AN, Najafi F. Silibinin encapsulation in polymersome: a promising anticancer nanoparticle for inducing apoptosisanddecreasing the expression level of miR-125b/miR-182 in human breast cancer cells. *J Cell Physiol* 2019;234:22285–98, https://doi.org/10.1002/jcp.28795.
57. Simon J, Flahaut E, Golzio M. Overview of carbon nanotubes for biomedical applications. *Materials* 2019;12:1–21, https://doi.org/10.3390/ma12040624.
58. Liu Z, Tabakman SM, Chen Z, Dai HJ. Preparation of carbon nanotube bioconjugates for biomedical applications. *Nat Protoc* 2009;4:1372–82, https://doi.org/10.1038/nprot.2009.146.
59. Hassan HAFM, Diebold SS, Smyth LA, Walters AA, Lombardi G, Al-Jamal KT. Application of carbon nanotubes in cancer vaccines: achievements, challenges and chances. *J Control Release* 2019;297:79–90, https://doi.org/10.1016/j.jconrel.2019.01.017.

60. Raphey VR, Henna TK, Nivitha KP, Mufeedha P, Sabu C, Pramod K. Advanced biomedical applications of carbon nanotube. *Mater Sci Eng C-Mater* 2019;100:616–30, https://doi.org/10.1016/j.msec.2019.03.043.
61. Mohseni-Dargah M, Akbari-Birgani S, Madadi Z, Saghatchi F, Kaboudin B. Carbon nanotube-delivered iC9 suicide gene therapy for killing breast cancer cells in vitro. *Nanomedicine-UK* 2019;14:1033–47, https://doi.org/10.2217/nnm-2018-0342.
62. Sundaram P, Abrahamse H. Effective photodynamic therapy for colon cancer cells using chlorin e6 coated hyaluronic acid-based carbon nanotubes. *Int J Mol Sci* 2020;21:1–15, https://doi.org/10.3390/ijms21134745.
63. Juzeniene A. Chlorin e6-based photosensitizers for photodynamic therapy and photodiagnosis. *Photodiagn Photodyn* 2009;6:94–6, https://doi.org/10.1016/j.pdpdt.2009.06.001.
64. Chen B, Li DY, Wang F. InP quantum dots: synthesis and lighting applications. *Small* 2020:1–20.
65. Huang YM, Singh KJ, Liu AC, Lin CC, Chen Z, Wang K, et al. Advances in quantum-dot-based displays. *Nanomaterials-Basel* 2020;10:1–26, https://doi.org/10.3390/nano10071327.
66. Zheng XT, Ananthanarayanan A, Luo KQ, Chen P. Glowing graphene quantum dots and carbon dots: properties, syntheses, and biological applications. *Small* 2015;11:1620–36, https://doi.org/10.1002/smll.201402648.
67. Manivannan S, Ponnuchamy K. Quantum dots as a promising agent to combat COVID-19. *Appl Organomet Chem* 2020:1–6.
68. Li ZJ, Xu H, Shao JD, Jiang C, Zhang F, Lin J, et al. Polydopamine-functionalized black phosphorus quantum dots for cancer theranostics. *Appl Mater Today* 2019;15:297–304, https://doi.org/10.1016/j.apmt.2019.02.002.
69. Tan LJ, He CY, Chu XJ, Chu YQ, Ding YM. Charge-reversal ZnO-based nanospheres for stimuli-responsive release of multiple agents towards synergistic cancer therapy. *Chem Eng J* 2020;395:1–11, https://doi.org/10.1016/j.cej.2020.125177.
70. Barreto JA, O'Malley W, Kubeil M, Graham B, Stephan H, Spiccia L. Nanomaterials: applications in cancer imaging and therapy. *Adv Mater* 2011;23:18–40, https://doi.org/10.1002/adma.201190041.
71. He C, Lin W. Hybrid nanoparticles for cancer imaging and therapy. In: Mirkin, C, Meade, TJ, Petrosko, SH, Stegh, AH, editors. *Nanotechnology-based precision tools for the detection and treatment of cancer*. Geneva: Springer International Publishing; 2015, vol 166. pp. 173–92.
72. Pan Y, Xue PJ, Liu SP, Zhang LC, Guan QB, Zhu JL, et al. Metal-based hybrid nanoparticles as radiosensitizers in cancer therapy. *Colloid Interface Sci* 2018;23:45–51, https://doi.org/10.1016/j.colcom.2018.01.004.
73. Thorat ND, Townley HE, Patil RM, Tofail SAM, Bauer J. Comprehensive approach of hybrid nanoplatforms in drug delivery and theranostics to combat cancer. *Drug Discov Today* 2020;25:1245–52, https://doi.org/10.1016/j.drudis.2020.04.018.
74. Li C, Wang KB, Li JZ, Zhang QC. Recent progress in stimulus-responsive two-dimensional metal-organic frameworks. *ACS Mater Lett* 2020;2:779–97, https://doi.org/10.1021/acsmaterialslett.0c00148.
75. Javanbakht S, Hemmati A, Namazi H, Heydari A. Carboxymethylcellulose-coated 5-fluorouracil@MOF-5 nano-hybrid as a bio-nanocomposite carrier for the anticancer oral delivery. *Int J Biol Macromol* 2020;155:876–82, https://doi.org/10.1016/j.ijbiomac.2019.12.007.
76. Karimi S, Namazi H. Simple preparation of maltose-functionalized dendrimer/graphene quantum dots as a pH-sensitive biocompatible carrier for targeted delivery of doxorubicin. *Int J Biol Macromol* 2020;156:648–59, https://doi.org/10.1016/j.ijbiomac.2020.04.037.

Łukasz Kaźmierski and Małgorzata Maj

9 3D tumor model – a platform for anticancer drug development

Abstract: While still attractive, the currently available 2D cell culture models present several limitations and if possible should be supplemented with their 3D counterparts, that is with spheroids/organoids or bio-printed structures. Those alternatives can sometimes show widely different results compared to the simpler 2D cell culture, especially during cytotoxicity testing that is often used during cancer drug development and in the rising field of personalized medicine. Although some of the methods like spheroid formation and basic alginate based bio-prints were already available for years, they still require huge amounts of optimization and troubleshooting to be used effectively. Proficient use of dedicated tools and software can help to overcome some of the difficulties associated with those seemingly well described models. In this article we compare the most popular and currently available methods of acquiring 3D bio-models while describing their limitations and shortcomings as well as technical hurdles that one has to overcome to succeed in the use of this complex model.

Keywords: bio-models, 1; cell culture, 1; cytotoxicity, 1; organoids, 1; spheroids, 1

9.1 Need for 3D tumor model (personalized medicine and drug development)

First of all, we can distinguish two types of 3D tumor models being used in current medicine that are far apart in use and methods. The first type is used solely as a representation of the actual tumor and often its surroundings, formed via one of the available 3D printing technologies. It is used by doctors during diagnosis or surgeons before surgery to better visualize the tumor in space. The second type of 3D tumor model can be used in *in vitro* experiments and is acquired via complex cell culture technics. Currently, because of low cost, wide knowledge and fine-tuned methods 2D *in vitro* cell culture-based assays are still the most commonly used ones. Drug efficiency tested only on a 2D model can have decreased efficiency in real-world applications, sometimes as crucial as cancer treatment [1]. Inherently animal biological structures are much more complex than those achieved via 2D culture modeling, therefore differences in *in vitro* and *in vivo* studies can occur. Those might be attributed to cells grown in a 2D culture lacking the complex 3D tissue architecture,

---

This article has previously been published in the journal Physical Sciences Reviews. Please cite as: Kaźmierski, L., Małgorzata, M. 3D tumor model – a platform for anticancer drug development *Physical Sciences Reviews* [Online] 2021, 6. DOI: 10.1515/psr-2019-0061

interactions between cells and between cells, and the extracellular matrix [2]. Although the 3D *in vitro* model is attractive, we have to keep in mind it is still relatively young, with some discrepancies in the available methods and vastly more complex than the alternatives as it will be described in the following sub-chapters.

9.1.1 3D printed tumor models

Tumor models created using modern FDM (fused deposition modeling), SLA (stereolithography) and MSLA (masked stereolithography) technologies are growing in popularity. Rapid prototyping is one of the main uses of 3D printer technologies, those are tailored toward creating personalized items in low volumes. In this case, it is being used to create an exact replica of a tumor and often surrounding organs of patients. It is possible to create a 1:1 replica within hours from acquiring a proper digital model.

9.1.2 Need for new *in vitro* models

Owing to the growing need for new, more accurate models representing human physiology, novel *in vitro* testing methods are being developed. Unfortunately, owing to the complex nature of biological processes that take place in tissues and the interactions between cells on a 3D plane, standard 2D *in vitro* models often fail to mimic them in an adequate matter [3]. As seen in the literature, 3D modeling technics are getting more and more popular, between 2009 and 2019 there has been three times increase in research using *in vitro* 3D modeling technics. Transitioning from the 2D to the 3D model might have the benefit of fine-tuning the model to suit the specific experiment application, mainly thanks to the multiple methods available to achieve 3D culture [4]. Although still irreplaceable, the expensive and ethically burdened, animal model can be in part phased out by its *in vitro* 3D counterparts, it plays nicely into the three Rs principle (Replacement, Reduction, and Refinement) proposed by Russel and Burch [5].

9.2 General understanding of 3D printing technology

Before getting into the use of 3D printer technology in medicine and diagnostics, it is absolutely necessary to have a good understanding of the basic technology. 3D printing is a type of additive, computer-assisted manufacturing (CAM) method widely used to produce models designed using a variety of computer-assisted modeling (CAM) methods. There are actually multiple methods of 3D printing in the industry. One of the most popular methods is FDM 3D printing (Fused Deposition Modeling), this method is known in the industry especially for prototyping capabilities and low material cost. Another pair of popular methods are SLA and digital light processing 3D printing. Both are valued for their accuracy but are more costly and produce fewer functional parts compared to FDM methods. Industries requiring high precision and strength from parts often use SLS 3D printing methods (Selective laser sintering) or a variant of this method using metal powder instead of powder polymers SLM (Selective Laser Melting) or DMLS (Direct Metal Laser Sintering).

9.2.1 FDM 3D printing

Although the concept of 3D printing was first mentioned in 1974 [6], this manufacturing method was not developed till the early 1980s. During those years a multitude of patents have been granted, those limited the growth of the technology in a major way and locked the use of it for the general public and small innovating companies. One of the major patents regarding the FDM manufacturing process expired in 2009, its expiration immediately enabled for prices of FDM 3D printers to drop from 10,000 USD to 1000 on average [7] Nowadays prices of this technology have fallen even further with the popularization of standard NEMA 17 stepper motors and their stepper drivers. The market is also pushed forward by the popularization of private and hobby use of this technology, a multitude of open-source projects help the industry growth in a huge way. Other major patents also expired between 2013 and 2015 regarding SLS and DLP 3D printing, further opening innovation opportunities for small companies, including companies developing complex bioprinters [8–11].

FDM is being widely accepted as the most popular method of 3D printing in the industry. In its original form, it uses thermal plastics to reproduce the desired 3D models. In the general construction of an FDM 3D printer, we can distinguish several key systems and device groups that work synchronized with each other during the print (Figure 9.1, Table 9.1). Based on knowledge about the basics of FDM printer construction it is easy to deduce the advantages and limitations of this technology. It is well suited for rapid prototyping, low volume manufacturing, and customized parts. It is especially versatile in manufacturing intricate internal structures if those can be printed by a layer-by-layer basis with or without support. The technology, unfortunately, does not scale well when large quantities of items are required, producing two identical items will require twice the time of a single item. Compare that to injection molding technics, where manufacturing multiple copies of an item takes about the same time as a single item.

9 3D tumor model – a platform for anticancer drug development

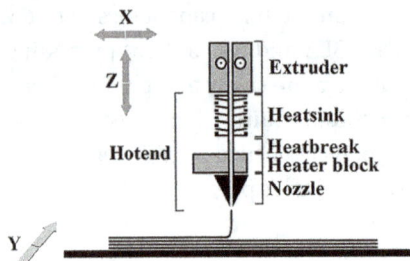

Figure 9.1: Simplified scheme of an FDM type 3D printer.

Table 9.1: Generalized FDM 3D printer components divided into functional groups.

System/device group	Components	Main function
Motion system and frame	Frame	A construction is usually based on metal rails, sheet metal, or acrylic on which the printer is built on. It is very important to ensure the squareness of the frame.
	Motors	Stepper motors that provide the torque required by the motion system (X, Y, and Z axis) and extruder (often referred to as the E axis). Their gearing and step amount play a big role in the final resolution of the printer.
	Motion transition system	Elements required for transmitting the motion of the stepper motors onto the axis such as belts, linear rails, wheels, and bushing bearings.
	Enclosure	The enclosure of the printer mainly serves the purpose of separating the print volume from the outside environment. It helps to prevent from drastic temperature changes, drafts, and in certain cases from light. It can be equipped with a heater to increase the print area temperature and a fume extractor to prevent harmful compounds from entering the work area.
Hotend	Nozzle	Provides an extrusion point for the molten filament, is the actual area of melting the filament.
	Heater block	Made of high conductivity metals such as aluminum or copper. Provides adequate thermal mass and houses the heater cartridge and thermal sensor.
	Heat break	Physically and thermally separates the "hot" and "cold" zones of the hotend. Usually in the form of a thin metal tube.
	Heat sink	Usually, a block of aluminum or other light metal with high thermal transition properties. It ensures that the heat transmitted from the heater block through the heat break will be properly dissipated. Usually has a form of active cooling attached such as a fan or a liquid cooling solution.

Table 9.1 (continued)

System/device group	Components	Main function
Extruder	Extruder gearing	The gearing usually provides a 1:1 torque transition from the extruder stepper motor but it is not uncommon for the extruder to have a different gearing ratio providing higher torque. Torque can affect the final print speed and reliability.
	Filament drive gear	Gear that comes in contact with the filament and drives it through the system, usually those are custom made to ensure compatibility with certain filament diameters. It is common to use dual gear drives to ensure higher grip levels and reliability.
	Idler arm	A lever that has an idle tension teethes gear or a gear attached indirectly to the extruder motor that provides tension to the filament. It enables tension adjustment on the filament.
	Extruder stepper motor	The motor that provides the motion to the gearing of the extruder.
Mainboard and sensors	Motherboard	The motherboard is responsible for controlling all print parameters and other components such as the motion system, temperatures, and fan speeds. Printer-specific firmware present on the motherboard usually provides possibilities of fine-tuning parameters such as acceleration, speeds, and PID (proportional–integral–derivative) tuning.
	Stepper drives	Stepper drives are separate (or integrated) of the motherboard, they are required on the motherboard to control the stepper motors.
Print bead	Print surface	The surface on which the print is going to be located on. Depending on the application it can provide different levels of adhesion for the first print layer.
	Base	The base provides a flat, even, and often heat conductive base on which the print surface is located on. The base is attached via rails, bushings, or wheels to the frame of the printer.
	Heater	The bed heater is either integrated into or located underneath of the print base. It ensures even heating of the print surface.

9.2.1.1 Nozzle

Although being the last part of the hotend assembly it is nevertheless the part that plays the most important role in the final extrusion of the molten filament. In most systems, the nozzle is firmly screwed or integrated into the heater block to provide adequate heat transfer from the heater cartridge to the melted filament. By design, nozzles are in fact the main melting zone for the filament entering from the heat break. Nozzles used in machines that require precise temperature control of the melting zone often have their own integrated temperature sensor. This solution is also beneficial when dealing with prints that are subject to high print speed variations that in effect might dissipate heat from the nozzle faster than both the heater block sensor can sense and the heater cartridge of the heater block can provide. The proper internal geometry of the nozzle and especially the tip of the nozzle is crucial. At the point of entry, the diameter of the nozzle is the same as the heat break, and only at the tip the diameter is reduced to a required, smaller one. Commercially available nozzle tip diameters range from 0.15 mm up to 1.20 mm, with 0.4 mm being the most popular choice.

Manufacturing accuracy and wear level of this part is a key factor in the success rate and quality of the final print, the inner walls of the nozzle need to be smooth and devoid of burrs that may obstruct the molten filament path or may unfavorably increase the adhesion of the filament to the walls. Improper filament path in the nozzle may result in issues such as clogs the external finish of the nozzle, especially close to the tip is also required to be smooth and devoid of burrs to prevent the molten filament to stick to it when the material gets out of the nozzle or the nozzle moves over print layers. Multiple materials are chosen to manufacture nozzles, they all present benefits and cones.

Brass is the most popular type of material in nozzle production. It is usually used with thermoplastics whose printing temperature does not exceed 300 °C (PLA, PETG, and ABS). It is also not suited for abrasive materials such as carbon fiber enhanced filaments because it is a relatively soft metal. Brass has relatively good thermal conductivity but the low thermal capacity to other materials used in nozzles.

Plated copper nozzles are used for high-temperature materials that require up to 500 °C such as Nylons, PEEK, PC, or PEI. They obtain superior thermal conductivity and thermal capacity compared to other nozzles. These nozzles are also suitable for use with abrasive filaments for example containing carbon fiber particles.

Stainless and hardened stainless steel can be used for the same materials as plated copper nozzles, unfortunately, steel has lower thermal conductivity and a possibility of cracking and loos of hardness when using them at temperatures over 400 °C. They are used mainly in applications where plated copper is not recommended and might react adversely with the filament.

Ruby insert nozzles provide superior abrasive filament wear protection, because of the ruby that replaces the actual tip of the nozzle. They are well suited for applications requiring work with carbon fiber modified filaments. The nozzles themselves should be treated as consumable, in case of damage or severe wear, they should be replaced to avoid inaccuracy and print failure.

9.2.1.2 Heater block

The heater block serves multiple purposes. It is the part of the hotend assembly that is responsible for connecting the heat break with the nozzle. It is also responsible for transferring the heat to the nozzle and temperature control (unless the temperature sensor is integrated into the nozzle). During the process of the print, it is inevitable to have deviations in the speed of the print and therefore differences in the required federate of the filament that is melted. During the melting process and the actual extrusion of molten filament, heat is removed from the system. Therefore, one of the most important features of the heater block is its increased thermal mass to compensate for those possible fluctuations. It is worth noting that heater blocks are usually quite compact compared to the rest of the hotend assembly, that is a conscious decision aimed at decreasing the total mass of the hotend and decrease the time of initial heater block warm-up.

There are two most commonly used materials for heater block construction, aluminum, and plated copper. The latter is used for high-temperature applications because of its increased softening temperature and deformation within the required temperature and increased thermal mass and superior thermal conductivity. Unfortunately, it also introduces a mass increase to the hotend assembly which is unfavorable for the final print quality.

Modern heater blocks house ceramic heater cartridges (30–80 W) that are the source of heat for the hotend. Depending on the temperature needed to achieve or the required speed of print different heater cartridges may be used for the application.

Because precise temperature regulation of heater block is very important to the print quality, a thermocouple or a thermistor (such as industry standard pt100 or pt1000) has to be used. The feedback from this sensor determines the power output of the heater cartridge and helps achieving the requested temperature during the print.

Printing particular high-temperature materials such as PEEK or PEI often requires a special printer setup. The main issue is the high hotend temperature required (330–350 °C) and the need for a heated chamber (110–120 °C). Those temperatures often demand the use of water cooling of certain motion system components such as stepper motors and electronics. Depending on the setup also the heat sink will require water cooling.

9.2.1.3 Heat break

The heat break connects the heater block with the heat sink. It is usually a thin metal tube. The main purpose of this part is to separate sections of the hotend into a "Hot" and "Cold" section. Because it is mandatory to keep as much heat as possible in the heater block and the nozzle, at the same time ensuring little heat fluctuations, the heat break is produced to have thin walls and to have a low thermal mass and low thermal conductivity. It is usually made out of stainless steel because of favorable thermal parameters and high part strength. The internal finish of this part has to be very smooth, to ensure an unobstructed filament path. It is also important to ensure a proper thermal transfer between the heat break and the heat sink because the filament inside the heat can't exceed its softening temperature, otherwise, it might cause a clog and a failure in the print process. The connection between the heat break and the heat sink usually contains a thermal compound to ensure proper heat transfer or the heat break is integrated into the heat sink.

9.2.1.4 Heat sink

The heat sink of the hotend is supposed to dissipate any heat that may creep up the heat break, it is the "Cold" side of the hotend assembly. Most heat sinks contain an integrated fan to ensure proper heat dissipation. Usually, the fan is running at maximum rpm and has no pwm control to ensure that the amount of heat dissipation from the heat sink remains constant. Because the heat sink is usually the largest part of the hotend it is also the part via which the assembly is mounted to the frame of the printer.

9.2.1.5 Extruder

Extrusion of filament is precisely controller by the extruder that consists of a stepper motor that with the use of dedicated gearing and filament fixation mechanisms guides it toward the hotend. Extruders often use extra gearing to increase the torque required to force the filament through the hotend. The part that actually comes in contact with the filament is usually a single or dual goblin drive gear that bites into the filament and ensures firm movement. We can distinguish two types of extruders based on the mounting position, direct-drive extruders and indirect type extruders (often called Bowden type extruders).

<u>Direct type extruders</u> are mounted just above or are integrated into the hotend. This setup minimizes the filament path and ensures maximum control over the flow of filament because it does not introduce flex or rebound to the system. The downside to this type of extruder is a big mass increase to the axis that it is mounted and can introduce artifacts into the print or strain the motion system when printing fast.

<u>Indirect type extruders</u> (Bowden type extruders) are connected to the hotend via a PTFE (Polytetrafluoroethylene) Bowden tube. The main benefit of this setup is a significant mass decrease on the extruder axis that reduces the mass required to move via the stepper motors. Unfortunately, owing to a certain amount of flex and possible compression of the PTFE Bowden tube this setup is more prawn to extrusion inaccuracy and is often not suited to print filaments that are flexible.

9.2.2 SLA and MSLA 3D printing

SLA and MSLA are resin-based, photomechanical 3D printing processes. To simplify, photocurable resin is used in a bath on which bottom of a laser source or LCD (Liquid Crystal Display) matrix is placed to work as a light source with a specific wavelength that will cure the resin. Those are usually UV-based and have to be treated with caution because they can easily damage eyes if proper shielding is not used. The bottom of the bath is created with a smooth material that the cured resin will not stick to easily. The cured layer of resin will stick to a z-axis build platform mounted above the resin bath. To allow the next layer of resin to cure the print is lifted via the z-axis build platform, and so the process continues till the print is finished. The main advantage of this method is the print resolution that is dependent on the laser accuracy or the LCD resolution and the z-axis pitch and precision. The print time is not dependent on the X and Y dimensions of the printed item when using an LCD matrix because it cures the entire layer at the same time. A thin and long item will always take longer to print than a thick but shallow item. Unfortunately, resins used for this method are expensive, have lesser mechanical properties, and are usually highly toxic when not cured. The prints themselves are also not entirely cured after the print and require post curing using dedicated UV curing stations and post-processing. Owing to those reasons, SLA and MSLA printing are still not dominant, except when producing high precision small items.

9.2.3 SLM and DMLS 3D printing

SLM and DMLS are industry-specific methods. They are used most prominently in the aviation and automotive industry for manufacturing parts with high geometrical complexity and with extraordinary mechanical requirements. Those parts are often not possible to manufacture by any other method except 3D printing. One of the fastest-growing industry branches for powder metal 3D printing is the medical market. There is a growing interest in the dental and prosthetic sectors with this method. It enables the manufacturing of highly personalized bone structures that can be based on actual patient data. The major drawback to those technologies is the fact that a certain, inherent surface roughness will be present and some post-processing will be required before attaining a final product. This fact does not disqualify those methods as a whole, on the contrary, it incentivizes manufacturers to further develop those methods to suit medical industry standards [12].

In the aforementioned methods, a high-power laser is used to melt a powder containing fine metal particles from the top. After acquiring a melted layer, a very fine fresh layer of metal powder is added on top and the process repeats till the end of the print [4].

9.3 3D *in vitro* model

Cells on their own are complex 3D structures but as a simplification for a cell model to be considered a 3D model, cells need to have freedom of dividing freely in the 3D space or be forcefully placed in the 3D space with the use of scaffolds forcing a certain shape or formation. Cells on the borders of 3D *in vitro* models usually have access to the surrounding media, but those that are within the 3D structure have both restricted access to fresh media and restricted possibilities of growth, just like cells in real-world tissues. The fact of restricted access to media by the cells residing within the structures can be mediated by creating vascularization of those models but it is a complex and difficult task that is not needed in most applications.

9.3.1 Classical 2D cell culture

In classical 2D *in vitro* cell culture regarding most mammalian species cells are grown on dedicated TC-treated (Tissue Culture – treated) surfaces. Most commonly plastic (polystyrene) or in special cases glass is used as the main material of those surfaces. By definition, cells grow on a 2D plane with limited options to form complex structures. It is unlikely under standard conditions and often unwanted for the cells to overgrow the first layer, of cells. In most cases, overgrowing of a 2D culture would cause cell death of the lowest layer because of lack of fresh media access and inability to transport waste metabolites outside. Growing multiple layers of cells proficiently requires the use of specialized flow bioreactors with pH, glucose, and CO_2/O_2 levels control. Those retrocessions are needed because of the high metabolic level of cells cultured in such a way.

Cells grown on plastic surfaces in standard 2D cell culture are limited from the bottom by the vessel growth surface, and if they reach the side of the vessel, they are also limited by the vessel walls. To ensure proper cell adhesion to polystyrene TC treatment is required, most commonly manufacturers use methods such as energetic plasma activation, liquid surface deposition, and emerging functionalization. TC modifications have two main goals in mind, increasing the hydrophilic properties of the surface and changing the topography of the surface of the material [13].

9.3.2 Spheroid culture

One of the most popular methods of 3D cell culture especially popular when modeling tumor – drug dependence, it is the so-called spheroid culture. In recent years spheroid culture was popularized among scientists performing experiments on stem cells, cancer treatment, and regenerative medicine. In this type of culture, cells grow on a 3D plain and can expand in all directions. If not obstructed by physical means they will generate a formation of a spherical shape. Spheroids are used to artificially create a multi-layer cell culture with different properties than the standard 2D approach. They are used in *in-vitro* studies, especially regarding cytotoxicity, drug development, cancer research, and stem cell research. They provide a model in many ways more accurate to the one encountered in real physiological environments.

One of the most characteristic features – multiple layers of cells – provides a great research model for drug penetration into the tissue, insight to organ formation, and invasion of cancerous cells. Spheroids can be formed using multiple methods and the quality, quantity, properties, and size of spheroids are going to be dependent on them. Because multiple factors affect spheroid growth it is harder to control the formation and growth of spheroids, assays based on spheroids require normalization and careful planning.

Unfortunately, spheroids cannot fully represent the physiological picture, because there is no vascularization present, and no real growth obstructions, as such they are not ideal to use as model organs or vascularized tumors.

9.3.3 Classical 2D versus 3D cell culture approach

Although classical 2D cell culture was considered a standard in the field of life sciences and is still the most widely used method for drug screening, biomaterial cytotoxicity testing, and cancer research, it might not always be the best-suited method. Currently 3D cell culture methods are being applied to a great extent in all areas of life sciences research. There are some key differences that prove this 3D model type so popular. Compared to the 2D cell culture the spheroids are usually not attached to the bottom of the growth surface although most often they consist of adherent cells. This property paired with the fact that the internal structure of spheroids is complex and overall spheroids react in different ways than cells in 2D cultures gives spheroids an opportunity to become the leading model for cytotoxicity research and personalized medicine.

Compared to 2D culture it is much more difficult to perform assays using spheroids, not only are they rather thick samples compared to 2D culture but using some formation methods there are multiple spheroids per well (when using multiwell plates). If performing assays that require the use of optical means of analysis (such as microscopes) specialized software has to be used to analyze the data.

Owing to the unpredictable nature of spheroids regarding their positioning in the culture vessel of choice, it might be very difficult to acquire images using live imaging of live spheroids or even fixated spheroids without user intervention. Without software capable of finding spheroids before image acquisition on the X, Y plane of the microscope the system would dictate the acquisition of images from the entirety of the plane, in result producing enormous amounts of waste data and increasing assay time by multiple times. The problem intensifies when there are multiple spheroids per culture vessel and are of different sizes. Image acquisition time without spheroids fixated in place is also a big problem, performing a Z-stack image (especially using high magnifications) requires time, and if the spheroid moves while the acquisition takes place, subsequent images will be shifted on the Z plane, rendering the final image unusable.

9.3.4 Spheroid formation methods

Over the years multiple methods of acquiring spheroids have been developed. All of them are still being used depending on the assay type and the experiment setup. Those of them that will be discussed often have many variations and are highly dependent on the cell type used in the experiment (Table 9.2). Looking at those methods with tumor modeling angle in mind we can put in place some assumptions. Most cell types used are going to be f cancerous cell lines that on average have a higher metabolic rate than normal cells. Those cells will also have a higher average Time for Cell Doubling – time for a certain population of cells to double in number. That means that the cell culture will demand more frequent medium changes than a spheroid culture of normal cells. As such the time between spheroid formation and the first medium change needs to be shorter, favoring methods with lower formation times like Magnetic spheroid 3D printing or using a U-bottom plate with centrifugation. On the other hand, depending on the assay type one would need to change the method to acquire uncontaminated spheroids like those generated by standard F-bottom cell repellent surface spheroid formation or Hanging drop formation. For qualitative analysis, the number of spheroids per replication might not be as important as in HCS (High Content Screening) methods but we have to keep in mind that large amounts of spheroids per well (if a multiwall plate is used) will affect the quality of medium in a different way, the same is true when testing cytotoxic effects of drugs.

9.3.4.1 Spheroid formation and assay-specific problems
As described in previous chapters, the spheroid culture process is more complex than standard 2D cell culture. As in all assays, more complexity may provide more room for error and increase the deviation of obtained results. We shall discuss the problems arising from assays based on spheroids and solutions to those problems in specific stages of the workflow.

9.3.4.2 Spheroid formation
The spheroid formation is the most critical process that is also most prawn to error. Depending on the assay type and the equipment used to evaluate the results the user might prefer different experiment setups. In the most common experiment setup using multiwell plates having a single spheroid per experiment well is usually ideal. Unfortunately, achieving a single spheroid per well is rather difficult using popular methods such as F-bottom cell repellent surface spheroid formation or U-bottom cell repellent surface spheroid formation. Having multiple spheroids formed per well overall negatively affects drug development experimentation because of the inability of assay well to well normalization. Spheroids obtained from those methods are going to range in size and numbers. Those methods also introduce stray cells, unbound to any spheroid, and are difficult to remove from the medium

Table 9.2: Table presenting a comprehensive comparison of different methods of spheroid formation.

Formation method	Formation method complexity	Spheroid size reproducibility	Control over spheroid number	Spheroid optical clarity and contamination level	Speed of spheroid acquisition	Spheroid circularity and shape irregularity	Average price per spheroid formed	Assay incompatibility and problems
F-bottom cell repellent surface spheroid formation	Low	Low	Low	High clarity with low contamination level	Low	Slightly nonspherical	Low	Problems with HCS systems
U-bottom cell repellent surface spheroid formation	Low	Medium	Medium	High clarity with low contamination level	Low	Slightly nonspherical	Low	Optical based assays
U-bottom with centrifuge	Medium	Medium	High	High clarity with low contamination level	Medium	Irregular shape	Low	Optical based assays
Magnetic spheroid 3D printing	High	High	High	Low optical clarity with medium/high contamination level				
Hanging drop	Medium	High	High	High clarity with low contamination level				
Spheroid formation in matrix	Medium	Low	Low	High optical clarity with medium/high contamination level				
Spinner flask spheroid formation	Medium	Low	Low	High clarity with low contamination level				

(continued)

Table 9.2 (continued)

Formation method	Formation method complexity	Spheroid size reproducibility	Control over spheroid number	Spheroid optical clarity and contamination level
Magnetic spheroid 3D printing	Fast	Spherical and regular	High	Some optical based assays
Hanging drop	Medium	Spherical and regular	Medium	Compatible with most assays
Spheroid formation in matrix	Medium	Spherical and regular	Medium	Incompatibility based on matrix used
Spinner flask spheroid formation	Low	Spherical and with medium regularity	Low	Most assays requiring a precise number of spheroids and a specific size of spheroids

without extracting actual spheroids by accident or damaging them. Owing to multiple spheroids present per well it is also difficult to perform live imaging-based assays that track spheroid growth because spheroids are not stationary and often change position for example because of convection liquid movement. For those reasons especially when testing drugs cytotoxicity, it might be beneficial to use other formation methods like Magnetic spheroid 3D printing or Hanging drop methods. Of course, there are possibilities of performing HCS experiments using simpler spheroid formation methods. Some HCS systems such as Perkin Elmer Opera are capable of performing complex assays using such assay configurations. This is a growing branch of life science software development often using Artificial Intelligence and Machine Learning to aid scientists with their research.

9.3.4.3 Spheroid fluorescence visualization

One of the industry standard methods for determining multiple cell parameters such as growth, morphology, and presence of specific proteins is fluorescence visualization. The method uses the phenomenon of fluorescence, in short emission of light by a compound that has absorbed light of another specific wavelength. In life sciences, this method is used for qualitative and quantitative analysis using fluorometers and microscopes. Most unmodified mammalian cells do not exhibit autofluorescence and have to be stained or modified for researchers to observe fluorescence.

There are three widely used methods of visualizing cells using fluorescence:
- Use of chemical compounds that exhibit fluorescence properties and specifically or nonspecifically bind to certain organelle of the cell.
- Immunofluorescence; use of antibodies specific toward a certain protein that contain a fluorescent compound or will be attacked via a secondary antibody containing this type of compound in their structure.
- Use of cells that were modified with the use of molecular biology methods such as transfection or transduction and produce fluorescent compounds on their own specifically or nonspecifically during their growth.

Not all of those methods can be directly applied to visualize and analyze spheroids with the same effect as with standard 2D cell cultures.

There are two main issues overcome, first is the penetration of agents used for visualization. The molecular weight of frequently used antibodies is about 150 kDa and is not suited for penetrating even a few layers of cells, in effect not being able to stain the entire spheroid. A proper stain is only obtained on the surface of the spheroid but not within it. The same problem is present while staining spheroid directly with fluorescent agents such as DAPI (4′,6-diamidino-2-phenylindole) or Hoechst 33342 (Trihydrochloride). The problem is exaggerated while using more complex methods that require the metabolism of the agents via cells, like in the case of apoptosis assays using the DEVD method. Often it might be a better approach to stain the cells before spheroid formation in a non-stable or stable matter using accumulating agents such as PKH days (for example PKH26, PKH67), quantum dots, or performing transfection or transduction of cells. It is worth noting that transduction is considered a more stable method than transfection but is fraught with more possible negative effects on cell metabolism and properties. It is considered good practice to assess the effects of any genetic modification on the cell line before performing any spheroid assays. It is also possible to buy ready-to-use transfected and transduced cell lines or order a modification of an already owned cell line.

The second issue is based on the physical limitation of light penetrating into the spheroid structure to excite the fluorescent dye and the emitted light exiting the spheroid structure. Classical epifluorescence methods lack both the power to properly excite all the fluorescent days and to detect the signal coming from the deeper parts of the structure. In theory, increasing the power of the excitation light source could ultimately improve the visualization but because we usually use live imaging methods for spheroid visualization then this would cause phototoxicity issues (toxicity from light). Therefore, the optimal method of spheroid visualization would be to use confocal or multiphoton systems [14–17].

9.4 Bioprinting

Bioprinting is based on the principles of polymer-based 3D printing but uses for tissue engineering purposes. Bioprinting excels at creating structures, virtually impossible to obtain using standard tissue engineering methods. Theoretically, it's possible to obtain complex structures such as tissues or organs containing multiple cell types intricately intertwined with each other just like they would in the original. Unfortunately, because the technology merges aspects of 3D modeling and machine engineering with biology, succeeding in bioprinting even simple models is an extremely difficult feat.

The holy grail of cancer treatment diagnosis in personalized medicine would surely be achieving a 1:1 tumor model constructed from cells acquired from the patient. Such a model would allow to, pinpoint the best method of treatment for any particular patient.

9.4.1 Bioprinter types

The bioprinting technology is quickly evolving and applying both, well-established methods and cutting-edge technologies from different industries. In the passing years not only have we seen significant improvements in existing bioprinter types but also novel types of bioprinters emerging.

9.4.1.1 Dispensing-based bioprinters (extruder bioprinters)

Dispersion-based bioprinters are among the most popular and well-refined machines. They are heavily based on standard FDM 3D printers and expect the modified extrusion system to share most of the parts with them. The extrusion system moves in the X and Z axis or X and Y axis depending on the frame type and lays lines of filament (a full plane of lines is considered to be a layer). Those are solidified fully or partially by chemical, or physical means before the next layer of filament is added on top. Dispensing-based bioprinters work best when extruding a continuous stream of bioink, similar to plastic filament-based 3D printers, and it is a good practice to prepare the files so that there are as few unnecessary retractions and travel moves as possible. It is also possible to use multiple extruders with different cell types or scaffolding materials for printing structures requiring multiple cells and structures that need to be printed at steep angles. As with classical 3D printing, those support materials can be removed afterward by chemical means. One of the main drawbacks of extruder bioprinters is the low cell viability after the print which is caused by the printing process.

The extrusion system had to be modified from dispersing molten plastic to dispersing bioinks. Compared to molten plastic bioinks need to be treated much more gently to guarantee cell survivability at acceptable levels during and after print. For various reason, the most optimal type of extruder setup, in this case, is the direct type extruder. An indirect (Bowden) type extruder does not offer enough control over retractions and federates acceleration, it would require a complex heating system of the capillary and would be much more difficult to calibrate for different bioink densities. Direct type extruders in bioprinters can be separated into three main groups based on the mechanism used to force the bioink through the nozzle: pressure, piston, or rotary screw.

9.4.1.1.1 Pressure and piston-based extruders

Pressure extruders (pneumatic extruders) use a source of pressure like a compressor to move the air piston above the bioink. Those are similar in concept to an automatic pipette and are characterized by the worst retraction characteristics of all extruder bioprinters. One of the advantages is that this type of extruder can be mounted semi-directly, decreasing the mass of the axis on which it is situated. Moving mass decrease is always beneficial, it can increase axis travel speeds and decreases the amount of surface artifacts. The main issue is that air is compressible and as such, any pressure change in the system is going to be dampened slightly by it.

Piston-based extruders have a piston that comes into contact directly with the bioink without air between. It ensures more control over the flow and gives the possibility to work with denser bioinks. Because there is no unintentional air dampening the retraction behavior is much more manageable and the system is able to print much more complex geometries in a shorter amount of time. This type of extrusion is the most popular because of the lack of complexity and a good balance between speed and resolution obtained from it.

9.4.1.1.2 Rotary screw-based extruders

This type of extrusion is made possible by a rotating screw connected to the motor and is fitted inside a cylindrical container (Figure 9.2). The rotating screw is used to extrude the bioink instead of the piston in the previous systems. Overall, this type of system provides a good level of volume control and can extrude the bioink at higher pressures than the previous systems. It can also cope with more viscous bioinks with higher cell densities. Unfortunately, pressure changes can damage cells more easily, decreasing cell viability after and during the print. The entire portion of the bioink can be placed inside at once or there can be an inlet present to supply the extruder with the bioink. Rotary-based systems are also the most complex of the extruder bioprinters, that means more assembly and disassembly time and more maintenance between prints.

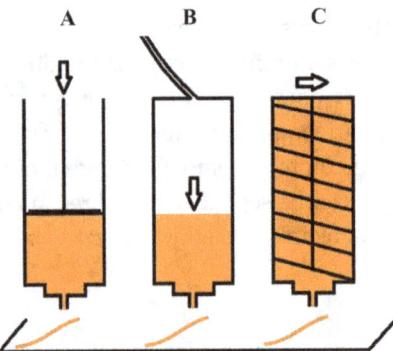

Figure 9.2: Types of dispensing-based bioprinter extruders, A – piston-based type, B – pressure-based type, C – rotary screw type.

Media used during a bioprint are most often referred to as bioinks. They contain a mixture of cells and the printing media but sometimes are deprived of cells and are just used to print an interfacing layer or support structures that can be dissolved after printing via chemical agents or biological processes (bio-degradable).

By design, most bioinks have different properties during and after print. Those are highly dependent on the application and print method but we can generalize that most of bioinks are in liquid form during the print and are supposed to become solid soon to after they exit the nozzle.

9.4.1.2 Droplet bioprinters

Droplet-based bioprinters do no use a consistent stream of bioink while printing, rather than that they use single droplets of bioink introduced one after the other. That gives those printers higher possible resolution and more control over the addition process. In theory, it is possible to modify every droplet's composure and print with multiple cell types in fast succession. There are a few technologies used for droplet bioprinters: inkjet, electrohydrodynamic jetting, laser-assisted, and piezoelectric.

9.4.1.3 Multi-photon bioprinters

Multi-photon (also called 2-Photon) is the most advanced, youngest, and most expensive method of bioprinting based heavily on the principles of multi-photon microscopy. The method uses photocured bioinks and resins. The curing light source consists of at least two lasers of a higher wavelength than normally needed to cure the bioink. If decoupled, the lasers on their own would not be able to cure the bioink, only when perfectly collimated and timed will they be able to serve their function, on this principle, it is possible to print hollow spaces, and structures not obtainable by other methods of bioprinting. The increased wavelength gives the benefit of higher penetration than the use of short wavelengths; it also has a less phototoxic effect. Overall, during an hour print cell viability throughout the final product is retained at a level of 80% using this method. This method introduces the highest resolution available to date, giving researchers a $1 \times 1 \times 3$ μm print resolution (there are four dimensions because the printer uses voxels as its primary metric).

A downside to this method is the build volume, because of complicated optics and high precision needed, the build area is only 12×12 cm and the maximum depth depending on the print conditions oscillates between 1 and 10 mm. Although it is possible that the build platform dimensions might increase, the penetrating power of the laser is not likely to improve because of the laser not being able to penetrate deeper into the uncured bioink [18, 19].

9.4.2 Bioprinter media (bioinks)

Owing to the increased market need for novel materials used for bioprinting with different types of cells the diversity of those materials has increased drastically. Bioinks are complex materials that rarely consist of only a single compound. They need to support the cells that are suspended in it, maintain a favorable pH, osmotic potential, and include some nutrition value for cells, they also need to maintain some permeability for substances to enter and exit cells. Bioinks need to be produced being sterile because any contamination with microorganisms would affect negatively the growth of cells. Bioink content mostly determines the viscosity, density, and potential for solidification. Those properties as a result determine the federate and speed of print. They can affect also the maximal layer height, overhangs maximal angle, and the build surface type needed. We can divide bioinks into two main groups, synthetic and natural [20].

Synthetic bioinks are in general less popular, there are two main kinds used:
- PEG (Polyethylene glycol) – Bioinert, used because of its mechanical properties, it is obtained by polymerization of ethylene oxide. In real life use, it is often combined with additives [21].
- Pluronics – It characterizes with specific physical properties; it changes its structure and increases viscosity when heated up to physiological temperatures. To achieve full solidification a cross-linking agent has to be used. It is also considered being a biocompatible biomaterial [22].

Natural bioinks are widely used and modified by scientists, the most common ones are:
- Gelatin – Protein-based, usually modified because of low viscosity in targeted temperatures. Methacrylation can improve its properties when in physiological temperatures [23].
- Collagen – Protein-based, used because of its natural occurrence in organisms and proven usage in the medical industry in other applications. Depending on the species from which it is harvested it can pose slightly different properties [24].
- Agarose – A common polysaccharide of marine origin. Often used in molecular biology and other industries such as the food industry. Depending on the concentration it can form semi-solid structures in physiological temperatures [25].
- Gellan Gum – Polysaccharide, similar to alginate and often used as an additive to other compounds when used in bioprinting applications. It is synthesized by Sphingomonas elodea bacteria [26].
- Alginate – Polysaccharide, commonly it is cross-linked via calcium ions present in the cross-linking solution [25].

9.4.3 Workflow and software

The workflow for acquiring a bioprint is more challenging than for standard plastic 3D print but shares most of the steps. The main differences come from the model you are going to work on and its acquisition process. There are also differences coming from the application of the model and its complexity. If the aim of the study using bioprinted constructs is to show changes in cell morphology we must understand the limitations of visualizing technologies available on the market. If cells are label in some way (using fluorescent markers or being genetically modified beforehand) the best available method would be visualization using a multi-photon microscope, that gives you a 2–3 mm depth of visualization, or a confocal system that can acquire images up to 1–1.5 mm. Those depths are estimates assuming ideal conditions. Obviously observing cell morphology without prior staining even with the use of phase-contrast microscopes is impossible in this case.

9.4.3.1 Model acquisition

For most *in vitro* research applications, scientists chose a simpler approach than modeling the entire organ or tissue, omitting the model acquisition and designing the simplified model in CAD (Computer Assisted Design) software. However, it is possible to source a dimensionally accurate digital 3D model of an organ, tumor, or body part using commercially available acquisition methods. Some of them need to be performed as standalone, and some can prove extractable data from already existing data such as CT, MRI, or X-ray scans. The latter can be especially beneficial for the patient because he does not have to undergo an extra medical procedure.

9.4.3.2 CAD (Computer Assisted Design)

Software plays a major part in the printing workflow; it enables model creation and correction and controls the process of printing. First, we need to separate CAD and CAM (Computer-aided manufacturing) software that is used during the workflow. CAD is used during designing the actual model using design tools, some also consider software used for 3D sculpturing to be CAD software. Using only CAD, it is possible to create a suitable model for our application. For example, while performing an invasion assay with the use of tumor cells we can design a simple tumor model represented via mesh and a vascular system which walls consist of normal cells [27]. Of course, it is possible to design more complex tumor models that can be placed in bioreactors but in most cases, simplified models are good enough. While designing complex models, with advanced vascular structures to mimic tumors within which angiogenesis occurred it is important to place attachment points to incur flow and make the model hydraulically tight [28]. Most of the complex models are usually acquired beforehand and were not developed in CAD software, those usually need their mesh to be cleaned and dimensioned and calibrated. Depending on the print method, some features need to be added to the model. A ready 3D model can be exported in a .obj or .stl format, both are compatible with most CAM software solutions.

9.4.3.3 CAM (Computer-aided manufacturing)

The basic tool for translating a 3D model into movements of the 3D bioprinter is called a "slicer". It is a piece of software compatible with printer firmware environments and can prepare proper tool-paths for the printing process. All 3D printers are so-called "dumb printers", which refers to them performing any movement that will be present in the generated GCODE, even if this code is harmful for the equipment. Slicer software in essence translates .obj or .stl files into GCODE (code understandable by the machine). During the preparation in the slicing software, almost all print parameters can be tweaked and corrected to suit your need. Such parameters as speed, federate, wall thickness and amount, retraction, and temperatures can be adjusted. The slicing process can have a greater effect on the final print than the print model design itself. During this process, support structures can be added, and the model itself can be rotated and repositioned for an optimal print position. During preparing digital models for printing is worth noting that most bioprinter manufacturers have an optional or integrated CAM software package that simplifies the printing process in major ways but at the same time can take away the flexibility offered by proprietary software solutions. The firmware of a printer plays a major role in the printing process, unfortunately almost all manufacturers of bioprinters use locked firmware that cannot be modified by the user.

9.4.3.4 Print post-processing

Any actions carried out to enhance, improve, or finalize the print after its completion are considered to be post-processing. Those are almost always needed by items produced using FDM-based technologies. For instance, if using support structures there is a need to remove them after the print is finished by physically detaching them or dissolving them in dedicated agents or water. In the case of bioprints it is less usual to deburr the product but can happen when working with biocompatible polymers.

9.4.4 Typical concerns in bio-printing

Printing with complex compounds and especially living cells is burdened with a range of problems that require novel solutions and a deep understanding of the issues. Most of the problems will arise because of the fragile nature of mammalian cells that are most often used for printing.

9.4.4.1 Cell viability

Cell viability should always be the primary concern during any bioprint operation. The entire printing process should be tuned in such a way that maximizes the number of viable cells in the final print and increases the chances of cell survival after the print. It is worth noting that regardless of efforts, there is going to be some cell viability decrease present both during, and after the actual print. There are many causes to this phenomenon, to name a few, a standard passage used for cell preparation for the bioprint on itself causes cell viability to decrease. Another parameter that will also affect viability in a negative way is the pressure present in the nozzle and pressure changes during the print. There are many other factors at play such as temperature changes, pH changes, or toxicity of the agents used for crosslinking.

All of the above are highly dependent on cell type, the actual methodology used and the skill of the operator performing the procedures.

One of the main factors at play considering cell viability during a bioprint that is often omitted is the pressure and changes in pressure that the cells experience. At high pressures, the delicate cell membrane is going to be destroyed causing cell death. The same effect is going to be observed if a sudden pressure change is going to be present. Even at low nozzle pressures (5 psi) cell viability can fall by 25%, at higher pressures it will deteriorate even further, for a value of 20 psi cell viability can fall by even 50% (compared to unprinted, control cells). The pressure also affects the recovery time of cells after print, a higher pressure will cause longer recovery times (time for cells to increase their viability) [29].

Cell density has no significant effect on overall viability after printing at concentrations between 2 and 8 million cells/ml of printing medium (for 3T3 cells with alginate printing medium used) [29].

9.4.4.2 Uneven cell density through the print

Maintaining a constant cell density during the bioprint might prove challenging. In most cases, the user has no control over regulating the concentration of cells through the print unless a flow change is imposed. Therefore, it might be difficult to correct changes in cell density in the final print.

At the core of the problem lies a force beyond our control that constantly affects the cells; gravity. Almost all of the media used while bioprinting require some kind of solidifying agent or factor, and before exiting the nozzle cells are suspended in a printing medium that is far less viscous than in the final product. This simply means that cells are in a suspension and are prone to sedimentation toward the bottom of the container that they are in before they exit the nozzle. If a direct type extruder is used (most popular in bioprinters) that means that cells are going to sediment at the bottom of the vessel where the nozzle entrance is located. Whereas the cells are still uniformly suspended in the solution at the beginning of the print the cell concentration of the medium exiting the nozzle will be correct. When cells will start to sediment at the bottom of the container the concentration of cells will increase but at some point, it is going to fall below the starting concentration. Figure 9.3 shows a typical representation of this problem when using direct type extruders without a mixing device. It is worth noting that the problem is especially visible when using media with low viscosity (for example with a viscosity close to water).

The problem is usually smaller in scale during small prints or when using very viscous media as presented in Figure 9.4.

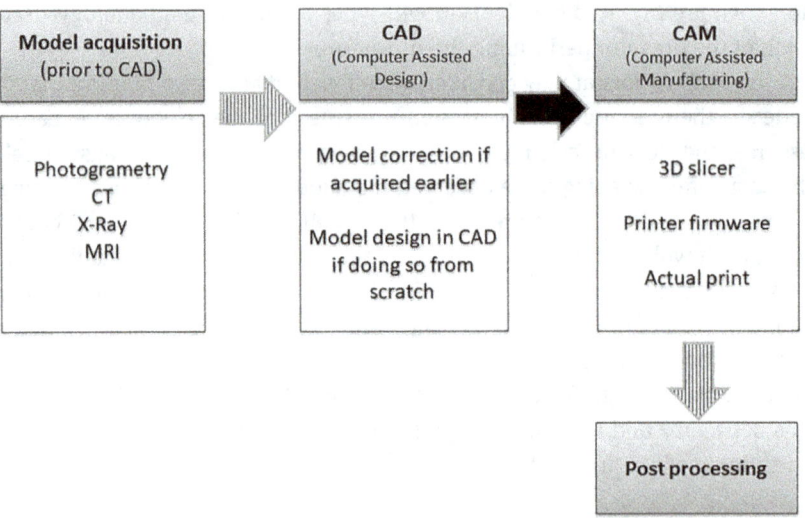

Figure 9.3: Graphical representation of a workflow for exemplary 3D bioprint.

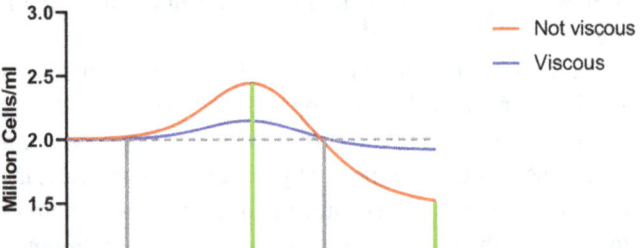

Figure 9.4: Example cell concentration of media exiting the nozzle during ta bioprint using different viscosity media with a starting concentration of two million cells per milliliter of media. The data was exaggerated to increase visual clarity.

9.4.4.3 Layer adhesion and layer sagging

Inadequate or improper solidification of bioinks after exiting the nozzle will have a negative effect on the success rate of bioprints and their quality. There are multiple causes of this phenomenon and because of this fact it is often difficult to troubleshoot.

One of the most common defects in bioprinting is layer sagging, visible when printing molten substances (Figure 9.5). Sagging is appearing when the solidification of a layer is not happening fast enough and the molten layer is flowing down on the layer's underneath "sagging" on them. For example, when printing with an alginate-based bioink the crosslinking agent used (in this case calcium ions) might

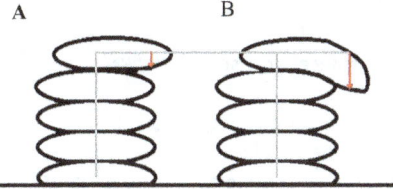

Figure 9.5: Representation of layer "sag" in FDM method used for bioprinting. A. example of layer properly solidified, B. example of layer improperly solidified, "sagging."

not be in a proper concentration. The amount of alginate in the bioink used might also be too low for proper solidification. When using UV-curated bioinks, the problem might be caused by a UV intensity set too low or parts of the print might obstruct the light path and in effect decrease the effectiveness of UV curing. Sagging might also occur if the project itself is carried out in such a way that many walls are printed at excessive angles. Usually, angles that are harsher than 45° are going to cause sagging problems. In bioprints, this problem might be more dangerous than in classical plastic-based FDM 3D printing because the solidification of layers is not as fast as in that method. In certain bioinks the solidification process is also slow compared to plastics and cannot be accelerated. In some cases, there is only a narrow printing speed that strikes a balance between the nozzle not clogging and the bioink being properly solidified. An extreme decrease in print speeds is not recommended, print speed below 5 mm/s (mm/s of filament flow given fixed nozzle diameter) can cause dripping, or clogs when using chemical solidifying methods such as calcium-based solutions. In effect decreasing print speeds will not result in a major quality improvement of the print and the percentage increase in cell viability is not worth risking the chance of a print failure during the process. There are more extreme examples of improper media solidification. In the case of problems in the solidification of the first layer (the layer adhering directly to the print bed) during the print process, the entire print might get loose and detach from the print bed. The first layer is considered the most crucial layer in any bioprint. During the curing process of successive layers, the print might contract slightly and, in the process, exert some force causing detachment from the print surface if the first layer was not attached properly [30, 31].

Author contributions: All the authors have accepted responsibility for the entire content of this submitted manuscript and approved submission.
Research funding: None declared.
Conflict of interest statement: The authors declare no conflicts of interest regarding this article.

References

1. Shoemaker RH. The NCI60 human tumour cell line anticancer drug screen. *Nat Rev Canc* 2006;6:813–23. https://doi.org/10.1038/nrc1951.
2. Chen L, Xiao Z, Meng Y, Zhao Y, Han J, Su G, et al., The enhancement of cancer stem cell properties of MCF-7 cells in 3D collagen scaffolds for modeling of cancer and anti-cancer drugs. *Biomaterials* 2012;33:1437–44. https://doi.org/10.1016/j.biomaterials.2011.10.056.
3. Gardner E. *3-D printing models, augmented reality images help surgeons visualize tumors*. USA: RSNA Daily Bulletin; 2017.
4. Jensen C, Teng Y. Is it time to start transitioning from 2D to 3D cell culture? *Front Mol Biosci* 2020;6:7–33. https://doi.org/10.3389/fmolb.2020.00033.
5. Tannenbaum J, Bennett BT. Russell and Burch's 3Rs then and now: the need for clarity in definition and purpose. *J Am Assoc Lab Anim Sci* 2015;54:120–32.
6. Ariadne. *New Sci* 1974;64:80.
7. Scott Crump S. U.S. patent no. 005121329; 1989.
8. Almquist TA, Smalley DR. U.S. patent no. US5569349A; 1995.
9. Smalley DR, Vorgitch TJ, Manners CR, Hull CW, VanDorin SL. U.S. patent no. US5597520A; 1994.
10. Childers CM, Charles WH. U.S. patent no. US5609812A; 1993.
11. Charles WH. U.S. patent no. US5762856A; 1995.
12. Revilla León M, Klemm IM, García-Arranz J, Özcan M. 3D metal printing – additive manufacturing technologies for frameworks of implant-borne fixed dental prosthesis. *Eur J Prosthodont Restor Dent* 2017;25:143–7. https://doi.org/10.1922/ejprd_revillaleon05.
13. Lerman MJ, Lembong J, Muramoto S, Gillen G, Fisher JP. The evolution of polystyrene as a cell culture material. *Tissue Eng B Rev* 2018;24:359–72. https://doi.org/10.1089/ten.teb.2018.0056.
14. Leary E, Rhee C, Wilks B, Morgan JR. Accurate quantitative wide-field fluorescence microscopy of 3-D spheroids. *Biotechniques* 2016;61:237–47. https://doi.org/10.2144/000114472.
15. Claudia M, Kristin Ö, Jennifer O, Eva R, Eleonore F. Comparison of fluorescence-based methods to determine nanoparticle uptake by phagocytes and non-phagocytic cells in vitro. *Toxicology* 2017;378:25–36. https://doi.org/10.1016/j.tox.2017.01.001.
16. Smyrek I, Stelzer EH. Quantitative three-dimensional evaluation of immunofluorescence staining for large whole mount spheroids with light sheet microscopy. *Biomed Opt Express* 2017;8:484–99. https://doi.org/10.1364/boe.8.000484.
17. Lazzari G, Couvreur P, Mura S. Multicellular tumor spheroids: a relevant 3D model for the in vitro preclinical investigation of polymer nanomedicines. *Polym Chem* 2017;8:4947–69. https://doi.org/10.1039/c7py00559h.
18. Naghieh S, Sarker M, Izadifar M, Chen X. Dispensing-based bioprinting of mechanically-functional hybrid scaffolds with vessel-like channels for tissue engineering applications – a brief review. *J Mech Behav Biomed Mater* 2018;78:298–314. https://doi.org/10.1016/j.jmbbm.2017.11.037.
19. Zeming G, Jianzhong F, Hui L, Yong H. Development of 3D bioprinting: from printing methods to biomedical applications. *Asian J Pharm Sci* 2019;15:529–57. https://doi.org/10.1016/j.ajps.2019.11.003.
20. Gopinathan J, Noh I. Recent trends in bioinks for 3D printing. *Biomater Res* 2018;22:11. https://doi.org/10.1186/s40824-018-0122-1.

21. Gungor-Ozkerim PS, Inci I, Zhang YS, Khademhosseini A, Dokmeci MR. Bioinks for 3D bioprinting: an overview. *Biomater Sci* 2018;6:915–46. https://doi.org/10.1039/c7bm00765e.
22. Nie S, Hsiao WL, Pan W, Yang Z. Thermoreversible Pluronic F127-based hydrogel containing liposomes for the controlled delivery of paclitaxel: in vitro drug release, cell cytotoxicity, and uptake studies. *Int J Nanomed* 2011;6:151–66. https://doi.org/10.2147/IJN.S15057.
23. Wang X, Ao Q, Tian X, Fan J, Tong H, Hou W, et al., Gelatin-based hydrogels for organ 3D bioprinting. *Polymers* 2017;9:401. https://doi.org/10.3390/polym9090401.
24. Lee A, Hudson AR, Shiwarski DJ, Tashman JW, Hinton TJ, Yerneni S, et al., 3D bioprinting of collagen to rebuild components of the human heart. *Science* 2019;365:482–7. https://doi.org/10.1126/science.aav9051.
25. Köpf M, Campos DF, Blaeser A, Sen KS, Fischer H. A tailored three-dimensionally printable agarose-collagen blend allows encapsulation, spreading, and attachment of human umbilical artery smooth muscle cells. *Biofabrication* 2016;8:025011. https://doi.org/10.1088/1758-5090/8/2/025011.
26. Wu D, Yu Y, Tan J, Huang L, Luo B, Lu L, et al., 3D bioprinting of gellan gum and poly (ethylene glycol) diacrylate based hydrogels to produce human-scale constructs with high-fidelity. *Mater Des* 2018;160:486–95. https://doi.org/10.1016/j.matdes.2018.09.040.
27. Albritton JL, Miller JS. 3D bioprinting: improving in vitro models of metastasis with heterogeneous tumor microenvironments. *Dis Model Mech* 2017;10:3–14. https://doi.org/10.1242/dmm.025049.
28. Satyavrata S, Nikhita J. 3D printing for the development of in vitro cancer models. *Curr Opin Biomed Eng* 2017;2:35–42. https://doi.org/10.1016/j.cobme.2017.06.003.
29. Yu Y, Zhang Y, Martin JA, Ozbolat IT. Evaluation of cell viability and functionality in vessel-like bioprintable cell-laden tubular channels. *J Biomech Eng* 2013;135:91011. https://doi.org/10.1115/1.4024575.
30. Liliang O. *Study on microextrusion-based 3D bioprinting and bioink crosslinking mechanisms*. Singapore: Springer; 2019.
31. Mishbak HH, Cooper G, Bartolo PJ. Development and characterization of a photocurable alginate bioink for three-dimensional bioprinting. *Int J Bioprint* 2019;5:189. https://doi.org/10.18063/ijb.v5i2.189.

Index

(FA@MSNPs-[Cu(L)(dppz)]$^+$) NPs 196
2,3-Dimethylmaleic anhydride (DMMA) 203
2-Photon 231
3-aminopropyltrimethoxysilane (APTMS) functionalized QDs 204
3D cell cultures 97
3D modeling 214
3D platform 86
3D printer technologies 214
3D printer technology 214
3D structures 221
3D tissue architecture 213
3D tumor models 213
4T1 breast cancer cell lines 199
5-fluorouracil (5-FU) 204
5-fluorouracil 32

"Hot-spot" MRI 180
A375 melanoma cells 203
A549 lung cancer cells 197
ABS 218
absorbed dose 59
acetylation 124
adjuvant treatment 28
adjuvants 158
ADME (Administration-Distribution-Metabolism-Excretion) 88
adoptive transfer of CIK 151
advanced 3D cell culture metabolomics 98
Afatinib 2
afterloading 59
Agarose 232
aggressive metastasis 194
agonist antibodies 149
albumin protein 198
Alectinib 2
Alginate 232
algorithms 69
ALK (anaplastic lymphoma kinase) 5
ALK fusions 6
ALK rearrangement 5
alkylating drugs 29
Allogenic NK cells 152
Alpelisib 13
Alpha particles 59
American Cancer Society 194
amphiphilic block copolymers 199

annihilation radiation 54
antibody directed enzyme prodrug therapy 46
anti-cancer 119, 127
antimethabolites 31
antioxidant 126
anti-tumor response 156
apoptosis 52
applications 204
Artificial organs 87
AuNPs 175, 177
autologous LAKs 151
autologous vaccines 158
autophagy 123
barium sulfate 177
Base 217
BAX (Bcl-2-associated X protein) 201
Beam collimation 69
beam profile 67
Beams Eye View 69
bioavailability 37
biocompatible and biodegradable dendrimers 197
biocompatible and biodegradable QDs 203
biocompatible nanocarriers 202
bio-degradable 230
biodegradable aliphatic polycarbonates 199
biodegradable polymeric nanoparticles (NPs) 155
biodevices 200
bioink 230
Bioluminescence Imaging 183
Biomarkers 1
biopolymeric NPs 194
Bio-printer media 231
Bio-printing 228
bispecific killer engagers (BiKEs) 149
bleomycin 35
*Bletilla striata*polysaccharide 199
BLI 183, 186
BMDMs 181, 186
Boron-Neutron Capture Therapy 59
bortezomib (BTZ) 197
brachytherapy 58
BRAF 8
BRAF V600E mutation 9
Bragg Curves 65
Bragg peak 66

BRCA 14
breast cancer 77, 114, 124
bremsstrahlung 52
Brigantinib 2
brigatinib 6
build-upeffect 66

Cabozantinib 7
cancer 81, 193
cancer cells resistance 86
cancer diagnosis 195, 202
cancer photodiagnosis 202
cancer radiotherapy 196
cancer therapy 140, 204
cancer treatment 228
cancer treatment approaches 207
capecitabine 32
Capmatinib 2, 8
carbon nanotubes (CNTs) 194
carboplatin 36
carboxylatemethylcellulose 204
carcinogenesis 118, 120
carmustine 46
CAR-NK cells 152
caspase 9 201
caspase-9 (iC9) suicide gene 202
caspase-9 enzyme 202
Caucasian colon adenocarcinoma human cell lines (Caco-2 cells) 202
CCL2-CCR2 axis 154
CD39 148
CD73 148
CD94/NKG2A 145
CD96 147
cell co-cultures 85
Cell culture analog (CCA) system 88
cell death 126, 235
Cell density 235
cell imaging 202
Cell migration 85
cell repellent surface 225
cell tracers 173
cell tracking 173, 186
Cell viability 234
Cell-to-cell interactions 81
cellulose 198
ceritinib 6
CEST MRI 182

charge-altering releasable transporters (CARTs) 155
Checkpoint receptors 144
chemoprevention 118
chemopreventive 122
chemotherapy 27, 112, 202
chimeric antigen receptor (CAR) 152
Chip-based body systems 89
Chitin 198
chitosan 198
chlorin e6 (Ce6) 202
cisplatin 36
classical monocytes 139
Clinical target volume 62
clinical tests 204
cobimetinib 9
coelenterazine 183
COL1A1/PDGF-B 11
Collagen 232
Colorectal 114
colorectal cancer 122
Compton effect 54
Computer Assisted Design 233
computer assisted manufacturing 214
conditioned medium 83
confocal 228
conventional DCs (cDCs) 140
corpuscular radiation 52
COVID-19 disease 193
cremophor 39, 40
critical micelle concentration (CMC) 199
Crizotinib 2, 5, 6
CT 174, 176, 177, 186
CT contrast agents 174
CT26 murine colorectal carcinoma cell lines 199
CT-based immune cell tracking 176
CTLA-4 147
culture techniques 82
curcumin (CUR) 198
Curcumin 122, 123
CXCL12/CXCR4 axis 154
Cyber Knife 74
cyclophosphamide 29
cytarabine 32
cytokines 151
cytostatic drugs 28
cytotoxicity 227

Dabrafenib 2
dacarbazin 31
Dacomotinib 2
dactinomycin 35
DAPI 227
DC-based vaccine 156
DCVAC 157
DEC-205 158
delineation of target 62
delivery of anticancer drug 204
dendrimer 197
dendrimers 194
dermatofibrosarcoma protuberans 11
DEVD 227
Dextran 198
diagnose cancer 204
Diagnostic biomarker 2
diagnostic imaging 173
direct ionization 52
Direct Metal Laser Sintering 215
Direct type extruders 220
Dispersion-based bioprinters 229
DLP 3D printing 215
d-Luciferin 183
DMLS 221
DNA 117
docetaxel 34
Dose distribution calculation 64
dose fractionation 74
DOTMA 181
doxorubicin (DOX) 199
doxorubicin 34
Droplet based bioprinters 231
drug delivery 202
drug delivery applications 195
Drug development 87
dysprosium 180, 181
D-α-tocopheryl polyethylene glycol succinate (TPGS) 199

effective dose 60
EGFR mutations 4
Einstein's theory 54
electrochemiotherapy 46
electron energy conversion 57
Enclosure 216
encorafenib 10
endo-metabolome 95

enhanced permeability and retention effect (EPR) 199
Entrectinib 2, 6, 8
environmental control 84
environmental factors 117
epigenetic modification 118
epirubicin 34
equivalent dose 60
Erlotinib 2
escape from immune surveillance 86
estrogen receptor 120
Estrogen receptors 13
etoposide 33
exo-metabolome 95
experimental design 95
external photon beams 57
extracellular metabolome 95
extraction method 96
Extruder gearing 217
Extruder stepper motor 217
extrusion system 229

F-bottom 225
ferrihydrite 179
ferritin 179
filament 218
Filament drive gear 217
first layer 237
flattening filter free 68
flattening filters 67
FLI 184, 186
Fluorescence imaging 184
fluorescence visualization 227
fluorescent protein (FP) 184
Fluorescent proteins 186
fluorine-19 180
fluorophore 184
Fluorophore 185
folate receptor (FR) 196
FOLFOX4 10
form of anticancer drug 37
Frame 216
fulvestrant 13
fused deposition modeling 214
fusion proteins 150

gadolinium (Gd) chelate 200
gadolinium 178

gallic acid 197
Gamma Knife 74
gamma radiation 56
Gas chromatography (GC) 84
Gastrointestinal stromal tumors 12
Gefitinib 2
Gelatin 232
gellan gum 198, 232
gemcitabine 32
gene directed enzyme prodrug therapy 46
gene therapy 202
Genetically modified NK cells 152
Genomics 92
GFP 185
Gold NPs 196
Gregory Gregoriadis 200
Gross tumor volume 62
gum acacia (GA) microspheres 198
gum arabic 198
gut 104
GVAX 158

Hanging drop 225
Heat break 216
heat shock proteins 126
Heat sink 216
Heater 217
Heater block 216
heater cartridges 219
Heavy particles 66
HeLa cervical cancer cells 204
HER2 gene amplification 16
High-Content Screening 227
highly-shifted proton MRI 180, 181
High-performance liquid chromatography (HPLC) 84
HIPEC 43
histones 117
Hoechst 33342 227
hotend assembly 219
HT-29 127
HTLA-230 and HTLA-ER neuroblastoma cell lines 197
human gastric carcinoma drug-resistant SGC7901/ADR cells 203
human gastric carcinoma SGC7901 cells 203
hyaluronic acid (HA)-SWCNTs 202
hyaluronic acid 198
hyaluronidase 43

hybrid NPs 203
hydrogel matrices 91
hyperbranched polymers 196
hypofractionation 74

Idler arm 217
ifosfamide 29
IL-1β 142
immature DCs (iDCs) 140
immobilizing accessories 62
Immune cell tracking 176
immune checkpoints 82
immune system 115
immunosupressive cytokines 142
immunotherapy 105
in vitro 213
in vivo conditions 86
indirect cell labeling 176
indirect ionization 53
Indirect type extruders 220
indocyanine green 200
indoleamine 2,3-dioxygenase (IDO) 148
inductive treatment 28
infusion solutions 39
Inhibitory killer cell immunoglobulin-like receptors (KIRs) 145
innovative approach 87
inorganic NPs 194
Intensity Modulated Radiation Therapy 72
intercellular communication 83
intestinal 104
intestine 125
Intracellular metabolome 96
Intraoperative Radiotherapy 77
intraperitoneal chemotherapy 43
intrathecal chemotherapy 45
intratumoral NK cells 141
intrinsic toxicity 196
inverse planning 72
ionizing radiation 51
irinotecan 35
Iron oxide nanoparticles 175
Iron oxide NPs 196

Jurkat T 127

KERMA 60
KIT 12
KRAS 10

Lapatinib 16
larotrectinib 8
Larotrektinib 2
late-stage diagnosis 194
Layer adhesion 236
limitations 82
liposomes 41, 194, 198, 204
liquid-based chromatography (LC) 93
Lirilumab 145
loaded anticancer drug 204
Lorlatinib 2
low volume manufacturing 215
luciferase 183
Lung cancer 112
Lymphocyte-activation gene 3 (LAG3) 147

M1 macrophages 142
M1 pro-inflammatory/anti-tumor 139
M2 anti-inflammatory/pro-tumor 139
M2 macrophages 142
macrophages 177, 179, 182, 184, 185
macrophages 180
macrophages polarization 139
Magnetic spheroid 3D printing 225
mantle cell lymphoma 197
Margetuximab 17
mass spectrometry (MS) 93
mature DCs (mDCs) 140
MDA-MB-231 breast cancer cell lines 196
medical research 193
mesoporous silica nanoparticles 196
metabolic quenching 96
metabolic reprogramming 95
metabolites 93
metabolome profiling 93
Metabolomics 93
metabolomics in 2D cell culture 95
metabonomics 93
metal-organic frameworks (MOFs) 203
METex14 skipping mutations 8
methotrexate 32
methoxypoly(ethylene) glycol polycaprolactone diblock (mPEG-SS-PCL-OH) polymersomes 201
methylation 118
$Mg_{(1-x)}Cu_xFe_2O_4$ superparamagnetic particles 196
micellar NPs 204
micellar system 40

Michigan Cancer Foundation-7 (MCF-7) breast cancer cell line 196
microbiome 104, 105
Microfluidic 2D models 84
microfluidic co-culture device 86
Microfluidic devices 84
microfluidic systems 84
Microfluids 3D models 85
microsatellite instability, MSI 18
Model acquisition 233
Modified antibody molecules 149
molecular cytogenetics analysis 5
monalizumab 145
monitoring 1
Monitoring biomarker 3
monoclonal antibodies 155
mononuclear phagocyte system (MPS). 138
Motherboard 217
Motion transition system 216
Motors 216
MRI 178, 186
MRI images 180
MS-based metabolomics 94
MSLA 221
multidrug resistant cancer 194
multi-leaf collimator 69
multiphoton 228
multiple myeloma 197
Multi-wall carbon nanotubes (MWCNTs) 201
multiwell plates 223
myeloid-derived suppressor cells (MDSCs), 142

nanomedicine 194
nanoparticle 174, 175
nanoparticles (NPs) 194
nanospheres 203
nanosuppression 41
Natural bioinks 232
Nature Cancer 194
NEMA 17 215
neoadjuvant treatment 28
Neutrons 59
neutropenia 116
NF-κB 123
NF-κB protein 120
nitric oxide (NO) 203
NK cell lines 152
NK cell-based adoptive cellular immunotherapy 151

NKG2A expression blockers 146
non-classical monocytes 139
none 18
nonpegylated liposomes 41
Non-small cell lung cancer 2
novel NPs 204
Nozzle 216
NTRK gene fusions 7
Nuclear Magnetic Resonance (NMR) 94
nucleic acids 155
Nylons 218

of activating and inhibitory receptors 138
Olaparib 14
Omic 92
onclolytic viruses (OVs) 153
oncobiome 104, 116
oncogenesis 106
oncogenic driver 5
operating room 77
Optical imaging 183
Organoid technology 92
Organoids 90
Organs at risk 62
Osimertinib 2
oxaliplatin 36

p21 gene 124
P53 (tumor protein p53) 201
p53 gene 120
paclitaxel (PTX) 198
paclitaxel 34, 39, 40
pair production 54
PARACEST MRI 182
Paramagnetic contrast agents 178
PBPKs - physiologically based pharmacokinetic model 89
PC 218
PC-3 prostate tumor cells 200
PD-1 143, 147
PD1/ PD-L1 115
PDGFRA 12
PEEK 218
PEG-PAMAM dendrimer 197
pegylated liposomes 41
PEI 218
Pembrolizumab 2, 17, 147
penumbra 67
Percentage Depth Doses 65

perfluorocarbons (PFC) 181
pertuzumab 15, 16
PETG 218
pH 222
Pharmacodynamic (response) biomarkers 3
pharmacodynamic biomarkers 1
pharmacological agents 150
phenotype of monocytes 142
phenylsulfonyl furoxan (PSF) 203
phospholipid bilayer 200
photodynamic therapies 202
photoelectric effect 54
photosensitizer 202
pH-sensitive acetal bond 199
physicochemical stability 39
PIK3CA 13
PIPAC 45
PLA 218
Planning target volume 62
plant alkaloids 33
plasmacytoid DCs (pDCs) 140
Pluronics 199, 232
polifeprosan 46
Poly(D,L-lactic acid) (PLA) 199
poly(ethylene glycol$_{400}$)-oleate polymersome NPs 201
Poly(α-amino acid) 199
poly(γ-glutamic acid) (γ-PGA) 198
polyamidoamine (PAMAM) dendrimers 197
Polycaprolactones (PCL) 199
polydopamine (PDA) 198
polyethylene glycol (PEG) 199
Polyethylene glycol 232
polyethylene glycols 41
poly-L-Lisine (PLL) dendrimers 197
polymeric micelles 194
polymersomes 194
polyphenol concentration 121
poly-propylene imine (PPI) dendrimers 197
post processing 234
Predictive biomarkers 1
Pressure extruders 229
print speed 218
Print surface 217
Probiotic 105, 114
Probiotics 105, 127
prognostic 1
Programmed death 1 receptor (PD-1) 17
protect healthy tissues 79

protein NPs 204
proteomics 92

quality factor 60
quantum dots 194
Quercetin 125, 126

radioactive decays 54
Radiosurgery 74
Radiotherapy planning 61
Rapid prototyping 214
RAS–RAF–MEK–ERK–MAP kinase pathway 8
redox stimulus 201
Renal 113
repair 52
reporter gene 184
reprogramming TAMs 155
Resveratrol 119, 120, 122
RET rearrangements 7
Risk (screening) biomarkers 3
RNA delivery 155
ROS proto-oncogene 1 (ROS1 6
ROS1 rearrangements 6
Rotary screw-based extruders 230
route of administration cytostatic drugs 37
rucaparib 14

safety 1
Safety biomarker 3
sample preparation 94
scattered (secondary) radiation 52
scientific strategies 204
Selpercatinib 2, 7
semiconductor crystal particles 202
semimature DCs (smDCs) 140
silibinin 201
Silica NPs 196
Silver NPs 196
simvastatin 200
Single-wall carbon nanotubes (SWCNTs) 201
slicer 233
Soluble MICA/MICB 148
spheroid 222
Spheroid formation in matrix 225
Spinner flask spheroid formation 225
SPIONs 179
Starch 198
STAT3 143
statins 200

Stepper drives 217
stereolithography 215
Stereotactic radiotherapy 74
stimuli-responsive vehicles 203
strategies for restoring NK cell's 144
strategies to escape the immune response 141
subcutaneous administration 42
Superparamagnetic contrast agents 179
Superparamagnetic iron oxide
    nanoparticles 179
surgery 77
Synbiotic 115
synbiotics 105
synthetic polymer particles 204
synthetic polymers 201
synthetic vesicles 200

T cell immunoglobulin and mucin-containing
    domain (TIM)-3 146
T cell immunoreceptor with immunoglobulin
    and immunoreceptor tyrosine-based
    inhibitory motif domains (TIGIT) 146
$T1$ 178
$T2$ 179
TAA-targeting antibodies 148
talazoparib 14
Tamoxifen 13
targeted drugs 198
taxoids 34
temozolomide 31
temperature regulation 219
temperature sensor 219
TGF-β 142
The leukocyte immunoglobulin-like receptor
    (LIR), 146
theranostic 175
thermal conductivity 219
thulium 180, 181
Time for Cell Doubling 224
tip 218
Tissue Culture 222
tissue engineering 91
TLR agonists 155
tolerance doses 63
tomotherapy 73
topotecan 35
Total Body Irradiation 76
Total Marrow Irradiation 76
toxicity profile of cytostatic drugs 37

transarterial chemoembolisation 45
transduction 227
transfection 227
Transwell system 83
Trastuzumab 15, 200
Treatment planning system 67
Tregs 142
triazine-derived dendrimer 204
triple negative breast cancer cells (TNBC) 198
trispecyfic (TriKEs) and teraspecyfic (TetraKEs) killer engagers 149
tumor 193
Tumor microenvironment 82
tumor-associated macrophages (TAMs) 142
tumor-derived factors (TDFs). 154
turmeric plant 198
tyrosine kinase inhibitors (TKIs 4

U. S. Food and Drug Administration (FDA) 200
U-bottom 225
ultrasmall superparamagnetic iron oxide nanoparticles 179
USPIONs 179
UV 237

V600K mutations 8
vinblastine 33
vincristine 33
vinorelbine 33
viral vectors 157
viscous media 235
Volumetric Modulated Arc Therapy 73

water cooling 219
weighting factors 60
Wnt signaling pathway 121
World Health Organization (WHO) 194

xanthan gum 198

z-axis 221
ZnO QDs 203
Z-stack 223

γ-PGA NPs 198

www.ingramcontent.com/pod-product-compliance
Lightning Source LLC
Chambersburg PA
CBHW082322220526
45470CB00008B/2380